Mucosal Immunology and Virology

Stephen K. Tyring (Ed.)

Mucosal Immunology and Virology

Springer

Stephen K. Tyring
Departments of Dermatology, Microbiology &
    Molecular Genetics and Internal Medicine
The University of Texas Health Science Center
Houston, TX
USA

British Library Cataloguing in Publication Data
A catalogue record for this book is available from the British Library

Library of Congress Control Number: 2005935331

ISBN-10: 1-84628-201-2   e-ISBN 1-84628-206-3   Printed on acid-free paper
ISBN-13: 978-1-84628-201-0

Printed in Singapore.   (BS/KYO)

9 8 7 6 5 4 3 2 1

Springer Science+Business Media
springer.com

Preface

Mucosal immunity plays a very prominent role in protection against viruses and other infectious agents in the gastrointestinal tract, in the eyes, in the anogenital mucosa and in the respiratory tract. Because the gastrointestinal (GI) tract is the largest and most complex of these systems, the greatest attention in this text has been given to the GI tract. Therefore, separate chapters are devoted to the virology and the mucosal immunology of the gastrointestinal tract. A third chapter focuses on the proximal end of the gastrointestinal tract (i.e. the oral cavity). The mucosal immunology and virology of the distal end of the gastrointestinal tract is covered in the chapter on the anogenital mucosa. Mucosa-associated lymphoid tissue (MALT) plays a role in protection against all viral (and other) infections except those that enter the body via a bite (e.g. yellow fever or dengue from a mosquito or rabies from a dog) or an injection or transfusion (e.g. HIV, Hepatitis B). In these cases where a break in the skin is necessary for viral entry, peripheral (i.e. systemic) immunity plays a greater role. Peripheral immunity is primarily controlled by the bone marrow, lymph nodes and spleen. Both systems involve humoral and cell mediated immunity, but their immune functions also can be divided into innate and adaptive immunity. Innate immunity provides a rapid response to a new antigen and sets the stage for the adaptive response; innate immunity is pathogen nonspecific and has no recall. On the other hand, adaptive immunity is organism/antigen-specific, is involved in disease recovery, has

immunological memory, thus providing protection against reinfection.

Public health measures are of primary importance in preventing the immune system from being overwhelmed with infectious organisms. Such measures include sanitation, hand washing, use of masks, gloves and condoms, blood testing, discouraging sharing of needles, etc. Since 1797 vaccines have also played an important role in prevention of viral infection/disease. Interestingly, some vaccines against mucosal infection appear to work when administered via injection or by direct exposure to MALT. For example, both intranasal and injectable influenza vaccines are available, as are both oral and injectable forms of the polio vaccine. Other vaccines given by the nasal, oral or anogenital mucosal routes are under study. Not only can the route of administration be critical in maximum induction of the mucosal immune response, the adjuvant also plays an important role, as was seen in the marked difference in efficacy between protective and non-protective herpes simplex virus vaccines.

Three vaccines were recently approved by the United States Food and Drug Administration. Two of these vaccines, RotaTeq and Gardasil, stimulate mucosal immunity. RotaTeq is an oral pentavalent vaccine approved for prevention of rotavirus infection, the most important cause of severe infantile gastroenteritis worldwide. Gardasil is a quadrivalent (injectable) vaccine approved for prevention of human papillomavirus (HPV) types 16 and 18, the leading causes of cervical cancer worldwide, as well as for prevention of HPV types 6 and 11, the leading causes of condyloma acuminatum. The third, recently-approved vaccine is Zostavax which will be used for the prevention of herpes zoster and postherpetic neuralgia. Therefore, this vaccine will not be used in the classic sense as a prophylactic vaccine, but rather will be given to prevent the clinical reappearance of virus that infected the patient decades earlier. Future vaccines will be designed to safely boost mucosal immunity using innovative routes, adjuvants and carriers (e.g. fruits and vegetables) and may include new therapeutic as well as prophylactic vaccines.

Stephen K. Tyring

Contents

List of Contributors

Carlos A. Barrera, Ph.D.
Department of Pathology
University of Texas Medical Branch
Galveston, TX, USA

Dorsey M. Bass, M.D.
Department of Pediatrics
Stanford University
Palo Alto, CA, USA

David A. Bland, B.S., M.S.
Microbiology and Immunology
University of Texas Medical Branch
Galveston, TX, USA

Tibor Farkas, D.V.M., Ph.D.
Division of Infectious Diseases
Cincinnati Children's Hospital Medical
  Center
Cincinnati, OH, USA

David B. Huang, M.D., Ph.D., M.P.H.
Division of Infectious Diseases
Department of Medicine
Baylor College of Medicine
Houston, TX, USA

Xi Jiang, Ph.D.
Division of Infectious Diseases
Cincinnati Children's Hospital Medical
  Center
Cincinnati, OH, USA

Vandana Madkan, M.D.
Center for Clinical Studies
Houston, TX, USA

Denis P. Lynch, D.D.S., Ph.D.
Department of Surgical Sciences
Marquette University School of
  Dentistry
Milwaukee, Wisconsin, USA

Victor V. Reyes, Ph.D.
Departments of Pediatrics and
  Microbiology and Immunology
University of Texas Medical Branch
Galveston, TX, USA

Anthony Simmons, M.D., Ph.D.
Department of Pediatrics
University of Texas Medical Branch
Galveston, TX, USA

Karan Sra, M.D.
Center for Clinical Studies
Houston, TX, USA

Stephen K. Tyring, M.D., Ph.D., M.B.A.
Departments of Dermatology,
  Microbiology & Molecular Genetics
  and Internal Medicine
University of Texas Health Science
  Center
Houston, TX, USA

Richard L. Ward, Ph.D.
Division of Infectious Diseases
Cincinnati Children's Hospital Medical
  Center
Cincinnati, OH, USA

Mitchell P. Weikert, M.D., M.S.
Department of Ophthalmology
Cullen Eye Institute
Houston, TX, USA

Steven Yeh, M.D.
Department of Ophthalmology
Cullen Eye Institute
Houston, TX, USA

The Role of Mucosal Immunity in Protection against Viral Diseases

Vandana Madkan, Karan Sra, and Stephen K. Tyring

The study of mucosal immunity has provided incredible insight into the human body's complex and intricate system of battling the many viruses and other pathogens it encounters on a daily basis. Mucosal surfaces, which collectively form an area larger than one and one half football fields (1, 2), are mucus-secreting membranes lining all body cavities or passages that communicate with the exterior. They are located primarily in the gastrointestinal, urogenital, and respiratory tracts and are portals of entry for disease-causing organisms. In fact, the great majority of pathogens enter via the mucosa, with few exceptions such as infections that are introduced via arthropod or other bites, injections, or blood transfusions (1). The body is dependent on immune cells and antibodies present in the mucosal lining to protect it against the onslaught of organisms to which it is exposed; therefore, it comes as no surprise that the human gastrointestinal tract contains more lymphocytes than all other lymphoid organs (components of the peripheral immune system) combined (3).

Many of the world's most devastating diseases are spread via mucosal infection. According to the World Health Report of 2004 conducted by WHO, more than 90% of the world's deaths from infectious diseases are caused by the following six disease processes: lower respiratory infections, HIV/AIDS, diarrheal diseases, tuberculosis, malaria and measles. Combined, they kill greater than 13 million persons yearly (4). Of these six diseases, five are primarily spread by mucosal infection, with malaria being the only significant exception (3). Even certain diseases

once thought to be non-infectious, such as some cancers, have now been shown to have an infectious etiology. Cervical cancer, for example, is one of the most common cancers among women in the developing world. It is now known to be associated with human papillomavirus infection, a disease spread through infection of the mucosa of the anogenital tract (1).

Surprisingly, despite the fact that the overwhelming majority of infectious agents penetrate the human body at these mucosal surfaces, most vaccines developed thus far were designed to target the peripheral immune system. As the counterpart to the mucosal immune system, the peripheral immune system is made up of the bone marrow, lymph nodes, and spleen and relies on presentation of foreign antigens to lymph nodes via the lymph fluid. Peripheral immunity is known to fight an infection once it has entered the body and stops the infection from causing disease. The mucosal immune system, however, defeats the pathogen at the mucosal barrier and prevents infection from occurring (5). Although several vaccines, including all traditionally recommended childhood vaccines, are injectable and therefore would likely stimulate the peripheral immune system, it is unclear if they provide adequate mucosal immunity as measured by secretory IgA, etc. (6).

For instance, studies conducted comparing the immune response to inactive poliovirus vaccine (IPV) and oral poliovirus vaccine (OPV) revealed that IPV alone could not induce a sufficient mucosal immune response; however, if the IPV was given in individuals already "primed" with the oral vaccine, a sufficient mucosal response was generated (7). With both natural infection and OPV, secretory IgA is seen at the mucosal surfaces of exposed individuals. IPV has been shown to produce a strong IgA response in recipients, but only in those persons who have previously had a natural infection or oral vaccination. If there was no previous infection, IPV vaccinees would remain susceptible to mucosal infection with the polio virus. If exposed and infected, they may remain asymptomatic, but still able to spread disease to others who are not fully vaccinated. The overall significance of this finding is incredibly important if mankind is to truly eradicate poliomyelitis. In communities where IPV recipients, who can contribute to the transmission of the virus, remain in contact with those with low vaccine rates, and therefore have no mucosal or peripheral protection against the disease, epidemics of poliovirus could ravage human life. This very situation has been observed in the Netherlands in small religious communities in 1978 and 1992 that were not vaccinated, but still in contact with those populations who were (7). Since an entirely oral vaccination series may carry its own morbidity, a suitable solution may be to initially vaccinate with OPV and use IPV boosters (7).

As more insight is gained regarding the immune system and its complex pattern of response with regard to the interplay between peripheral and mucosal immunity, research is leading to improved vaccines. Due to the knowledge that mucosal defense can provide an advantage over peripheral immunity in preventing infection and that traditional vaccines may not provide mucosal immunity, several new vaccines targeting mucosal response are being studied. As mentioned above, enteric infections cause impressive morbidity and mortality, especially in the developing world, and are spread by mucosal transmission. Each year, 2 million children worldwide die from enteric infections with 400 of those deaths occurring in the United States (8). Possible etiologic agents include viruses such as rotavirus and noroviruses as well as invasive bacteria such as E.coli and shigella, or toxic bacteria including *Cholera* species. The injectable cholera vaccine, which was designed to prevent infection with *V. cholerae*, was judged largely ineffective (9). However, as this virulent pathogen is responsible for severe outbreaks and occasional epidemics of diarrhea, attempts to develop a mucosal vaccine were launched. Cur-

rently available oral cholera vaccines have been shown to be effective and may even provide community protection (10). The most commonly employed vaccine (Dukoral) is made of a recombinant toxin subunit and inactive whole cholera cells. The vaccine has been shown to stimulate IgA production in the intestinal mucosa (9).

Rotavirus is noted to be the most common cause of infectious diarrhea worldwide and is noted to be responsible for 20% of deaths due to infectious diarrhea (8). The virus kills almost half a million children globally each year (9). In the United States, its peak incidence is the winter months. Two new oral vaccines are being developed in attempt to decrease the number of people afflicted each year. RotaRix and RotaTeq are both attenuated oral viruses shown to be efficacious in certain populations and are currently undergoing further clinical trials (9).

Enteric infections are the second deadliest infections in children worldwide, behind only respiratory infections. Perhaps the most well known etiology of lower respiratory infections is the influenza virus, whose natural variability and genomic mutations can cause epidemics and even pandemics, killing millions each year. Statistics cite 20,000 deaths annually in the US alone and tens of thousands hospitalized (11) with a large financial burden placed upon even industrialized countries. Standard vaccinations against the influenza virus have been administered for years in the United States, with preference given to those individuals most likely to suffer significant morbidity and mortality from the disease, e.g. people over the age of 65 or those with chronic medical conditions, etc. The influenza vaccine used more commonly within the United States is an injectable vaccine targeting the peripheral immune system which produces a serum IgG antibody response. An occasional mucosal protective effect has also been documented (9) but this also includes more mucosal IgG than IgA. However, the newly licensed, more readily dispensed vaccine, FluMist, is intranasally administered and is shown to provide immunity more comparable to immunity gained from a natural infection. Natural infection is more likely to provide long term immunity, with an IgA response noted in nasal washings and an IgG response in the lower respiratory tract. It protects against reinfection of the offending virus as well as against antigenically similar strains (11). Similarly, live virus vaccinations have been shown to have both IgA and IgG mucosal responses, as well as serum IgA and IgG responses. However, it is noted that the serum response is much higher with the inactivated vaccination (11).

Additionally, the side effect profile for live virus vaccination for influenza is much lengthier and more serious than for the inactivated vaccination. Despite the large number of inactivated flu vaccines administered each year in the United States, there are very few adverse reactions noted besides local injection site reactions. The live, attenuated vaccination has also appeared to be safe and effective in studied populations (11), but short courses of symptoms including coughing, sneezing, nausea, vomiting and other systemic symptoms were uncommonly noted. There are also hypothetical concerns, including genetic alteration of the live virus due to human contamination and CNS side effects secondary to the close proximity of the anatomic site of administration of the vaccination.

Although further studies must be conducted in order to determine an accurate profile of the advantages and disadvantages of each, there are still substantial advantages to the live virus vaccination. For instance, with a virus such as influenza, antigenic drifts and shifts cause new virulent strains. Patients may require booster vaccinations with live vaccines over the course of a lifetime. This, however, is still less than the yearly injections required currently by the inactive vaccine. In addition to providing more substantial mucosal protection, FluMist's administration would provide other benefits as well. The shortage of injectable flu vaccines in 2004 caused lines of elderly and chronically ill patients to form outside public health clinics once shipments of vaccine were received. One major advantage of mucosal vaccines is ease of administration. In the case of the influenza virus and its corresponding mucosal vaccine, FluMist, patients would be able to administer it themselves. Even more beneficial is the number of patients who may have originally refused vaccination due to fear of needle involvement but who may reconsider once needles are eliminated.

Few diseases have the dubious distinction of wiping out whole villages and destroying the economic infrastructures of countries by killing much of its work force; in this sense, HIV is the largest threat facing underdeveloped countries today. In some estimates, 60% of 15 year olds in Sub-saharan African countries hit hardest by the epidemic will not see their 60th birthday (4). While new drugs have shown promise in controlling disease progression, it can do so only for those who can afford it. In countries like Uganda, India, and Thailand, the fastest growing HIV-seropositive populations are among sex-workers and their immediate partners, a population that can hardly afford costly anti-retroviral therapy. Although there is currently no vaccine that is near availability, prevention of the disease by stopping its spread via anogenital mucosal contact may serve as the best method of curbing the disease's ravages. Efforts to produce a mucosal vaccine are targeted at the oral or nasopharyngeal mucosa, as early studies show that immune response in these areas confers protection at other mucosal sites by mechanisms that are not well understood (12). Additionally, these studies performed on a group of mice in Japan showed that once the animals were stimulated with a nasal immunization with gp160-HIV-liposome, effective HIV-specific immunity was noted at both mucosal and systemic sites. HIV-specific antibody titers were found by ELISA in serum, saliva, fecal extracts, and vaginal washings. The mice used in the study were deficient in Th1 or Th2 cells; also included were "wild-type" mice (those with no deficiency in immune cells); each type was noted to have appropriate mucosal responses (12).

Mucosal vaccinations are but one aspect of pioneering medical technology against HIV. Topically applied mucosal microbicides are currently undergoing research; certain compounds may have the ability to adsorb HIV-1 from physiologic fluids, preventing viral contact with target cells. For instance, cellulose acetate phthalate (CAP) is usually used in the pharmaceutical industry as a topcoat of capsules and may have a role as a topical microbicide. When confronted with HIV-1, the chemical causes the shedding of envelope glycoprotein and therefore, a loss of infectivity (13). Other compounds, such as sodium dimandelic acid ether and even the host's own beta-defensins may also provide protection against HIV as well as HSV (14, 15).

Although human herpesvirus (HSV) 1 and HSV 2 have less economic detriment than HIV and are not commonly the cause of death in healthy adults, a vaccine to help decrease transmission would reduce the psychological trauma of otherwise healthy adults and may well decrease the number of neonatal deaths from herpes encephalitis. A previous study showed that an injected recombinant glycoprotein (gD) vaccine could provide protection against HSV 2 in females who had no antibodies to either HSV 1 or HSV 2 prior to vaccination but could not protect females who had antibodies to HSV 1 prior to vaccination nor protect men (regardless of their antibody status). The observed discordant immunity, although not fully explained, led to hypotheses that the vaccine had stimulated mucosal immunity which provided more protection to females due to their larger mucosal surfaces, i.e. in the vagina, versus their male counterparts with less mucosal surface in the genital area (16). Further analysis of the study also showed that prior infection with HSV 1 can be protective against HSV 2; however, this may also depend on the site of infection for both viruses as stimulation at certain mucosal surfaces may or may not provide protection at other mucosal surfaces. Bernstein et al. also studied the proposed herpes vaccine and found that the vaccination did not decrease or increase the number of herpes recurrences in people with nongenital HSV disease (17).

Clearly, the mechanism of mucosal immunity is complex and not fully understood. However, significant benefits are to be gained as we endeavor to regulate the mucosal immune system to prevent infection and disease. Not only can we prevent common and devastating communicable diseases, but mucosal immunity may also play a role in preventing autoimmune diseases (5, 9). Although no studies have been published to date regarding such activity in humans, researchers showed that oral stimulation with antigen may actually lead to tolerance of that substance. In certain rat populations, immune tolerance to myelin basic protein developed after the protein was fed to rats who previously were afflicted with experimental autoimmune encephalomyelitis, an animal model of multiple sclerosis (18). It is thought that repeated high levels of antigen exposure at the gut mucosa could lead to several mechanisms that would suppress potentially damaging immune responses (19). Some of these responses include apoptosis of autoreactive T cells as well as anergy, depending on the dose administered (9). Tolerance can also be induced by repeated antigen exposure at other sites, although with varying responses (9). Hence the study of mucosal immunology could not only be used for the reduction of infectious disease, but also for the prevention of autoimmune related disorders.

Conversely, inducing an immune response against certain non-infectious antigens may also be desirable. In an effort to develop an anti-fertility vaccine, researchers have attempted to develop a vaccine that would cause an immune response against human sperm and thereby cause the development of anti-sperm antibodies in men and/or women (2). This would mimic what is observed naturally among 30% of those couples who are infertile, the development of anti-sperm antibodies in mucus secretions. While controversial, the research involved in this endeavor emphasizes the breadth of possibilities that would be provided by better understanding the mucosal immune system.

References

1. Hoft DF, Eickhoff CS. Type 1 immunity provides both optimal mucosal and systemic protection against a mucosally invasive, intracellular pathogen. Infect Immun 2005;73:4934–4940.
2. Seipp R. Mucosal Immunity and Vaccines. 2005. Retrieved from the World Wide Web July 02, 2005: http://www.bioteach.ubc.ca/Biomedicine/mucosalimmunity/.
3. Russell MW. Mucosal immunity. Chapter 5: 64–80. In New Bacterial Vaccines (Eds. Ellis, R.W. and Brodeur, B.R.) Eurekah.com, Georgetown TX and Kluwer Academic/Plenum, New York. 2003.
4. Davey S. Infectious Diseases are the biggest killers of the young. 1999, World Health Organization. Retrieved from the World Wide Web July 08, 2005: http://www.who.int/infectious-disease-report/pages/ch1text.html.
5. Anderson AO. Peripheral and Mucosal Immunity: Critical Issues for Oral Vaccine Design. 2004. Retrieved from the World Wide Web July 14, 2005: http://www.geocities.com/artnscience/crossregulation.html.
6. Ogra PL. Mucosal immunity: Some historical perspective on host-pathogen interactions and implications for mucosal vaccines. Immunol Cell Biol 2003;81:23–33.
7. Herremans TM, Reimerink JH, Buisman AM, Kimman TJ, Koopmans MP. Induction of mucosal immunity by inactivated poliovirus vaccine is dependent on previous mucosal contact with live virus. J Immunol 1999;162:5011–5018.
8. Frye RE, Tamer MA. Diarrhea. 2005. Retrieved from the World Wide Web July 28, 2005: http://www.emedicine.com/ped/topic583.htm.
9. Holmgren J, Czerkinsky C. Mucosal immunity and vaccines. Nat Med 2005;11:45–53.
10. Lucas ME, Deen JL, Seidlein L, et al. Effectiveness of mass oral cholera vaccination in Beira, Mozambique. N Engl J Med 2005;352:757–767.
11. Cox RJ, Brokstad KA, Ogra P. Influenza virus: immunity and vaccination strategies. Comparison of the immune response to inactivated and live, attenuated influenza vaccines. Scand J Immunol 2004;59:1–15.

12. Sakaue G, Hiroi T, Nakagawa Y, et al. HIV mucosal vaccine: nasal immunization with gp160-encapsulated hemagglutinating virus of Japan-liposome induces antigen-specific CTLs and neutralizing antibody responses. J Immunol 2003;170:495–502.

13. Neurath AR, Strick N, Li YY, Debnath AK. Cellulose acetate phthalate, a common pharmaceutical excipient, inactivates HIV-1 and blocks the coreceptor binding site on the virus envelope glycoprotein gp120. BMC Infect Dis 2001;1:17.

14. Scordi-Bello IA, Mosoian A, He C, et al. Mandelic acid condensation polymer: novel candidate microbicide for prevention of human immunodeficiency virus and herpes simplex virus entry. Antimicrob Agents Chemother 2005;49:3607–3615.

15. Quinones-Mateu ME, Lederman MM, Feng Z, et al. Human epithelial beta-defensins 2 and 3 inhibit HIV-1 replication. AIDS 2003;17:F39–48.

16. Stanberry LR, Spruance SL, Cunningham AL, et al. Glycoprotein-D-adjuvant vaccine to prevent genital herpes. N Engl J Med 2002;347:1652–1661.

17. Bernstein DI, Aoki FY, Tyring SK, et al. Safety and immunogenicity of glycoprotein D-adjuvant genital herpes vaccine. Clin Infect Dis 2005;40:1271–1281.

18. Whitacre CC, Gienapp IE, Orosz CG, Bitar DM. Oral tolerance in experimental autoimmune encephalomyelitis. III. Evidence for clonal anergy. J Immunol 1991;147:2155–2163.

19. Wu HY, Weiner HL. Oral Tolerance. Immunol Res 2003;28:265–284.

2

Anogenital Mucosal Immunology and Virology

Anthony Simmons

The immune system is now regarded as being divided operationally into two separate compartments (1), namely peripheral (sometimes imprecisely called systemic) immune tissue and mucosa-associated lymphoid tissue (MALT). This segregation is justified by the ability of each compartment to operate independently of the other as well as in concert. Understanding this fact is congruent with the notion that mucous membranes not only are a physical barrier to potentially harmful pathogenic organisms but also represent a vital first line of defense caused by activity of a diverse range of innate and adaptive host responses. The physiologic importance both of innate and adaptive host defenses at mucosal surfaces has made mucosal immunity a research priority.

The realization that optimal protection against some of the most prevalent viral pathogens may require mobilization of the mucosal immune system in addition to peripheral immunity is clearly germane to the design of effective vaccines. The essence of each compartment of the immune system is reviewed briefly here, to demonstrate that each arm of a host's defenses against pathogenic organisms differs in anatomic organization, mode of induction, and the balance of effector cells responsible for the eventual outcome. The overall message conveyed is that the most effective response to a pathogenic virus at mucosal surfaces is mediated by cells induced within the same compartment. Subdivision of the immune system into peripheral immune tissue and MALT should not be confused with the distinction between humoral

and cellular responses or the differences between innate and adaptive immunity (Figs. 2.1 and 2.2). In fact, the mucosa, including that of the anogenital tract, is a major site both of innate and adaptive host defenses involving cellular and soluble effectors.

However, substantive differences between the two immune systems are apparent, and these variations have profound implications for selective stimulation of the most appropriate responses for protection against any specific pathogen. This chapter documents important ways in which the mucosal immune system, particularly as it relates to the anogenital tract, differs from peripheral immunity.

Challenges for the Immune System at Mucosal Surfaces

Mucous membranes comprise the largest organ in the body with a surface area of about the size of one and a half tennis courts, about 200 times the area of the skin. To gain access to their hosts, the vast majority of pathogens must cross a mucous membrane barrier before causing an infection. In fact, blood-borne transmission, involving insect bites, injections, or transfusions, is virtually the only other way for viral pathogens to be transmitted between hosts. The mucous barrier includes the entire gastrointestinal, respiratory, and urogenital genital tracts. Here, the term *anogenital tract* is used to functionally describe an important mucosal compartment that is the site affected commonly by

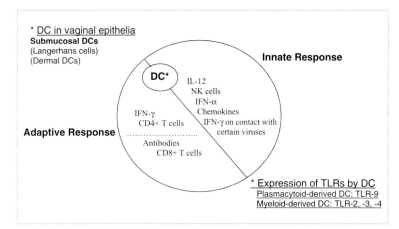

Figure 2.1. The interplay between adaptive and innate immune systems. This diagram highlights the importance of the dendritic cell (DC) (in its various guises in different tissues) in influencing both innate and adaptive immunity and some of the mechanisms by which the innate response might influence the environment in which the adaptive response develops in the vagina. Recent evidence suggests that CD11b(+) submucosal DCs but not Langerhans cells are responsible for inducing protective T_H1 responses against herpes simplex virus 2 (HSV-2) (7). Sub-mucosal DCs are rapidly recruited to infected vaginal epithelium followed by the appearance of interferon-γ (IFN-γ) secreting CD11c(+) DCs in local lymph nodes. This results in stimulation of CD4+ T cells. No other cell-type appears to present HSV peptides in the context of class II major histo-compatability complex (MHC) in draining lymph nodes. IL-12, inter-leukin-12; NK, natural killer; TLRs, Toll-like receptors.

sexually transmitted infections, including herpes simplex virus (HSV) and human papilloma virus (HPV). Human immunodeficiency virus (HIV) is not considered here because the immune response to this agent does not specifically involve mucosal immunity; HIV or HIV-infected cells usually enter the host via a breach of anogenital mucosa, and infection is systemic. However, what is dealt with is the devastating effect that compromised immunity caused by HIV has on other viral infections of the anogenital tract.

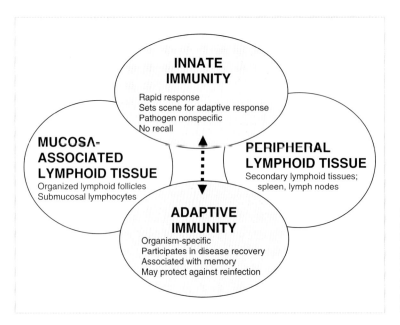

Figure 2.2. Anatomic and functional divisions of the innate and adaptive immune systems and the different roles played by each compartment. The innate response is very rapid due to TLRs on dendritic cells in particular, leading to secretion of chemokines and cytokines that attract lymphocytes to the infected area.

The Peripheral Immune System

The immune response operates throughout the body, and specific structures have evolved to be the first point of contact between immune cells derived from the bone marrow and the foreign antigens that they must encounter in order to protect the host. Here, in anatomic terms, the peripheral immune system refers to the bone marrow, lymph nodes, and spleen. A critical component of peripheral immunity is delivery of antigens to lymph nodes by the lymphatic system. The central nervous system, which does not have a conventional lymphatic system, is clearly distinct from other tissues and is often referred to loosely as an immunologically "privileged" site. It is certainly a specialized site with respect to immunity but it is not considered in detail here.

Historically, approaches to immunization against viruses have been focused on stimulation or peripheral adaptive immunity, which is a perfectly viable strategy for dealing with infections that have a prominent systemic phase in their pathogenesis, such as mumps, measles, rubella, and poliomyelitis. This approach, however, has generally been unsuccessful for protecting against primary infections involving the mucosa as a major site of disease, prominent examples being herpes simplex, papillomavirus infections, and molluscum contagiosum. It is now understood that separate populations of lymphocytes circulate and reside preferentially in one compartment or another (i.e., peripheral or mucosal).

The Mucosal Immune System

Mucosal immunity comprises a network of tissues, cell types, and soluble effector molecules that is responsible for protecting the host against infection at mucous membrane surfaces. The importance of mucosal immunity is highlighted by the fact that about 80% of activated B lymphocytes are resident not in traditional secondary lymphoid tissues, such as spleen and lymph nodes, but rather congregate in MALT (2). Mucosa-associated lymphoid tissue comprises a collection of anatomic units, and the immune tissues associated with each site are represented by their own terms (Table 2.1).

The purpose of dividing MALT into different compartments is not to cause confusion but rather to correlate function with anatomy and aid scientific communication. Prominent among the different subdivisions are gut-associated lymphoid tissue (GALT), discussed in Chapter 3, and nose-associated lymphoid tissue (NALT), both of which have been studied extensively. Salivary gland– (and salivary duct–) associated

Table 2.1. Diversity of the mucosal immune system: proteins and main subdivisions

	Entity	Label	Description
Proteins	Immunoglobulin	Ig	Antibody; three types, IgA, G, and M, are all found in
	Secretory immunoglobulin A	sIgA	The mucus cornerstone of mucosal immunity secreted into mucus
	Secretory immunoglobulin M	sIgM	IgM, an Ig that is also secreted across membranes
	Polymeric immunoglobulin receptor	pIgR	Polymeric immunoglobulin receptor found on epithelial cells
	Polymeric IgA/IgM molecules	pIgA/IgM	A complex of Ig molecules that binds to pIgR
	Secretory component	SC	Attaches to Ig enabling it to pass across mucous membranes
Compartments	Peyer's patch	PP	A prominent secondary inductive site in the wall of the gut
	Mucosa-associated lymphoid tissue	MALT	The collective name for mucosal immune tissues
	Intraepithelial lymphocyte	IEL	A population of cells commonly found amidst epithelial cells
	Gut-associated lymphoid tissue	GALT	Includes PP and appendix (secondary lymphoid tissues)
	Nasopharynx-associated lymphoid tissue	NALT	Includes tonsils and adenoids (secondary lymphoid tissues)
	Salivary duct–associated lymphoid tissue	SALT	IEL prominent at this site; a part of innate response
	Conjunctiva-associated lymphoid tissue	CALT	Forms a functional unit with LDALT to protect the eye
	Lacrimal drainage–associated lymphoid tissue	LDALT	Plays a major role in preserving integrity of ocular surface
	Larynx-associated lymphoid tissue	LALT	Presumed to be a respiratory inductive site in children
	Bronchus-associated lymphoid tissue	BALT	Defends against inhaled pathogens

MALT has been identified anatomically both in humans and nonhuman primates. In humans, the existence of salivary gland–associated MALT is disclosed by the occurrence of lymphomas arising in salivary glands, often in association with autoimmune disorders like Sjögren's syndrome. The lungs are clearly sites of common exposure to pathogens, yet bronchus-associated lymphoid tissue (BALT) is identifiable in only about 40% of normal lungs from children and adolescents, but BALT is not evident in normal adults. However, the importance of BALT is highlighted by the formation of organized lymphoid structures when the lungs become inflamed.

Inductive and Effector Phases of Mucosal Immune Responses

The host response to foreign antigens is conveniently divided into inductive and effector phases. The inductive phase of immunity refers to priming of lymphocytes by contact with cognate antigen, presented to them in so-called inductive sites, which are classically equated anatomically with lymph nodes and spleen. However, at mucous sites induction of the immune response occurs not only in draining lymph nodes but also in local collections of lymphoid cells, which are often located in histologically recognizable lymphoid structures such as Peyer's patches in the gut wall or solitary organized lymphoid follicles, which can be identified in many mucous membranes, particularly those associated with the gut. Following the inductive phase, primed lymphocytes migrate to regional lymph nodes where they proliferate and mature. The priming of lymphocytes at mucosal sites profoundly influences their subsequent behavior owing to the participation of certain mucous membrane–selective cell-adhesion molecules, which confer an ability to reaccess membranous sites. This process is referred to as "homing," and the molecules responsible are sometimes called homing molecules. Consequently, when mucosally primed lymphocytes have been expanded in lymph nodes and reenter the bloodstream they have access to mucous membranes at *effector sites* such as the lamina propria of most mucosae. Indeed, mucous membranes at many sites contain organized lymphoid tissue that can be recognized readily as lymphoid follicles on histologic examination. In the uninfected genital tract, however, no such follicles are recognized. Therefore, the obvious question that arises is: Is a more effective immune response in the anogenital mucosa engendered by immunization at a distant mucosal site such as the nose (Fig. 2.3)?

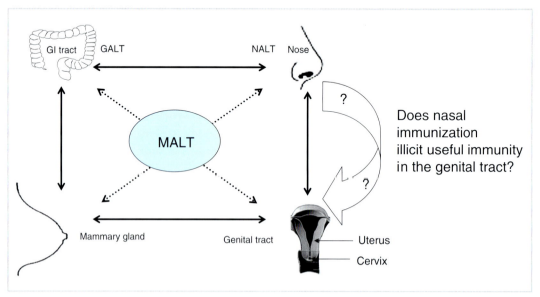

Figure 2.3. Major sites of mucosa-associated lymphoid tissue (MALT). There is a potential for immunologic interactions between mucosal tissues. This diagram emphasizes the possibility that immunization in the common mucosal immune system may be an effective means of pro-tecting distant mucosal sites. For instance, it has been demonstrated in principle that nasal immunization protects efficiently against genital herpes (20) and papilloma virus infections (18). NALT, nose-associated lymphoid tissue.

Peyer's patches in the wall of the gut are one of the first specialized collections of lymphoid tissue to be recognized. The organized mucosal lymphoid system has a follicular structure reminiscent of lymph nodes, but is additionally equipped with a specialized cell-type called an M cell that is specialized for transepithelial transport (3) and thus they play a critical role in delivery of foreign antigens to MALT, particularly in the gut.

Other mucosal sites contain diffuse lymphoid tissue that was not recognized until specialized techniques became available for its identification, and there is still much to learn about the intricacies of immunity at many mucosal sites. The genital tract falls into this category.

Both Innate and Adaptive Immunity Operate at Mucous Sites

It has become increasingly apparent that the early adaptive immune response to foreign antigens is dependent on the innate environment encountered during the inductive phase of that response (Fig. 2.1; Table 2.2). However, this may involve specific alterations in the expression of co-stimulatory molecules by antigen-presenting cells, resulting in secretion of cytokines and other ancillary molecules.

This scenario lowers the threshold for stimulation of T-cell responses (4). The overall principle illustrated here is that, in the case of primary viral infections, resistance and recovery are governed by a delicate balance between innate (germ-line encoded) host factors and adaptive immunity (Fig. 2.1).

The first line of innate resistance to a pathogenic organism is usually the physical barrier comprising skin and mucous membranes, which generally must be breached for an infection to be established. Direct infection of mucosal cells, is a prominent way in which this barrier is compromised. For instance viral infections are frequently established in mucosal cells of the nasopharynx, bronchi, gastrointestinal tract, conjunctivae, salivary glands, skin, and, of particular relevance to this chapter, the anogenital tract. It follows that developing effective vaccines and other immunotherapies for sexually transmitted infections requires a detailed understanding of immunity at mucosal surfaces. There are only a few exceptions where the body's outer tegument does not present a physical barrier to infection. For instance, arthropod vectors may directly inoculate pathogenic viruses into the bloodstream; alternatively, virally contaminated blood may be infused directly, either by transfusion or as a result of sharing needles during intravenous drug abuse.

Table 2.2. Elements of innate and adaptive immunity

	Key players	Key features	Key properties
Innate responders	Immature dendritic cells Epithelial/endothelial cells Natural killer (NK) cells Granulocytes Monocyte/macrophage lineage Mast cells and basophils	Rapid response (minutes)	Molecular pattern recognition Germ-line encoded Receptor rearrangement not necessary Nonclonal Results in cross-talk by causing expression of chemokines, co-stimulatory molecules and cytokines
Interplay between innate and adaptive responses	Mature dendritic cells γδ T cells Chemokines Cytokines Complement proteins	Cross-talk enabling innate environment to influence adaptive response	↕ Overlap (communication)
Adaptive responders generate	B-lymphocytes (e.g., mucosal IgA antibodies) T-lymphocytes (helper, suppressor, cytotoxic) Complement Chemokines Cytokines	Memory; a unique feature of acquired immunity is the requirement for several days or prior exposure to antigen	Specific antigen recognition Rearrangement of gene segments necessary to generate specificity Delayed

Principles of the Immune Response to Viruses

The Innate Immune System: Pathogen-Associated Molecular Patterns and Toll-Like Receptors

Innate immunity has developed in order to recognize a wide variety of pathogens, including viruses, without prior exposure, which rapidly engenders a number of antimicrobial and inflammatory responses. It appears that structurally conserved motifs on pathogens known as PAMPs (pathogen-associated molecular patterns) are recognized by a family of specialized receptors present on cells like macrophages. This can be likened to identification of supermarket goods using bar codes. The concerned receptors (bar-code readers) are known as Toll-like receptors (TLRs), so named because of their similarity to the Toll receptors identified more than a decade ago for their role in embryonic development of the fruit fly, *Drosophila melanogaster*. They play a critical role in host defenses against pathogens and other noxious stimuli. Ten TLRs have been identified and cloned in mammals. Each TLR appears to be involved in the recognition of a unique set of PAMPs (Table 2.3). Upon ligand binding, this family of receptors has been shown to activate a variety of signaling pathways involved in antiviral, antibacterial, antiinflammatory, and antitumor activities. As a result, several new targets for potentially useful therapeutic inventions have been identified (e.g., TLR-7 and -8).

Table 2.3. Toll-like receptors and some of their most important ligands

Toll like receptor	Prominent ligands
TLR-1	Microbial lipoproteins
TLR-2	GPI anchors
TLR-3*	Double-stranded RNA
TLR-4	Lipopolysaccharide of gram-negative bacteria
TLR-5	Bacterial flagellin
TLR-6	With TLR-2, a co-receptor for mycoplasma lipoproteins
TLR-7 and -8*	Unknown (but interacts with imidazolquinoline drugs)
TLR-9*	Unmethylated CpG
TLR-10	Unknown

* Of particular relevance to viruses.

Only a few molecular structures unique to viruses have been identified that cause activation of the innate immune system. One is double-stranded RNA (dsRNA), which is present in the genomes of some viruses and occurs transiently during replication of many others. In mammals, the receptor for dsRNA has been identified as a TLR family member, TLR-3. However, the interaction between TLR-3 and its ligand is incompletely understood. Another member of the same family of receptors, TLR-9, is known to respond to nucleic acids having unmethylated CpG motifs, including HSV DNA (4,5). In a mouse model of genital herpes, HSV-2 has been shown to interact with mucosal plasmacytoid dendritic cells via TLR-9 (6). The widespread presence of CpGs in HSV DNA appears, therefore, to mean that CpG "islands" are the PAMPs responsible for recognition of HSV-2 by innate immune cells. However, it appears that CD4+ T cells recognize specific antigenic peptides presented by submucosal dendritic cells (7), revealing an interesting difference in the roles of different dendritic cell populations in the genesis of innate and adaptive responses. Apparently HSV DNA is recognized by TLR-9 much more efficiently in the context of the virus rather than when extracted. Signaling through different TLRs appears to trigger distinct but overlapping cellular responses (8). For example, TLR-3 leads mainly to production of interferon-β and various chemokines, whereas TLR-4 stimulates to interferon-β secretion, accompanied by phagocytosis and inflammation.

Upon ligand binding, TLRs have been shown to activate a variety of signaling pathways involved in antiviral responses. For instance, recognition of HSV-2 by TLR-9 on plasmacytoid dendritic cells triggers secretion of high levels of type I interferons, which have powerful antiviral functions. Toll-like receptor–mediated response to HSV-2 infection in vivo may require a cooperative cascade of responses by, first, infected stromal cells, and, second, uninfected dendritic cells (9). Both appear to be required for TLRs to steer the antiherpes immunity toward a T_H1 response.

Adaptive Immunity

The adaptive immune response differs from innate immunity in two major respects. First, the adaptive response is antigen-specific and, second, it has memory. Hence, adaptive immunity may participate not only in recovery from primary contact with microbial and viral antigens but also in protection against reinfection.

Classically, recovery from established infection involves cytotoxic T lymphocytes that express CD8 molecules on their surfaces (10–13). CD8[+] T cells recognize virally infected cells by interacting with virally encoded proteins that have been cleaved by special machinery (proteosomes) in the infected cell into small peptides before being displayed on cell surfaces in association with class I molecules encoded by the host's major histocompatibility complex (MHC-I). This strategy has evolved to allow perusal of a site of infection or neoplasia by circulating lymphocytes and provides a unique system for distinguishing normal from abnormal (e.g., infected) cells. However, this dependence on MHC-I restricted CD8[+] T cells for recovery from infection has resulted in coevolution by many viruses of several strategies to evade cell-mediated immune surveillance.

Viruses probe the limits of immune responses and immunity because resistance, recovery, and immunity of vertebrate hosts against them comprise important elements both of natural and innate resistance combined with adaptive immune responses involving T and B cells. Therefore, viral infections have provided excellent opportunities to assess the biology, physiology, and molecular aspects of immune responses and help in characterizing the three basic parameters of immunity, namely specificity, tolerance, and memory.

The Mucosal Immune System in the Genital Tract

Although the genitourinary tract is considered to be a component of the mucosal immune system, the genital tracts both of males and females have been studied in detail only relatively recently. The genitourinary immune system displays characteristic features that are distinct from those of other typical mucosal sites or the peripheral compartment. For example, antibodies in female genital tract secretions are derived not only locally from resident plasma cells and also from the blood, as reflected in their structural heterogeneity. For instance female genital tract secretions contain a significant proportion of monomeric immunoglobulin A

(IgA), which is derived by extravasation of immunoglobulin directly from the bloodstream. Furthermore, many aspects including the distribution and properties of immunocompetent cells, and the proportions of Ig isotypes and their molecular forms, are under hormonal influence, especially in females (14–16).

It might be reasonable to assume that these factors explain the otherwise surprising degree of protection against genital warts provided by immunization with HPV virus-like particles (VLPs), which cannot replicate and would thus be expected to stimulate primarily an antibody response against capsid antigens (17). However, it was shown by Dupuy et al. (18,19) that intranasal immunization with HPV-16 VLPs or the HPV *L1* gene elicit cellular responses that could be detected in vaginal and splenic lymphocytes capable of cytotoxicity and interferon-γ (IFN-γ) secretion.

Cross-Talk Between Mucosal and Peripheral Immunity

The inductive phase of mucosal immunity leads to trafficking of primed lymphocytes and antigen-presenting cells through regional lymph nodes, providing opportunity for interactions (cross-talk) between the peripheral and mucosal compartments of the immune system (Table 2.2). An important and topical issue that arises is whether mucosal immunization is the most effective way to protect against mucosal infections. The well-documented circulation and homing of lymphocytes primed in the mucosa to mucosal effector sites has led investigators to approach the question of whether nasal rather than systemic immunization (Fig. 2.3) is more protective against genital herpes (20,21) and papillomavirus infections (18).

Cells and Molecules of Mucosal Immunity

Secretory antibodies are the cornerstone of adaptive mucosal immunity. Immunoglobulin A, and to a lesser extent IgM and IgG, antibodies secreted into mucus adherent to mucosal surfaces are the first specific line of defense against many potentially invasive microbes. If this barrier is breached, antigens meet corresponding serum-derived IgG antibodies within the mucosa itself, resulting in formation of immune complexes that activate complement with resultant formation locally of mediators of inflammation. Fortunately, development of persistent inflammation that would be detrimental to the host is regulated by competition for antigen in the mucosal stroma by serum-derived IgA and locally produced dimeric or monomeric IgA. Selective expression of adhesion molecules by vascular endothelial cells in the mucosa regulates the preferential extravasation of B and T cells belonging to the mucosal immune compartment.

Mucosal secretory IgA (sIgA) antibodies generally lack complement-activating properties (22–24) and cause exclusion of invading pathogens by noninflammatory mechanisms. The cellular basis for this major first-line specific defense is the fact that exocrine glands and secretory mucosae contain most of the body's activated B cells, particularly the gut lamina propria where at least 80% of all Ig-producing immune cells are found (2). At all exocrine sites, IgA secreting cells produce mainly dimers and larger polymeric IgA (pIgA) that can be transported actively through secretory epithelial cells by a receptor-mediated mechanism. It has been estimated in humans that more IgA is synthesized and secreted each day than IgG and IgM combined (25).

Prominent Viral Infections of the Anogenital Region

Herpes Simplex

Genital herpes (GH) is generally (but not exclusively) caused by infection with HSV type 2 (HSV-2) (26–28). It is estimated from type-specific serologic studies that 22% of adults in the United States are infected with HSV-2. All infected persons can shed HSV periodically from their genitalia or other dermatomes below the waist irrespective of recognizable symptoms, resulting in a vast reservoir of unrecognized and therefore undiagnosed GH in the human population. The pathogenesis of herpes simplex is complicated by frequent shuttling of virus between mucocutaneous surfaces and the sensory nervous system. When skin or mucous membranes first come into contact with HSV,

virus not only causes a primary infection, which may be inapparent, at the site of acquisition, but also simultaneously is shuttled in a retrograde direction along axons to sensory nerve ganglia. In sensory neurons a dormant, nonreplicating, infection (referred to as latency) is established for the life of the host. The significance of latency is that it represents a reservoir of HSV genomes that reactivates periodically. Spread of virus between neurons of a ganglion, followed by reactivation of HSV, causes infectious virus to be returned to mucocutaneous surfaces by anterograde transport along nerves, which may cause recurrent disease anywhere in the same dermatome as that of the primary infection (Figs. 2.4 and 2.5). Alternatively, HSV may be shed from the mucosa and transmitted to others without causing overt symptoms.

Hence, mucosal immunity is presented with unique challenges: it is required not only to promote recovery from primary disease but also to respond rapidly to virus released from cutaneous nerve endings of an infected host.

To summarize, herpes simplex has two discrete phases, namely primary infection of the immunologically naive host and recurrent infections in the immune host.

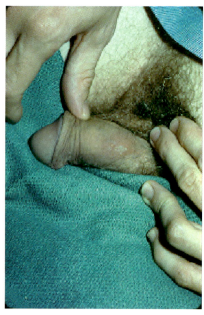

Figure 2.4. Typical circumspect appearance of recurrent genital herpes (GH) lesions on the penis, kept in check by mucosal host responses. More often recurrences are so mild that they are unrecognized by the patient as GH.

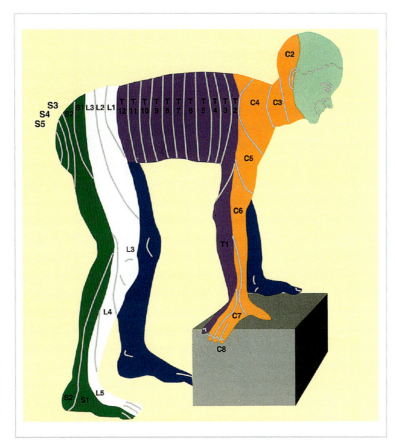

Figure 2.5. Diagram showing the dermatomes of the human body, which illustrates the areas of skin innervated by each sensory nerve ganglion. The genitalia are innervated by S2 and S3, from which it can be seen easily from this diagram how latent infection of these ganglia readily causes recurrent lesions on the buttocks, thighs, and many other sites below the waist.

An HSV-2 infection causes avid innate and adaptive immune responses in the female genital tract. It results in cervical antibody responses to the virus in females, but whereas the specificity profiles of cervical and serum IgG antibodies are similar, those of cervical and serum IgA differ, suggesting a local origin for cervical IgA but not for IgG (29). Anamnestic (memory) cervical antibody responses have been detected in HSV-1–seropositive women newly infected with genital HSV-2 (30). Antibodies might be expected to be beneficial both at promoting recovery from primary infection and at preventing recurrence, yet neither appears to be the case. In the extreme case of congenital agammaglobulinemia, humans generally recover normally from primary infections, and only a few cases of recalcitrant herpes lesions are apparent in the literature (31). Recurrences in these individuals are generally neither unduly frequent nor severe. The reasons behind this apparent anomaly are becoming clearer and they involve more than the simplistic model of a virus hiding in neural tissues out of reach of antibodies. At the same time that it was realized that activated lymphocytes, regardless of antigen specificity, readily breach the blood–brain barrier (32), the molecular basis for the inability of antibodies to attack HSV-infected cells has been delineated as yet another specific example of an immune evasion strategy coevolved by a large DNA virus and its host.

Herpes simplex virus has evolved stealth mechanisms to evade attack of infected cells by antibodies. The fact that viruses such as HSV can persist in the immune host and cause recurrent infections of skin and mucous membranes (Fig. 2.4), despite a florid systemic virus-specific antibody response (33), highlights the point that mechanisms have evolved by which HSV is able to evade antibody-mediated defenses. In contrast, symptomatic primary infection tends to be more severe (Fig. 2.6). The molecular basis of these mechanisms includes the facts that two of

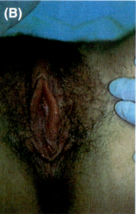

Figure 2.6. A: Primary GH in an immunocompetent male showing moderate swelling (balanitis). B: Symptomatic primary GH in a female patient illustrating extensive bilateral distribution of lesions. (*Source:* Courtesy of S. Tyring.)

the glycoproteins encoded by HSV-1, gC and gE, interact with complement and immunoglobulin Fc, respectively (34,35). These properties have profound effects both on innate and acquired immunity, including interfering with complement components C1q, C3, C5, and properdin, and blocking antibody-dependent cellular cytotoxicity. Systemic antibodies, however, may help to prevent life-threatening dissemination of HSV in neonates (36).

Recovery from Genital Herpes Simplex Virus Infections and Its Dependence on T Lymphocytes

Observations made in humans immunocompromised by diseases or therapeutic agents causing selective depression of cell-mediated immunity made it clear decades ago that cell-mediated immune mechanisms are paramount for recovery from herpes simplex (28). Subsequently, studies in experimental animals led to the same conclusion and have gone a long way toward uncovering some of the mechanisms involved (37). Graphic examples have been provided by the global pandemic of AIDS, where HSV can be recalcitrant and life-threatening (Fig. 2.7). Patients with prolonged disease who are treated but cannot resolve acute infections tend to develop HSV strains that are resistant to antiviral compounds such as acyclovir and its derivatives (Fig. 2.8).

Recurrent Herpes Simplex

Herpes simplex virus causes recurrent blistering lesions of the skin and mucous membranes in the same dermatome that was acquired, that is, the primary infection.

In contrast to patients with defects purely of humeral defenses, the common sexually transmitted infections that cause symptoms primarily in genital skin and mucosa are all significantly

Figure 2.7. Recurrent herpes simplex virus (HSV) infection on the buttock of an HIV-positive person, demonstrating the dependence of resolution of lesions on intact (mucosal) cellular immunity. (*Source:* Courtesy of S. Tyring.)

Figure 2.8. Persistent mucocutaneous lesions in the genital region are characteristic of resistance to antiviral drugs used in patients with compromised immunity. Shown here are recalcitrant lesions in the groin caused by acyclovir-resistant HSV, which cannot be eradicated by the immune system of an HIV-infected patient. Prolonged viral growth in the presence of drug leads to resistance usually by mutations in the gene encoding an enzyme known as viral thymidine kinase (TK), which must phosphorylate acyclovir before it can interfere with HSV replication. Fortuitously, TK mutants cannot reactivate from latency and therefore do not readily survive in the general population. (*Source:* Courtesy of S. Tyring.)

worse in hosts with impaired cell-mediated immunity. For instance, HSV has long been recognized as a cause of severe, progressive ulcers and life-threatening cutaneous ulcers in persons with leukemia, with other disorders (particularly HIV infection; Fig. 2.7), or using drugs that compromise the functions of T cells in mucous membranes.

Genital Warts

Papillomaviruses cause some of the most common sexually transmitted infections in the world. The immune system often fails to control papillomavirus infections of the anogenital tract, resulting in the appearance of unsightly genital warts. Of the 100+ HPV types identified to date, approximately 30 are spread by sexual contact. Most are harmless and reminiscent of HSV; many of those infected with HPV are asymptomatic. However, some HPV infections result in the appearance of warts in genital areas, including the vagina, cervix, vulva, penis, and rectum. A minority of HPVs lead to abnormal Pap smears and a handful (particularly HPV-16 and -18) are associated with cancers of the cervix, vulva, vagina, anus, or penis.

The American Social Health Association estimates that at least 20 million people in the U.S. are infected with HPV and approximately 5.5 million new cases of sexually transmitted HPV infections are reported every year. The prevalence of new infections appears to be rising (38). Genital HPV infections are spread by direct contact during oral, vaginal, or anal sex with an infected partner. Genital warts develop within 3 months in approximately two thirds of people who have sexual contact with an HPV-infected partner. In women, warts generally occur around the vagina, on the uterine cervix, or in the perianal region. In men, genital warts are less common but if present, they usually are found on the tip of the penis, the penile shaft, the scrotum, or around the anus. Warts may also develop rarely in the mouth or throat following oral sex with an infected person.

Though warts are characteristically devoid of inflammation, the inflammatory response observed histologically at the base of resolving warts and the uncontrolled behavior of warts in immunocompromised persons suggest that the host response may play an essential, albeit protracted, role in their control. In the setting of impaired immunity, HPV infections may also become severe and difficult to eradicate (Fig. 2.9).

Because HIV-infected and immunocompromised patients have an increased incidence of persistence and progression to neoplastic change, they need to be monitored more closely over time. For routine screening, at least yearly Pap smears and visual inspection of the external genitalia should be performed. For a number of reasons, including poor follow-up care after release from prison, the current standard of practice for incarcerated HIV-infected women is to perform Pap smears every 6 months. Most medical institutions have colposcopy available on site. Some authors recommend a baseline colposcopy for all HIV-infected women with presence of HPV infection. There appears to be a reduction of accuracy of Pap smears in this group secondary to obscuring inflammation from cervicitis. Colposcopy should be done on all women with abnormal Pap smears including atypia and low-grade dysplasia. All dysplasias should be treated aggressively.

The giant condyloma of Buschke and Löwenstein (GCBL) is most likely a florid example of the inability of the host immune response to control genital HPV infection. It is a locally destructive verrucous lesion that typically appears on the penis but may occur elsewhere in

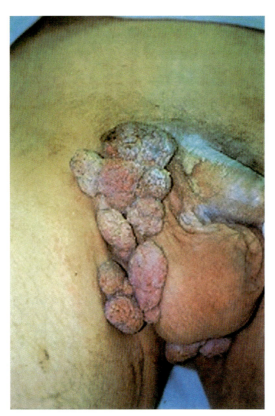

Figure 2.9. Uncontrolled growth of genital warts in a patient with compromised cell-mediated immunity caused by HIV. (*Source:* Courtesy of S. Tyring.)

the anogenital region (Fig. 2.10). It is most commonly considered to be a regional variant of verrucous carcinoma and oral papillomatosis.

The cause of GCBL is not known with certainty but the favored hypothesis is that papillomaviruses are involved because HPV types 6, 11,

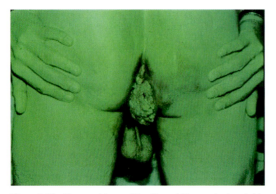

Figure 2.10. Giant condyloma of Buschke and Löwenstein. A probable manifestation of uncontrolled HPV infection, which might be considered a failure of mucosal immunity. (*Source:* Courtesy of S. Tyring.)

16, 18, and on one occasion type 54 have been shown to colocalize with the lesions (39). The E6 proteins of HPV-6 and HPV-11 bind to the p53 tumor-suppressor protein less efficiently than that of HPV-16 and HPV-18 but could in theory lead to accelerated degradation of the p53 protein. The E6 protein also inhibits p53 transcription. An alternative hypothesis for its pathogenesis is a spontaneous mutation in the p53 protein leading to clonal cell proliferation. Several reports have shown some overexpression of p53 in genital warts including one that studied GCBL specifically (40) and squamous cell carcinomas (SCCs), but a recent study concluded that, despite the overexpression, p53 mutations were not present. Other implicated agents are chronic chemical exposure, chronic irritation and poor hygiene.

Although it is slow growing and seldom metastasizes, GCBL is highly destructive to adjacent tissues. The most common site for GCBL is the glans penis, but it can be found on any anogenital mucosal surface, including the vulva, vagina, rectum, scrotum, and bladder. Frequently, it is mistaken for a recalcitrant condyloma. In the U.S., GCBL is fortunately rare, accounting for less than 24% of penile cancers, which, in turn, are 0.3% to 0.5% of male malignancies. Verrucous carcinoma, however, has been assessed as accounting for approximately 50% of all low-grade SCCs of the penis; GCBLs located outside the penis are much more infrequent. There have been less than 100 cases of GCBL arising in the perianal region, vulva, or bladder.

Our understanding of the role of mucosal immunity in HPV-related cervical diseases is in its infancy. Naturally produced serum anticapsid antibodies appear to be ineffective at protecting either HIV$^+$ or HIV$^-$ women from subsequent infection, presumably because of the low level of antibodies induced by natural HPV infection (41). Enigmatically, nonreplicating virus like particles, which induce a high level of antibody, have been shown to protect women against HPV-16 infection and subsequent HPV-16–related intraepithelial neoplasia (17), although it appears that nasal administration of HPV-16 VLPs, despite being unable to replicate, induce specific IFN-γ secreting CD4$^+$ T cells and cytotoxic CD8$^+$ cells in the vagina (18). This observation addresses two issues simultaneously. First, VLPs are able to induce cell-mediated as well as antibody responses, and second, nasal immunization may be an effective immunization strategy for protecting against HPV at distant mucosal sites, congruent with data on HSV.

Innate immunity may also be very important for protecting the genital mucosa against progression of papillomavirus infections because imiquimod, which is an immunomodulatory compound that stimulates the immune system through TLR-7, causes regression of warts (42), strongly implicating innate immune mechanisms in their control. However, it was shown recently that combining a cytotoxic T lymphocyte (CTL) epitope with epicutaneous administration of imiquimod causes avid priming of CD8$^+$ cells with a wide range of activities from proliferation to cytotoxicity and cytokine production (43). Further imiquimod is known to be able to interact with TLR-7 on the surfaces of epidermal dendritic cells, but the consequences of this interaction on development of antiviral adaptive immunity is unexplored. Memory correlates best with antigen-dependent maintenance of elevated antibody titers in serum and mucosal secretions, or with an antigen-driven activation of T cells.

Molluscum Contagiosum

Molluscum contagiosum virus (MCV), a member of the poxvirus family, causes one or more small skin lesions, which generally resolve without treatment in a matter of months. Once a disease primarily of children, MCV infection has surfaced as a genital infection in adults (44). Molluscum contagiosum is sexually transmitted by direct contact between skin or mucous membranes and active lesions. Unlike other viral sexually transmitted infections, MCV may be transmitted from inanimate objects such as towels and clothing that come in contact with lesions. Its transmission has been associated with swimming pools and sharing baths with an infected person. It also can be transmitted by autoinoculation. Hence, crops of lesions are quite characteristic. The incubation period ranges between 1 week and 6 months, with an average of 2 months.

Lesions are usually present on the thighs, buttocks, groin, and lower abdomen, and may occasionally appear on the external genital and perianal region. The lesions, which develop slowly, tend to be flesh-colored or gray-white, and generally cause few problems. Lesions may last for only 2 weeks or as long as 4 years, with an average duration of 2 years. Although they may cause itching or tenderness in the area, in most cases the lesions pose no significant problems in immunocompetent hosts. Lesions may recur, but it is not clear whether this is due to reinfection, exacerbation of subclinical infection, or reactivation of latent infection.

Molluscum Contagiosum and the Immune System

In people with HIV infection, molluscum contagiosum is often a progressive disease, and the clinical features may be atypical in this group of patients (45,46). The lesions often are large and may be verrucous and markedly hyperkeratotic (Fig. 2.11). In addition, immunostimulatory

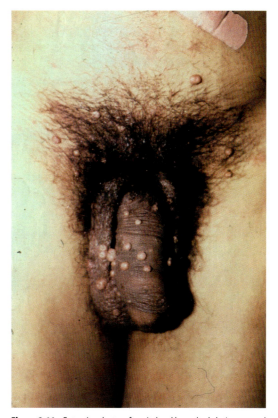

Figure 2.11. Extensive cluster of genital and lower body lesions caused by molluscum contagiosum in an HIV-positive patient demonstrating the importance of the immune system in its control. (*Source:* Courtesy of S. Tyring.)

compounds like imiquimod have been reported to be useful in the management of mollusca (47). Finally, MCV has been found to encode proteins that are homologous to CD150 (48), a major receptor involved in the pathway that activates interferon-γ, a key cytokine for viral immunity. This presumably represents yet another example of coevolution by a virus and its host to subvert a potentially protective mechanism that would otherwise lead to rapid elimination of the pathogen by cells known to be resident in, or quickly recruited to, mucosal sites. Hence the conclusion that the mucosal immune system is involved in control of MCV infections seems incontrovertible, but little is known about the specific mechanisms involved. The recent full sequencing of the MCV genome may result in a rapid change in this situation and the prospect of more effective therapies.

Conclusion

The mucosal immune system works largely autonomously from systemic immunity, protecting the largest organ in the body (skin and mucous membranes including the gastrointestinal tract and genital mucosa) from invading organisms. In total, this system occupies an area approximately equivalent to one and a half tennis courts. Both innate and adaptive immunity operate in the mucosa, but the cell types involved are subtly different. For instance, intraepithelial lymphocytes frequently use T-cell receptors comprising γδ rather than αβ chains and may have important regulatory roles. The bulk (80%) of activated lymphocytes (T and B) is found in mucosal tissues rather than classic secondary lymphoid organs like lymph nodes and spleen. Lymphocytes primed in the mucosal compartment expand in lymph nodes draining the site and are equipped with homing receptors that allow them to enter preferentially mucosal effector sites, once they leave the lymph node and enter the bloodstream via the thoracic duct. In the anogenital tract they reside primarily as solitary intraepithelial lymphocytes rather than organized lymphoid follicles. Although the primary defense mounted by the anogenital mucosa has been assumed to be antibodies specially adapted for secretion into mucus, it is now appreciated that innate and T-cell–mediated mechanisms are more important against viruses, such as HSV, HPV, and MCV.

References

1. Abraham R, Ogra PL. Mucosal microenvironment and mucosal response. Am J Trop Med Hyg 1994;50(5 suppl):3–9.
2. Brandtzaeg P, Halstensen TS, Kett K, et al. Immunobiology and immunopathology of human gut mucosa: humoral immunity and intraepithelial lymphocytes. Gastroenterology 1989;97(6):1562–1584.
3. Kraehenbuhl JP, Neutra MR. Epithelial M cells: differentiation and function. Annu Rev Cell Dev Biol 2000; 16:301–332.
4. Herbst MM, Pyles RB. Immunostimulatory CpG treatment for genital HSV-2 infections. J Antimicrob Chemother 2003;52(6):887–889.
5. Hochrein H, Schlatter B, O'Keeffe M, et al. Herpes simplex virus type-1 induces IFN-alpha production via Toll-like receptor 9-dependent and -independent pathways. Proc Natl Acad Sci U S A 2004;101(31):11416–11421.
6. Lund J, Sato A, Akira S, Medzhitov R, Iwasaki A. Toll-like receptor 9-mediated recognition of herpes simplex virus-2 by plasmacytoid dendritic cells. J Exp Med 2003;198(3):513–520.
7. Zhao X, Deak E, Soderberg K, et al. Vaginal submucosal dendritic cells, but not Langerhans cells, induce protective Th1 responses to herpes simplex virus-2. J Exp Med 2003;197(2):153–162.
8. Akira S, Sato S. Toll-like receptors and their signaling mechanisms. Scand J Infect Dis 2003;35(9):555–562.
9. Sato A, Iwasaki A. Induction of antiviral immunity requires Toll-like receptor signaling in both stromal and dendritic cell compartments. Proc Natl Acad Sci USA 2004;101(46):16274–16279.
10. Doherty PC, Christensen JP, Belz GT, Stevenson PG, Sangster MY. Dissecting the host response to a gammaherpesvirus. Philos Trans R Soc Lond B Biol Sci 2001; 356(1408):581–593.
11. Doherty PC, Topham DJ, Tripp RA, Cardin RD, Brooks JW, Stevenson PG. Effector CD4+ and CD8+ T-cell mechanisms in the control of respiratory virus infections. Immunol Rev 1997;159:105–117.
12. Doherty PC, Riberdy JM, Belz GT. Quantitative analysis of the CD8+ T-cell response to readily eliminated and persistent viruses. Philos Trans R Soc Lond B Biol Sci 2000;355(1400):1093–1101.
13. Stevenson PG, Belz GT, Altman JD, Doherty PC. Changing patterns of dominance in the CD8+ T cell response during acute and persistent murine gammaherpesvirus infection. Eur J Immunol 1999;29(4):1059–1067.
14. Richardson J, Kaushic C, Wira CR. Estradiol regulation of secretory component: expression by rat uterine epithelial cells. J Steroid Biochem Mol Biol 1993; 47(1–6):143–149.
15. Richardson JM, Kaushic C, Wira CR. Polymeric immunoglobin (Ig) receptor production and IgA transcytosis in polarized primary cultures of mature rat uterine epithelial cells. Biol Reprod 1995;53(3):488–498.
16. Kaushic C, Richardson JM, Wira CR. Regulation of polymeric immunoglobulin A receptor messenger ribonucleic acid expression in rodent uteri: effect of sex hormones. Endocrinology 1995;136(7):2836–2844.

17. Koutsky LA, Ault KA, Wheeler CM, et al. A controlled trial of a human papillomavirus type 16 vaccine. N Engl J Med 2002;347(21):1645–1651.

18. Dupuy C, Buzoni-Gatel D, Touze A, Bout D, Coursaget P. Nasal immunization of mice with human papillomavirus type 16 (HPV-16) virus-like particles or with the HPV-16 L1 gene elicits specific cytotoxic T lymphocytes in vaginal draining lymph nodes. J Virol 1999;73(11):9063–9071.

19. Dupuy C, Buzoni-Gatel D, Touze A, Le Cann P, Bout D, Coursaget P. Cell mediated immunity induced in mice by HPV 16 L1 virus-like particles. Microb Pathog 1997; 22(4):219–225.

20. Milligan GN, Dudley-McClain KL, Chu CF, Young CG. Efficacy of genital T cell responses to herpes simplex virus type 2 resulting from immunization of the nasal mucosa. Virology 2004;318(2):507–515.

21. Gallichan WS, Johnson DC, Graham FL, Rosenthal KL. Mucosal immunity and protection after intranasal immunization with recombinant adenovirus expressing herpes simplex virus glycoprotein B. J Infect Dis 1993;168(3):622–629.

22. Mestecky J, Russell MW, Jackson S, Brown TA. The human IgA system: a reassessment. Clin Immunol Immunopathol 1986;40(1):105–114.

23. Brandtzaeg P, Baklien K, Bjerke K, Rognum TO, Scott H, Valnes K. Nature and properties of the human gastrointestinal immune system. In: Miller K, Nicklin S, eds. Immunology of the Gastrointestinal Tract. Boca Raton, FL: CRC Press, 1987:1–86.

24. Kilian M, Russell MW. Function of mucosal immunoglobulins. In: Ogra PL, Mestecky J, Lamm ME, Strober W, McGhee JR, Bienenstock J, eds. Handbook of Mucosal Immunology. Orlando, FL: Academic Press, 1994:127–137.

25. Conley ME, Delacroix DL. Intravascular and mucosal immunoglobulin A: two separate but related systems of immune defense? Ann Intern Med 1987;106(6):892–899.

26. Ashley R, Benedetti J, Corey L. Humoral immune response to HSV-1 and HSV-2 viral proteins in patients with primary genital herpes. J Med Virol 1985;17(2): 153–166.

27. Corey L, Adams HG, Brown ZA, Holmes KK. Genital herpes simplex virus infections: clinical manifestations, course, and complications. Ann Intern Med 1983; 98(6):958–972.

28. Simmons A, Osman MN, Stanberry LR. Genital Herpes. In: Gorbach SL, Bartlett JG, Blacklow NR, eds. Infectious Diseases. Philadelphia: Lippincott Williams & Wilkins, 2004:904–915.

29. Brandtzaeg P. Mucosal immunity in the female genital tract. J Reprod Immunol 1997;36(1–2):23–50.

30. Ashley R, Wald A, Corey L. Cervical antibodies in patients with oral herpes simplex virus type 1 (HSV-1) infection: local anamnestic responses after genital HSV-2 infection. J Virol 1994;68(8):5284–5286.

31. Kraemer CK, Benvenuto C, Weber CW, Zampese MS, Cestari TF. Chronic cutaneous herpes simplex in a patient with hypogammaglobulinemia. Skin Med 2004; 3(2):111–113.

32. Irani DN, Griffin DE. Regulation of lymphocyte homing into the brain during viral encephalitis at various stages of infection. J Immunol 1996;156(10):3850–3857.

33. Lubinski JM, Jiang M, Hook L, et al. Herpes simplex virus type 1 evades the effects of antibody and complement in vivo. J Virol 2002;76(18):9232–9241.

34. Kostavasili I, Sahu A, Friedman HM, Eisenberg RJ, Cohen GH, Lambris JD. Mechanism of complement inactivation by glycoprotein C of herpes simplex virus. J Immunol 1997;158(4):1763–1771.

35. Weeks BS, Sundaresan P, Nagashunmugam T, Kang E, Friedman HM. The herpes simplex virus-1 glycoprotein E (gE) mediates IgG binding and cell-to-cell spread through distinct gE domains. Biochem Biophys Res Commun 1997;235(1):31–35.

36. Ashley RL, Dalessio J, Burchett S, et al. Herpes simplex virus-2 (HSV-2) type-specific antibody correlates of protection in infants exposed to HSV-2 at birth. J Clin Invest 1992;90(2):511–514.

37. Simmons A, Tscharke D, Speck P. The role of immune mechanisms in control of herpes simplex virus infection of the peripheral nervous system. Curr Top Microbiol Immunol 1992;179:31–56.

38. Koshiol JE, Laurent SA, Pimenta JM. Rate and predictors of new genital warts claims and genital warts-related healthcare utilization among privately insured patients in the United States. Sex Transm Dis 2004; 31(12):748–752.

39. Haycox CL, Kuypers J, Krieger JN. Role of human papillomavirus typing in diagnosis and clinical decision making for a giant verrucous genital lesion. Urology 1999;53(3):627–630.

40. Pilotti S, Donghi R, D'Amato L, et al. HPV detection and p53 alteration in squamous cell verrucous malignancies of the lower genital tract. Diagn Mol Pathol 1993;2(4): 248–256.

41. Viscidi RP, Snyder B, Cu-Uvin S, et al. Human papillomavirus capsid antibody response to natural infection and risk of subsequent HPV infection in HIV-positive and HIV-negative women. Cancer Epidemiol Biomarkers Prev 2005;14(1):283–288.

42. Smith KJ, Hamza S, Skelton H. The imidazoquinolines and their place in the therapy of cutaneous disease. Expert Opin Pharmacother 2003;4(7):1105–1119.

43. Rechtsteiner G, Warger T, Osterloh P, Schild H, Radsak MP. Cutting edge: priming of CTL by transcutaneous peptide immunization with imiquimod. J Immunol 2005;174(5):2476–2480.

44. Laxmisha C, Thappa DM, Jaisankar TJ. Clinical profile of molluscum contagiosum in children versus adults. Dermatol Online J 2003;9(5):1.

45. Smith KJ, Skelton H. Molluscum contagiosum: recent advances in pathogenic mechanisms, and new therapies. Am J Clin Dermatol 2002;3(8):535–545.

46. Smith KJ, Skelton HG, Yeager J, et al. Cutaneous findings in HIV-1-positive patients: a 42-month prospective study. Military Medical Consortium for the Advancement of Retroviral Research (MMCARR). J Am Acad Dermatol 1994;31(5 pt 1):746–754.

47. Hengge UR, Cusini M. Topical immunomodulators for the treatment of external genital warts, cutaneous warts and molluscum contagiosum. Br J Dermatol 2003; 149(suppl 66):15–19.

48. Sidorenko SP, Clark EA. The dual-function CD150 receptor subfamily: the viral attraction. Nat Immunol 2003;4(1):19–24.

3

Gastrointestinal Mucosal Immunology

David A. Bland, Carlos A. Barrera, and Victor E. Reyes

It has been contemplated that if every bit of matter on the surface of the planet were made invisible *except* for the kingdom Monera, the shape and form of everything around us would still be seen in ghost form due to the ubiquitous presence of bacteria on Earth.

Thus, humans have developed a complex immune system that enables us to coexist with these microbes. Our skin and airways are constantly exposed to bacteria and have tools in place to block or neutralize this constant assault. However, no tissue is exposed to a more diverse array of foreign matter, including bacteria, than the mucosal surfaces of our gastrointestinal (GI) tract. From birth, the gastrointestinal mucosa is continuously challenged by antigens that include dietary antigens, normal flora, and pathogens. Because of its extremely large surface area, created by the complex involution of crypts and villi, and lined with epithelial cells, the GI tract is susceptible as a site of colonization and entry for many infectious agents. Some pathogens colonize the surface of the epithelium and others reside within or invade through the epithelial barrier. The mucosal immune system must be able to recognize these pathogens while at the same time it must ignore commensal bacteria and dietary antigens. Therefore, it is not surprising that the local immune system servicing the GI tract is highly specialized and of central importance to human health.

Like many viruses, several enteric bacteria have evolved mechanisms to actively evade or disrupt the host immune response. Furthermore, the commensal bacteria colonizing the human GI tract add complexity to the local immune response by acting as a basic defense against pathogenic species as well as the causation of disease should the host immune system initiate inappropriate immunity against it.

Although the structure and function of the GI mucosal immune system is similar to the greater immune system, there are features of both inductive and effector immune sites in the gut that allow the GI immune system to respond appropriately to the unique challenge of enteric bacteria. Furthermore, recent studies suggest that control of the mucosal immune system may be much more decentralized than previously thought. Specialized immune sites such as the gut-associated lymphoid tissue (GALT), working in tandem with immune cells distinct to the GALT, such as intraepithelial lymphocytes, are redefining many immunologic paradigms. Concepts such as the interaction of the innate and acquired immune systems, and centralized immunologic control are currently being questioned in the context of the mucosal immune system. Moreover, we now know that a distinct population of T lymphocytes homes specifically to the GI tract. These specialized immune cells work with the structural cells of the gut to shape the host response to enteric pathogens.

This chapter outlines the elements responsible for the innate and acquired immune responses within the GI tract and how these are called into play during representative infections. The mucosal response in the GI tract must be able to differentiate dangerous antigens from commensal organisms and food antigens through suppression of immune responses directed at the latter. Unfortunately, overt activity of these immune defenses leads to common clinical problems such as food allergies and inflammatory bowel disease.

Innate Immune Defenses Within the Gastrointestinal Mucosa

An important function of mucosal immunity is that the innate immune elements aid in protecting the host from the antigenic load bombarding the GI interface. The innate defenses include a physical barrier provided by mucus, which is secreted by goblet cells, and provides a protective cover to the epithelium. Potential pathogens may become trapped in the mucus and are thus prevented from accessing the underlying epithelium. In addition to the mucus, a battery of proteolytic enzymes such as trypsin, chemotrypsin, and pepsin together with bile salts and low-pH extremes aid in the protection against potential pathogens. There are other important noncellular, humoral factors that contribute to the innate defenses at mucosal surfaces whose role will be described. The mucosal epithelium has long been regarded as a physical barrier to invading pathogens, but studies over the last decade have shown that epithelium contributes to host protection more than initially thought. Evidence accumulated from several independent studies suggests that the epithelium may directly influence adaptive immune responses, as will be reviewed below.

The Gastrointestinal Epithelium

Structure and Function of Mucosal Epithelium

The architecture of a selectively permeable epithelial cell barrier is essential in preventing the uncontrolled passage into the host of partially digested food, bacteria, and bacterial products, and in regulating fluid and electrolyte absorption and secretion. Tight and adherens junctions, which are located near the apical surface of columnar epithelial cells, separate the paracellular space between those cells from the intestinal lumen. Cell–cell adhesion at these junctions is maintained by a complex of pro-

teins, such as occludin, zonula occludens-1 and -2 (ZO-1 and -2), and members of the claudin family. These junctional proteins can be regulated by products of inflammatory and immune responses. For example, proinflammatory cytokines such as interferon-γ downregulate the expression of the junction component ZO-1 (1) (Fig. 3.1). This time-dependent decrease corresponds with a significant decrease in transepithelial resistance as evidenced by an increase in mannitol flux. In humans, expression of the tight junction protein occludin appears to be diminished in inflammatory bowel disease, suggesting that downregulation of epithelial occludin may play a role in enhanced paracellular permeability and in the neutrophil transmigration that is observed during active inflammatory bowel disease (2).

When damage to the epithelial cell layer occurs, several epithelial cell factors are produced that are known to influence epithelial cell proliferation, migration, and wound healing. These include growth factors such as epidermal growth factor (EGF), which exerts its effect partially through an elevated secretion of transforming growth factor-β (TGF-β) (3), fibroblast growth factor (FGF), hepatocyte growth factor (HGF), and intestinal trefoil factor (ITF). Intestinal trefoil factor is secreted toward the apical side of the epithelium, and its function is to increase the migration of epithelial cells toward sites of injury (4).

In the course of infection or insult to the epithelial layer, there is an increased influx of immune and inflammatory cells into the subepithelial compartment at the site of the insult. For example, in response to bacterial infection, human intestinal epithelial cells produce mediators, such as chemokines, that are essential for the onset of acute mucosal inflammation. In this case, a number of enteroinvasive bacteria, and some noninvasive bacterial pathogens that interact with the epithelial cell membrane induce epithelial cells to upregulate the production and release of potent chemokines for attraction of neutrophils (e.g., CXCL1/GROα, CXCL2/GROβ, CXCL5/ENA78, CXCL8/IL-8) (5,6), monocytes/macrophages (e.g., CCL2/MCP-1) (6,7), and immature CCR6-expressing dendritic cells (CCL20/MIP-3α) (8). Furthermore, under inflammatory conditions, human intestinal epithelial cells can also produce interferon-γ (IFN-γ–inducible chemokines (e.g., CCL9/Mig, CCL10/IP-10, CCL11/I-TAC) (9,10) that

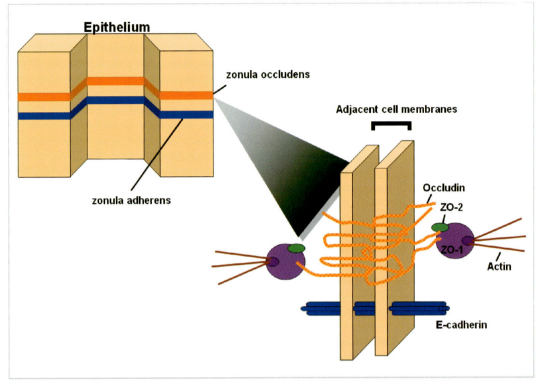

Figure 3.1. Junctional complexes within intestinal epithelial cells. The transmembrane protein occludin is the sealing protein of the zonula occludens (ZO) and it interacts directly with ZO-1. The ZO is formed near the apical surface and it forms linear arrays of ridges that fuse cells together. The zonula adherens provides enhanced mechanical strength to the epithelial barrier by forming a continuous belt-like connection though adjacent cells. The zonula adherens is composed of E-cadherin.

chemoattract CXCR3-expressing T cells that have a memory phenotype, and themselves produce IFN-γ. In contrast to the above-mentioned chemokines, the IFN-γ–inducible chemokines CCL9, CCL10, and CCL11 are not substantially upregulated in response to bacterial infection alone, but rather synergistically upregulate IFN-γ–induced expression of those chemokines. Moreover, differences in the kinetics of upregulated chemokine expression and production by intestinal epithelial cells, and differences in the biologic and functional properties of chemokines that have similar targets, may result in spatial and temporal chemokine gradients for the chemoattraction of target inflammatory cells within the intestinal mucosa (6).

Various studies have also begun to elucidate the role of the epithelium in providing signals important for the development of mucosal adaptive immunity. In this regard, small intestinal epithelial cells constitutively produce CCL25/TECK whose cognate receptor CCR9 is expressed on α4β7-expressing T cells that pref-

erentially localize to the small intestinal mucosa (11,12). In contrast, human colon epithelial cells do not produce CCL25 but do produce CCL28/MEC, whose cognate ligand is CCR10 (13,14). Thus, subpopulations of CCR10 expressing cells can preferentially localize to sites of CCL28 production in the gut (e.g., colon, salivary gland). The latter chemokines appear to be important for selective migration of lymphocyte subsets to the small intestine and colon, respectively.

In addition to chemokines, intestinal epithelial cells are capable of expressing a number of cytokines in response to bacterial infection. Some of the cytokines expressed by intestinal epithelial cells include EBI3, tumor necrosis factor-α (TNF-α), interleukin (IL)-12p35, IL-15, granulocyte-macrophage colony-stimulating factor (GM-CSF), and macrophage inhibitory factor (MIF) (5,15–18). Besides affecting gene expression patterns of immune cells in the mucosal tissue, it has been suggested that factors released by epithelial cells upon bacterial infection or under the influence of proinflammatory cytokines might directly induce endothelial expression of adhesion molecules such as intracellular adhesion molecule-1 (ICAM-1) and vascular cell adhesion molecule-1 (VCAM-1), which could lead to an increased mucosal influx of immune and inflammatory cells (19).

Epithelial Cell Receptors and Their Role

Consistent with their role as an integral component of the mucosal immune system, epithelial cells constitutively express or can be induced to express receptors important for host immunity. For example, gastrointestinal epithelial cells express major histocompatability complex (MHC) class II and nonclassical MHC class I molecules (e.g., CD1d, MICA) (20) as well as CD86 co-stimulatory molecules, suggesting that GI epithelial cells can function as antigen-presenting cells (APC). A conventional APC is able to internalize antigens, process them, and present them in the context of MHC class II molecules. The expression of co-stimulatory molecules such as CD80 and CD86 allows the APC to optimally stimulate T cells via CD28 engagement. Thus, GI epithelial cells may sample antigens that are partially processed by proteases in the lumen, and these antigens may be further processed by proteases within the epithelial cells (21) for presentation to T cells in the lamina propria.

Studies by Mayer and colleagues have suggested that under normal conditions intestinal epithelial cells may selectively stimulate CD8+ T cells with suppressor activity, which may aid in the control of inflammation in the intestine. This selective activation of CD8+ T cells is due to the expression by the epithelium of a carcinoembryonic antigen (CEA) family member known as gp180, which is a ligand for CD8 (22,23). Interestingly, intestinal epithelial cells from patients with inflammatory bowel disease lack gp180 and fail to expand these suppressor CD8+ T cells. Another recent study from the same group demonstrated the expression by intestinal epithelial cells of novel B7 family members, B7h and B7-H1, with modulatory activity on T cells (24).

In addition to the expression receptors involved in interactions with T cells, human intestinal epithelial cells have been shown to express a wide array of cytokine receptors. These include putative receptors for IL-1, IL-4, IL-6, IL-7, IL-9, IL-10, IL-15, IL-17, IFN-γ, GM-CSF, and TNF-α (4,7,19) as well as receptors for several chemokines including CXCR4, CCR5, CCR6, and CX3CR1 (19,25). Several of these receptors are expressed on the basolateral surface of the epithelial cells, whereas others are expressed apically or in a bipolar fashion. This indicates that epithelial cell signaling and function can be influenced not only by luminal antigens, bacteria, and bacterial products, but also by cytokines and chemokines released from local immune cell populations in the mucosa, thereby allowing epithelial cells to detect and subsequently respond to immunologic changes within the subepithelial compartment.

Although some pathogenic bacteria have been shown to invade the epithelial cells and alter intracellular signaling pathways either directly or through the secretion of bacterial products into the cell (e.g., *Salmonella*, *Shigella*, enteroinvasive *Escherichia coli*, *Yersinia*, and *Listeria*) (26,27), intestinal epithelial cells have been also demonstrated to express Toll-like receptors (TLR), which represent an evolutionarily conserved family of receptors that function in innate immunity via recognition of conserved patterns in bacterial molecules (28,29). These receptors are described below in detail. Briefly, epithelial cells have been shown to variably express TLR2, TLR3, TLR4, and TLR5 and it has been suggested that lipopolysaccharide (LPS) and other bacterial products like flagellin might exert their effect on epithelial cells through those receptors (30,31).

Adhesion molecules are another important class of membrane molecules that can play a central role in regulating the trafficking of immune and inflammatory cells through tissues. These include ICAM-1, lymphocyte function associated antigen-3 (LFA-3) (CD58), E-cadherin, and biliary glycoprotein (BGP) (32,33). Epithelial ICAM-1 expression, for example, can be upregulated in response to coculture of intestinal epithelial cells with invasive bacteria or agonist stimulation with IFN-γ or TNF-α (29,34), and its expression is polarized on the apical surface of intestinal epithelial cells, with its density being greatest in the area of the intracellular junctions. As elevated levels of ICAM-1 have been shown to correlate with an increased apical adhesion of neutrophils (34,35), it is possible that increased apical ICAM-1 expression in response to pathogenic bacteria functions to maintain neutrophils that have transmigrated across the epithelium into intestinal crypts within that site as an epithelial defense mechanism.

Nuclear Factor (NF)-κB: A Central Regulator of the Intestinal Epithelial Cell Innate Immune Response

Cell signaling events within intestinal epithelial cells are initiated by a broad array of bacterial pathogens with different strategies for epithelial entry and different intracellular lifestyles that converge to activate the transcription factor NF-κB and its target genes (36). This has led to the concept that NF-κB is a central regulator of epithelial cell signaling pathways essential for initiating host innate immune responses to microbial infection. Activation of the I kappa kinase (IKK) complex, and notably the IKKβ subunit is an essential step in the activation of NF-κB by a number of enteric pathogens and by proinflammatory cytokines (e.g., TNF-α, IL-1). Moreover, some pathogens (e.g., *Yersinia*), through their type III secretory proteins, have developed strategies to prevent the activation of NF-κB and other signal transduction pathways, and consequently to modulate the resultant host inflammatory response (37,38).

Over the past decade, many studies have revealed the immunologic importance of intestinal epithelial cells in maintaining a physical barrier to the external environment and in functioning alongside cells of the immune and inflammatory system to prevent infection and epithelial injury. Intestinal epithelial cells play a key role in intraluminal host defense by producing antimicrobial peptides and other antimicrobial products, by producing signals essential for activating the onset of mucosal innate immunity, and in setting the stage for host adaptive immune responses by ensuring that the appropriate cell populations are brought into the intestinal mucosa.

Innate Humoral Factors

Antimicrobial Peptides

The study of the antimicrobial properties of plants and insects revealed a family of evolutionarily conserved proteins that have since been discovered in virtually all multicellular organisms studied, including humans. Efforts to categorize these peptides have been only modestly successful, as their vast diversity enables only simple classifications based on secondary molecular structure and size. These molecules are known as CAMPs (cationic antimicrobial peptides) and have the capacity to target viral, fungal, and bacterial pathogens. Currently, the antibacterial properties of CAMPS are best understood. Perhaps the most intriguing aspect of the CAMP–bacteria interaction is the difficulty for bacteria to become resistant to CAMP action. In fact, there is currently a large body of research dedicated to pursuing antimicrobial peptides as the next generation of antibiotic therapy.

Antimicrobial peptides are effective in disrupting bacterial infection because they target a fundamental difference in the design of bacterial cell membranes as compared to those of multicellular organisms; most bacterial cell membranes have a large component of negatively charged phospholipids groups in the outer leaflet of the bilayer while the outer bilayer in plant and animal cells is constructed of lipids with zero net charge. Taking advantage of this property, CAMPs of differing primary structure are able to work in similar fashion due to the consistent organization of their hydrophobic and hydrophilic residues into discrete amphoteric secondary structures.

There are currently several hypotheses as to the actual method of bacterial killing by CAMPs; all begin with the interaction of the peptide with the lipid bilayer followed by a physical disruption of the bacterial membrane. From this point, it is believed that CAMPs might cause physical holes in the bilayer, enabling cell contents to leak out. Other theories suggest a fatal depolarization of the bacterial membrane, the activation of hydrolases that degrade the cell wall, or the disruption of cell activity by the internalized CAMP.

The contribution of CAMPs to GI mucosal innate immunity appears especially significant, as many of these factors have been discovered in saliva, breast milk, as well as epithelial and Paneth cells of the intestine.

Defensins

Like the other CAMPs, defensins are highly cationic proteins/peptides. They are rich in arginine and have a molecular weight of 3.0 to 4.5 kd. Defensins are distinguished by their β-sheet structures and cysteine-containing disulfide bridges. Although human defensins are grouped into two categories, α and β, based on the arrangement of these bridges, a novel third category of circular peptides, known as theta-defensins, has been isolated in primates. Defensins are synthesized as prepropeptides that undergo posttranslational modifications to give rise to active forms.

α-Defensins are 29 to 35 residues in length and contain a triple-stranded β-sheet structure. So far, six α-defensins have been identified: human neutrophil peptides 1 to 4 (HNP-1 to HNP-4) and human defensin-5 and -6 (HD-5, -6) (39). As the abbreviation suggests, HNPs are expressed primarily in neutrophils, but HNP-1, -2, and -3 have been isolated from T cells and natural killer (NK) cells grown in the presence of IL-2 (40). HD-5 and -6 are expressed primarily by intestinal Paneth cells (41).

β-Defensins differ from α-defensins in that they are up to 45 residues in length with a different disulfide bridge (cysteine) pairing. There are four β-defensins (HBD-1 to -4), which, with a few exceptions, are produced primarily by epithelial cells (42).

Defensins, in addition to their direct antimicrobial properties, contribute significantly to the innate immunity of the GI tract by recruiting and activating leukocytes. In fact, it was only recently that the study of chemokines (the classic category for soluble recruiters of leukocytes) and defensins merged; based on structural and functional similarities, several proteins from the defensin and chemokine families can act as both CAMPs and chemotactic activators of immune effector cells. Thus emerges an important link between the innate and adaptive immune systems that facilitates host defense against enteric pathogens (43).

Lactoferrin

With only a few exceptions, lactoferrin has been isolated from the milk of all mammalian species tested, including humans (44). It is also found in exocrine fluids such as tears, saliva, bile, and pancreatic fluid. Furthermore, polymorphonuclear cells (PMNs) are major producers of this iron-binding protein. Human lactoferrin (hLf) is a 692 amino acid protein in its mature form, and

is folded into two lobes: an N terminal and C terminal lobe. There are iron-binding sites in each lobe (45).

Besides its role as an iron transporter, lactoferrin has significant antimicrobial properties (46). Although the significance of lactoferrin's iron-scavenging properties on bacterial infections is still being debated, its direct-action antimicrobial properties have been demonstrated. Lactoferrin binds to the lipid A portion of LPS, resulting in the separation of LPS from gram-negative bacteria. Sequence alignments of lactoferrin with other known LPS-binding proteins have demonstrated significant structural homology. This activity is thought to critically destabilize the bacterial membrane.

A second direct effect of lactoferrin on bacteria results from this protein's cleavage products. Gastric pepsin cleaves lactoferrin into an N-terminal-derived peptide called lactoferricin H (47). Although the mechanism of action is less understood, lactoferrin pepsin cleavage products have been shown to bind to the LPS of gram-negative bacteria and to teichoic acid of gram-positive bacteria. It is currently postulated that, from there, it contributes to the disintegration of the bacterial cytoplasmic membrane.

Lysozyme

Lysozyme is a highly cationic protein that induces hydrolysis of the 1,4-β links between N-acetylmuramic acid (NAM) and N-acetylglucosamine (NAG) of the peptidoglycan cell wall, and is thus destructive to certain gram-positive bacteria. However, lysozyme is less effective against many pathogenic bacteria such as hemolytic streptococci, Listeria, and mycobacteria. This resistance is due to peptidoglycan constituents of the cell wall, including O-acetyl groups, that interfere with lysozyme's interaction with the NAM–NAG linkages. Interestingly, experiments performed with heat-inactivated lysozyme have shown that this protein actually kills bacteria indirectly by activating bacterial autolytic enzymes collectively known as muramidases.

Because many gram-negative bacteria have thick, anionic outer membranes, lysozyme is often less effective. However, lysozyme can work in tandem with the complement system in utilizing peroxide to punch holes in the walls of these bacteria. Furthermore, lysozyme is able to work synergistically with lactoferrin to induce bacteriolysis.

Interferons

Interferons (IFNs) play a crucial role in human disease and are subdivided into type I IFNs (IFN-α and IFN-β) and type II IFN (IFN-γ). The importance of type I IFNs in inflammation, immunoregulation, and T-cell responses has been recognized, and various cell types, including fibroblasts and epithelial cells as well as cells of hematopoietic origin, are known sources of IFNs. Type I IFNs are multifunctional immunomodulatory cytokines with profound effects on the cytokine cascade, including various antiinflammatory properties. The antiviral effects of these proteins were among the first properties identified. These proteins are produced by virus-infected cells and upon their release act on neighboring cells where they establish an antiviral state.

Type II IFN is produced by T cells, NK cells, and macrophages upon activation. Interferon-γ has an immunopotentiating effect and further stimulates macrophage activation. It also induces the enhanced expression of MHC class I and II molecules in macrophages, dendritic cells, and B cells. Interestingly, this effect is also mediated in nonimmune cells. This effect on the expression of MHC molecules has implications on processes associated with antigen presentation to T cells.

Angiogenins

Angiogenins (Angs) represent a novel type of microbiocidal protein important in innate host defense. Angiogenin messenger RNA (mRNA) expression increases rapidly during inflammation, and protein levels in the serum rise during the acute-phase response (48). Mouse and human Ang genes are in chromosome 14 (49). Recent studies have shown that different Angs have restricted tissue distribution. Hooper and colleagues (50) recently showed that Ang4 is expressed by the intestine, and that its expression is induced by gram-positive bacteria. More specifically, Ang4 is secreted by Paneth cells in the intestine, and the secreted form also has potent microbiocidal activity against gram-positive bacteria. This group of investigators

suggested that Ang4 is a mediator of epithelial innate host defense in the GI tract, and other members of this family, such as Ang1, very likely represent previously unappreciated members of the host systemic innate defenses.

Normal Flora

The classic description of the purpose of the immune system often referred to the differentiation between self and nonself. However, two factors prevent this definition from being completely accurate. First, one of the most important tasks of the immune system is to identify and destroy transformed cells before they result in cancerous lesions. Second, the mucosal immune system must differentiate between potentially harmful enteric bacteria and the normal, beneficial flora that permanently reside in the human GI tract. The normal human gut contains from 10 trillion to 100 trillion bacteria, from over 500 different species, consisting of both aerobic and nonaerobic organisms. The colonization of the gut begins immediately after birth.

These commensal bacteria play a formative as well as an active role in the innate mucosal immune system. It is well established that animals born into sterile, germ-free environments have a dysregulated immune system. Furthermore, the structure of the mucosa-associated lymphoid tissue, such as Peyer's patches, never properly develops. There are multiple contributions of the normal flora to the host's defenses. For instance, the normal flora prevents colonization by pathogens by competing for attachment sites or for essential nutrients. In addition, the normal flora may antagonize other bacteria through the production of substances that inhibit or kill nonindigenous species. The intestinal bacteria produce a variety of substances ranging from relatively nonspecific fatty acids and peroxides to highly specific bacteriocins that inhibit or kill other bacteria. Furthermore, the normal flora stimulates the development of certain tissues, i.e., the cecum and certain lymphatic tissues in the GI tract. The cecum of germ-free animals is enlarged, thin-walled, and fluid-filled, compared to that organ in a conventional animal. Also, based on the ability to undergo immunologic stimulation, the intestinal lymphatic tissues of germ-free animals are poorly developed compared to conventional animals. An additional contribution that warrants mentioning is that the normal flora stimulates the production of cross-reactive antibodies. It is known that the normal flora behaves as antigens in an animal and therefore induces immunologic responses. Conceivably low levels of antibodies produced against components of the normal flora could cross-react with certain related pathogens and prevent infection or invasion.

Toll-Like Receptors

The speed by which a bacterial infection can overwhelm a host necessitates a molecular sentinel receptor system that (a) is expressed by host cells that are among the first to encounter microbial pathogens, (b) can distinguish between pathogen and host, and (c) has the ability to initiate a greater response to the harmful microbe. As agents of early immune response, the innate immune cells are the ideal cells to incorporate such receptors. A landmark in the study of the innate mammalian immune system came about after the discovery of macrophage and dendritic cell surface receptors similar to the Toll family of antimicrobial receptors previously found in *Drosophila*. Named TLRs (Toll-like receptors), members of this expanding family of receptors have been found on many human hematopoietic cells. Furthermore, the argument that epithelial cells of the GI mucosa are immune cells was bolstered by the discovery of TLRs on several GI surfaces. The TLR family acts as pattern recognition receptors for pathogen-specific molecular patterns (PAMPs), including CpG DNA motifs, peptidoglycan, LPS, flagellin, and other bacterial surface or breakdown products.

TLR1 and TLR2

TLR1 was identified by the presence of a domain homology found in both *Drosophila* Toll and human IL-1 receptors. It is expressed in the spleen and in peripheral blood cells, including macrophages. TLR2 recognizes and induces signaling after contact with a variety of PAMPs, including bacterial lipoprotein/lipopeptides, peptidoglycan, and glycosylphosphatidylinositol (GPI) anchors. TLR2 signaling is strongly enhanced by CD14. Recent reports suggest that TLR1 associates with TLR2 in mediating the response to microbial lipoproteins and triacylated lipopeptides, thus making them

especially important sentinel receptors during gram-positive infections, when LPS is unavailable (51).

TLR3

TLR3 recognizes double-stranded RNA (dsRNA) (52), which is only found in viruses. Therefore, this PAMP receptor has minimal significance to mucosal bacterial immunity, but is highly relevant in the response to rotavirus infection, because rotaviruses are the most common cause of severe diarrhea worldwide. Rotavirus account for approximately one million deaths each year, which represent 20% to 25% of all deaths due to diarrhea and 6% of all deaths among children younger than 5 years old (53).

TLR4

One of the most intensely studied aspects of the antimicrobial innate immune system has been the mechanism of cellular activation by the gram-negative bacterial product LPS. After the discovery that the GPI-anchored membrane protein CD14 is a receptor for LPS, the search for a signaling co-receptor began. This is due to the fact that CD14 does not have a cytoplasmic domain and thus cannot initiate signaling. It has since been discovered that CD14 associates with TLR4 upon LPS binding, and that TLR4 provides the cytoplasmic signaling needed for cell activation. For TLR4 to functionally associate with LPS, a soluble secreted molecule called MD-2 is required. This is of greater significance in the gut, as CD14 is not expressed in intestinal epithelial cells.

TLR5

TLR5 recognizes flagellin from both gram-positive and gram-negative bacteria. Signaling by TLR5 mobilizes NF-κB and induces TNF-α and IL-8 (54). In dendritic cells, flagellin induces the increased surface expression of CD83, CD80, CD86, MHC class II, and the lymph node-homing chemokine receptor CCR7 (53).

TLR6

Like TLR1, TLR6 acts as a co-receptor with TLR2. The TLR2-TLR6 heterodimer recognizes peptidoglycan as well as diacylated mycoplasma lipoproteins. Mycoplasma lipoproteins upon binding TLR2 and TLR6 can induce NF-κB activation, which is partially mediated by MyD88 and FADD, and apoptosis, which is regulated by p38 MAPK as well as by MyD88 and FADD (55).

TLR7 and TLR8

TLR7 and TLR8 are related to TLR9, and have a higher molecular weight when compared with TLRs 1 to 6. The natural ligand(s) for TLR7 and TLR8 has not yet been identified. However, studies with TLR7-deficient mice have shown that TLR7 recognizes imidazoquinoline compounds, which are small synthetic antiviral molecules. TLR7 activation leads to inflammatory cytokine release. TLR8 is also reactive to imidazoquinolines (56).

TLR9

TLR9 is localized intracellularly and it is involved in the recognition of specific unmethylated CpG oligodeoxynucleotide sequences (57). The unmethylated form of these oligodeoxynucleotide motifs distinguish bacterial DNA from mammalian DNA, thus fitting the classic description of a PAMP. Upon activation, TLR9 engages an intracellular pathway that initiates NF-κB translocation.

The Intermediate Immune Step

An intermediate step important in filling the gap in our defenses shares some properties with the innate immune step and some with the adaptive immune step. Like the classic adaptive immune step, it is composed of both of T and B lymphocytes. However, the receptors it uses are of restricted repertoire and are germ-line encoded, as are the pattern recognition receptors (PRRs) of the innate immune system. In the B-cell compartment these receptors are germ-line immunoglobulin V genes, many of which form receptors that are capable of interacting with bacterial surface molecules (58). The B cells of this intermediate immune step require no T-cell help for activation and differentiation to plasma blasts, nor do they undergo the time-consuming process of affinity maturation. Therefore, the antibody response of this system is not ideal but it is rapidly available. Finally, both the B and T

cells of this intermediate immune step are then recruited at anatomic areas where an infection can be rapidly detected.

Role of B Cells During Intermediate Immune Step

Two major phenotypes of the B-cell subpopulation have been described. The first are B1 cells that are present in the peritoneum and other body cavities. They are self-renewing and express unmutated conventional B-cell receptors (BCRs) of restricted repertoire. These receptors respond to bacterial antigens in a T-cell–independent fashion. The second is a B-cell population that is located in the spleen, at the marginal zone (MZ) of the follicles, where they are ideally positioned to engage blood-borne pathogens. Like the B1 cells, they have a limited repertoire and are triggered in a T-cell–independent process. These MZ B cells are activated by myeloid dendritic cells (DCs) rather than lymphoid DCs, which are involved in the activation of conventional follicular B cells. In addition, LPS greatly accelerates the differentiation of MZ B cells into immunoglobulin M (IgM)-secreting plasma cells. As with the B1 cells, this rapidly activated T-cell–independent population is situated at a crucial checkpoint where blood-borne bacteria captured by circulating immature DCs make their first contact with the intermediate or adaptive immune system.

Role of T Cells in the Gut During the Intermediate Immune Step

Lymphocyte populations with a restricted receptor repertoire and specialized functions are not restricted to the B-cell compartment. A population of T cells with similar properties has been demonstrated in the gut (59). Murine gut-associated lymphoid tissue contains as many T cells as the entire central immune system, and about half of these are not of thymic origin. Cells of this system are, at least in part, seeded from gut-associated structures called cryptopatches, which are described later in this chapter. The analysis of these cells has been frustrating because of the difficulty in recovering and analyzing the small numbers of cells that an individual cryptopatch contains. Rocha (60) has described how single-cell polymerase chain reaction (PCR) made analysis of the markers carried by these lineages possible, and how it facilitated their isolation by cell sorting and further functional analysis in adoptive transfer experiments. As with the B1 and MZ B cells, these T cells are positioned at an important mucosal surface where they could be poised to ward off bacterial attack. Similar to the B cells of the intermediate system, they also express a restricted receptor repertoire. These gut-associated T cells can now be examined in detail.

Adaptive Immune Response

The adaptive mucosal immune system has in place organized lymphoid tissues to initiate antigen-specific responses to potential pathogens. The mucosa-associated lymphoid tissue (MALT) is located in anatomically defined compartments. In the gut it is collectively referred to as the gut-associated lymphoid tissue (GALT), and in the nasopharynx it is termed the nasopharyngeal-associated lymphoreticular

tissue (NALT). The GALT may be considered the central MALT because of its overall mass, the antigen load to which it is exposed, and its influence in the immune system. The GALT is functionally divided into inductive and effector sites. Although the Peyer's patches and the appendix represent inductive sites in the GI tract, the tonsils and adenoids fulfill a similar role in the NALT. In mice there are isolated lymphoid follicles that have properties of an inductive site (61,62). The effector sites of GALT include the mucosal lamina propria and the intestinal epithelium (Fig. 3.2).

Gastrointestinal Inductive Sites

Peyer's Patches

Peyer's patches (PPs) are organized areas of lymphoid tissue in the gut mucosa and contain follicle centers as well as well-defined cellular

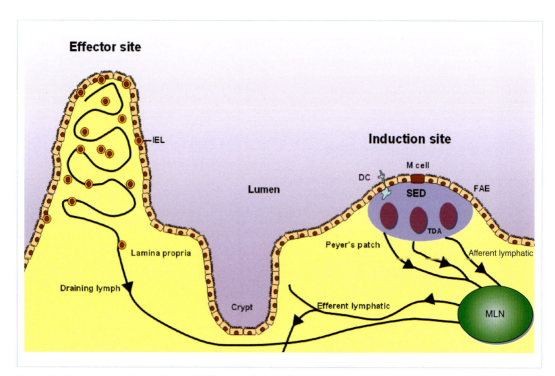

Figure 3.2. The gut-associated lymphoid tissue (GALT) contains both effector and induction sites. The subepithelial dome (SED) of the Peyer's patch contains B-cell follicles and a thymus-dependent area (TDA). Antigens are transported across the epithelium by M cells or by dendritic cells (DCs) that extend their processes through epithelial tight junctions into the luminal space. Effector sites are less organized and can be found throughout the lamina propria and epithelium of the GALT. Intraepithelial lymphocytes (IELs) are prevalent within these effector sites, which are drained by afferent lymphatics that supply the mesenteric lymph nodes (MLNs).

zones. The surface of PP is covered by the follicle-associated epithelium, which is a unique epithelial layer that differs from adjacent columnar epithelium in that it is cuboidal and does not contain secretory component. Instead, the follicle-associated epithelium contains specialized cells known as microfold (M) cells that have antigen-sampling capacity. The M cells are irregular and have microfolds in place of microvilli. Compared to absorptive epithelial cells, they have poorly developed brush borders and reduced enzymatic activity. M cells form a compartment at the basal membrane where T and B cells are clustered with some dendritic cells. M cells are able to internalize and transport antigens from the lumen to the underlying APCs. Soluble antigens as well as intact bacteria have been shown to be internalized by M cells (63). These findings suggested that M cells allow for controlled transport of antigens from the lumen into the PP as part of immune surveillance and eventual development of an adaptive immune response.

B cells in the PP are largely committed to IgA production, and those that are adjacent to the M cells are like germinal center B cells with a memory phenotype (36). The germinal centers under the dome of the PP contain dividing B cells undergoing affinity maturation and class switching to IgA. As discussed below, the IgA that is eventually produced contributes to the adaptive immune protection of mucosal surfaces. After antigenic stimulation in the PP, IgA$^+$ lymphoblasts migrate through the lymph and circulation to eventually reside in the lamina propria (LP). The high frequency in isotype switching to IgA by B cells in PP is very likely influenced by T cells, as suggested by various independent studies. For instance, T-cell clones of PP or GALT origin induced sIgM$^+$ sIgA$^-$ B cells to express sIgA$^+$ (64,65). The T cells that are implicated in this selective isotype switching produce TGF-β and have been referred to as Th3 or T regulatory (Tr) cells (66).

T-cell regions in PP are located adjacent to the B-cell follicle zone. Although all major T-cell subsets are present in PP, most of the T cells in PP are CD4$^+$ T cells and are phenotypically mature. Approximately two thirds of the T cells express αβ T-cell receptors (TCRs). The T cells include both T helper (Th)1 and Th2 cells. One third of the T cells are CD8$^+$ and include cytotoxic T lymphocyte (CTL) precursors (67). The presence of both small naive T cells (CD45RA$^+$)

as well as activated T cells (CD45RO$^+$) supports the notion that PPs are a route of lymphocyte recirculation. Also, immediately beneath M cells there are dividing CD45RO$^+$, CD69$^+$ T cells (68).

The T cells in PPs respond to antigens that are internalized and translocated by M cells. After translocation, those antigens then are processed by APCs. Antigen presentation in PPs is mediated by class II MHC$^+$ cells that include antigen-specific B cells, macrophages, and DCs. There are at least three different subtypes of DCs in PPs defined by their expression of the markers CD11b and CD8α. Those subsets are lymphoid DCs (CD8α), myeloid DCs (CD11b), and double-negative DCs (69). Each DC subset influences the induction of distinct T-cell subsets. Although myeloid DCs produce IL-10 and thus influence the differentiation of Th2 cells for humoral (i.e., IgA production) responses, lymphoid and double-negative DCs produce IL-12 needed for the induction of Th1 differentiation and eventual cell-mediated immune responses (70). Production of TGF-β as well as IL-10 by myeloid DCs may allow them to also influence the differentiation of Th3 or T regulatory (Tr) cells involved in the unresponsiveness to oral antigens, which is referred to as oral tolerance (66).

Cryptopatches

Cryptopatches represent a recently described primary lymphoid organ consisting of small clusters of lymphocyte precursors. Those lymphocytes express the IL-7 receptor, are c-kit$^+$, CD3$^-$, TCR$^-$, and RAG$^-$. These clusters of cells are present in mice, but not in the human intestine. Transfer of cryptopatch c-kit$^+$ lymphocytes from athymic nude mice into irradiated severe combined immunodeficiency (SCID) mice revealed that those lymphocytes have lymphopoietic capacity (61).

Gastrointestinal Effector Sites

Antigen-sensitized lymphocytes, both B and T cells, migrate from mucosal inductive sites (GALT or PP) to mucosal effector sites via the mesenteric lymph nodes, thoracic ducts, and bloodstream. Those lymphocytes enter the mucosal effector sites such as the lamina propria and epithelium, where they differentiate further. For instance, sIgA$^+$ B cells differentiate into IgA plasma cells. This differentiation is promoted by Th1 (i.e., IL-2) and Th2 (IL-5, -6, and -10) cytokines.

Lamina Propria

The lamina propria (LP) represents the basement membrane layer below the follicle-associated epithelium and around lymphoid follicles. It is located between the epithelium and the muscularis mucosa and consists of smooth muscle cells, fibroblasts, as well as cells of hematopoietic origin. Of these hematopoietic cells, in addition to lymphocytes, the LP also contains large numbers of macrophages and dendritic cells, which are responsible for the processing of antigens that cross the epithelium and are thus responsible for presentation of the resulting peptides to CD4[+] T cells. The CD4[+] T cells in the LP are predominantly $\alpha\beta$ TCR[+] and there is a smaller number of CD8[+] T cells that express the αEβ7 integrin, suggesting that they are destined for the epithelium (62). T cells in the LP express markers typically associated with activated T cells. They are α4β7[+], CD45RO[+], CD25[+], and human leukocyte antigen (HLA)-DR[+] (71). Normal, unstimulated LP lymphocytes show an increased level of apoptosis when compared with peripheral lymphocytes, which is perhaps attributed to their expression of Fas and FasL (72). These LP T lymphocytes also have an increased production of cytokines that are important in the eventual production of IgA.

B lymphocytes in the LP home there due to their expression of the α4β7 integrin that interacts with the mucosal addressin cell adhesion molecule-1 (MADCAM-1) expressed by mucosal endothelial cells. The recruitment of precursors of IgA[+] plasma cells is very likely mediated by the interaction of specific chemokines produced by cells in the LP or the intestinal epithelium and receptors on those B cells. Interestingly, a recent study showed that thymus-expressed chemokine (TECK), also known as CCL25, is a potent and selective chemoattractant for IgA antibody-secreting cells (73). In addition to the thymus, the intestinal epithelium is a source of this chemokine whose chemotactic effect is mediated via the CC chemokine receptor 9 (CCR9) expressed by IgA[+] plasma cell precursors (73). The cytokines (IL-2, IL-4, and IL-5) produced by LP T cells promote the generation of IgA plasma cells. Class switching and IgA[+] plasma cell differentiation may also occur in a T-cell–independent manner via LPS and stromal cell–derived TGF-β, IL-10, and IL-6 (74). IgA[+] plasma cells represent approximately one third of the mononuclear cells found in the LP. These IgA plasma cells actively produce dimeric and polymeric IgA. More than one half of the IgA plasma cells in the GI mucosa produce IgA2, whereas in lymph nodes and tonsils the predominant form of IgA produced is IgA1 (75). The characteristics of these antibodies are described later in this chapter.

Intestinal Epithelium

The gut epithelium is another effector lymphoid site, and it may play important immunomodulatory functions, as detailed above. In the epithelium there are populations of lymphocytes referred to as intraepithelial lymphocytes (IELs) that are found in spaces between epithelial cells above the basement membrane. They possess distinct features from those of LP and systemic lymphocytes. Populations of both $\alpha\beta$ and $\gamma\delta$ T cells are represented in the IEL. The ratio of $\alpha\beta$ to $\gamma\delta$ TCRs in IEL differs from one species to another. In mice, the $\gamma\delta$ TCR IELs predominate, whereas in humans the $\alpha\beta$ TCR IELs are more frequent (76). Most of the human IEL are CD8[+] and express the GI integrin αEβ7 that binds epithelial E-cadherin (77). The IELs also express markers of activated T cells, such as CD45RO[+] (78). The IELs are found at a frequency of one for every four to nine epithelial cells in the small intestine. It is estimated, based on immunohistology, that IELs represent almost half of the total T-cell numbers in all the lymphoid organs (79).

In comparison to the periphery, T cells with $\gamma\delta$ TCR are well represented in the gut, especially following infections. However, their functional significance is not fully characterized. It is suggested that $\gamma\delta$ T cells respond to bacterial antigens in the absence of antigen presentation, perhaps as part of an innate response mechanism (80). Some $\gamma\delta$ T cells bind directly to human class I MHC-like MIC molecules that are not loaded with peptides and that are inducible. MIC expression increases in enterocytes under stress conditions and some epithelial tumors, and these are recognized by $\gamma\delta$ IELs (81,82). Thus, $\gamma\delta$ TCR IELs may be responsible for the recognition of stressed cells.

Another unique population of IELs consists of CD8αα homodimer-expressing T cells. CD8αα-expressing T cells are rare outside of the intestinal epithelium. These T cells use the invariant FcεRIγ chain as part of their CD3 complex. These IELs are oligoclonal in their TCR repertoire (59). Various studies have suggested that CD8αα IELs are self-reactive and represent an extrathymic T-cell lineage. For instance, mice expressing the Mls-1^a allele are devoid of conventional T cells expressing the TCRs with Vβ6, Vβ8.1, and Vβ11, which bind the endogenous retroviral Mtv-7 superantigen, due to deletion. In contrast, the CD8αα TCR αβ IELs are present even at frequencies that are higher than in the nondeleting strain (79). To determine the requirement of class I MHC for the development of CD8αα TCR αβ IELs, mouse strains deficient in TAP, CD1, classic class I MHC, and β$_2$-microglobulin were examined and all except β$_2$-microglobulin knockout mice had extrathymic CD8αα αβ IELs (83). These observations suggested that these cells only require class Ib MHC molecules for their development. Multiple observations suggest that CD8αα does not function as a co-receptor for class I MHC, and these IELs may in fact be selected in a co-receptor–independent manner.

Cytotoxic T lymphocytes (CTLs) are largely responsible for the elimination of cells infected with intracellular pathogens, such as viruses. Most CTLs are CD8$^+$ T cells and they recognize viral peptides bound by class I MHC molecules. Because IELs are found at a frequency of one for every four to nine enterocytes and two thirds are CD8$^+$, they could represent an important first line of defense against intracellular enteric pathogens. Because IELs possess constitutive cytolytic activity (84), they are thought to play a role in the clearance of intracellular pathogens. Studies to support this notion showed that CD8$^+$ TCR αβ lysed *Listeria monocytogenes*—infected cells and their adoptive transfer into SCID mice resulted in clearance of rotavirus infection (84).

Immunoglobulin A

The discovery that mammals produce more IgA than all other antibody isotypes combined, that at least 80% of all plasma cells are located in the intestinal lamina propria (85,86), and that most IgA is secreted into the luminal spaces of the intestine has led to questions about how this acquired immune defense functions to protect against enteric bacterial and viral pathogens. These and related questions have now been investigated, including the sites of induction of IgA plasma cells, the migration patterns of these antibody precursors, the key elements involved in IgA class switching, and IgA responses to certain human pathogens. An interesting note is that the estimated daily amount of IgA that is secreted across the epithelium in the GI tract is approximately 3 g (87) (Fig. 3.3).

IgA1

Although there are small amounts of IgA in the serum, most exist as a form known as secretory IgA (sIgA). In fact, humans produce two isoforms of IgA. IgA1 is one of the IgA subclasses and exists in monomeric form in the serum (88). Although present in quantitatively less amounts than IgG or IgM, the role of serum IgA is currently being investigated. There is a receptor (FcαR1/CD89) for IgA1 on the surface of eosinophils, neutrophils, monocytes, and macrophages that has been shown to induce phagocytosis, antibody-depended cellular cytotoxicity, and secretion of inflammatory molecules (89–92). Studies involving mice transgenic for human CD89 have shown that, in the presence of inflammatory mediators, bacteria coated with IgA can be phagocytosed by Kupffer cells (93). This suggests that serum IgA might act as a backup defense against pathogens that escape the mucosal immune system without being opsonized locally.

IgA2

IgA2 is a dimer of two IgA molecules linked through their alpha chains by a J chain. As plasma cells in the lamina propria release sIgA, it is bound, endocytosed, translocated, and released apically by mucosal epithelial cells. The transmembrane cellular receptor that binds sIgA is called the polymeric immunoglobulin receptor (pIgR). The receptor-ligand union between pIgR and sIgA at the basolateral surface of the epithelial barrier is a covalent bond. After endocytosis of this complex and transport through the cell, there is a proteolytic cleavage event that releases IgA into the lumen with the N-terminal portion of pIgR still attached (88). This segment

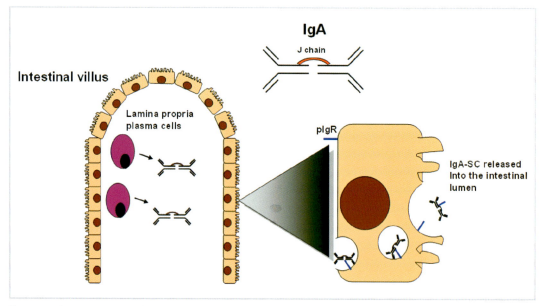

Figure 3.3. The polymeric immunoglobulin A (IgA) secreted by plasma cell binds to the polymeric Ig receptor (pIgR) expressed on the basolateral surface of the mucosal epithelium. The IgA–pIgR complex is transcytosed and trafficked to the apical surface. During this process, a disulfide bond is formed between the IgA and the pIgR. At the apical surface, an enzyme cleaves in between the ectoplasmic and transmembrane domains of the pIgR, releasing the IgA–secretory component complex (IgA-SC) into external secretions.

of pIgR is referred to as secretory component. Interestingly, it has been demonstrated that the pIgR-sIgA complex is capable of binding antigen both before and during its transport through an epithelial cell. This provides several potential benefits: First, antigens that leak through the mucosal barrier can be returned to the lumen. Second, IgG and IgM antibodies, with no other means of reaching the lumen, can be transported through the epithelial layer (94). Finally, it has been shown that IgA can bind and neutralize replicating viruses in epithelial cells during their intracellular transit (95).

Secondary IgA Access to the Gut

It is now known that IgA molecules can reach the intestinal lumen by two other means besides transepithelial endocytosis. IgA is also secreted into bile across biliary epithelia of the bile ducts and gallbladder. It then reaches the intestine via duodenal entry through the ampulla of Vater (96,97). A second alternative entry of IgA into the intestine is through maternal milk consumed by the infant. Investigation into the functional role of milk IgA on the developing immune system of the neonate has demonstrated that IgA from breast milk binds to commensal bacteria and restricts the full colonization of the young gut (98). Furthermore, breast-fed infants have a different intestinal flora composition than formula-fed infants (99). Studies using mice with various levels of immune competence revealed that the adaptive immunity conferred by milk is significant. It is known, for example, that antibodies to pathogenic bacteria exist in human breast milk (88).

IgA Induction

The GALT, representing the most functionally important site of the mucosal immune system, including both inductive and effector sites within PPs and isolated lymphoid follicles (ILFs), contains IgA+ plasma cell precursors. Within these immune centers, there are interactions between B cells, antigen-loaded dendritic cells, and local CD4+ T cells. It is important to note, however, that PPs are not an absolute requirement for IgA-producing B cells to exist in the gut (100). This microenvironment, unique in its constant exposure to and stimulation by foreign antigens, thus favors B-cell proliferation, class switching to IgA, and somatic hypermutation. Moreover, recent studies have shown two interesting requirements for IgA production by B2 cells in the GALT. The first is interaction with helper T cells and the second is a dependence on interaction with the commensal flora (101). B1 cells, however, are capable of producing IgA without help from T cells. (B2 cells are B cells of bone-marrow origin and stain heavily for IgM and weakly for IgD. B1 cells are of pleuroperitoneal origin and stain weakly for IgM and strongly for IgD.) The T-cell independence of the IgA production by B1 cells suggests an important role for B1-derived IgA molecules as a frontline defense against systemic invasion by intestinal bacteria. Evidence to support this idea is as follows: commensal bacteria bind more to B1-derived IgA than to B2-derived IgA. Second, normal mice that produce commensal bacteria-specific B1-derived intestinal IgA show no serum IgG or IgA with these specificities. Conversely, mice deficient in IgA have serum IgG specific for intestinal bacteria (102).

IgA Homing

From their induction sites in PP, IgA$^+$ B cells migrate to the draining mesenteric lymph nodes, undergo further proliferation, and differentiate into plasmablasts. These cells home back to their preferential targets in the intestinal lamina propria through the thoracic duct and blood (103). Specific interactions between lymphocyte receptors and their ligands on endothelial cells result in this homing of plasmablasts to their effector sites in the gut. One question that continues to intrigue investigators is why there is not parallel homing of IgG and IgM plasma cells to the gut lamina propria. The fact that only IgA-producing cells show this migratory pattern suggests the presence of specific chemokines produced by local gut cells. Investigation of this hypothesis has led to the discovery of a chemotactic factor in mice specific for IgA$^+$ B cells known as thymus-expressed chemokine (TECK/CCL25) (73). TECK is produced in the thymus as well as the epithelium of the small intestine. In addition to this chemoattractant, certain cells of the GI tract have been shown to be necessary for B-cell homing. Specifically, lamina propria stromal cells have been shown to be crucial for the presence of B cells in the gut lamina propria. Further studies have demonstrated that this finding is a result of a depend-

ence of B-cell homing to the gut on the lymphotoxin-β receptor (LTβR) found on lamina propria stromal cells (100,104). The mechanisms by which signaling through the LTβR selects for B-cell homing are not yet known.

IgA Class Switching

To address the questions about B-cell class switching to IgA phenotypes, B-cell cultures were grown in vitro. Because the role of cytokines in promoting and influencing the specificity of B-cell class switching is well documented, these B-cell cultures were supplemented with a nonspecific stimulant (LPS) and an array of cytokines. The results indicated that TGF-β and IL-4 promote the switch from IgM to IgA, and IL-10 can synergize with TGF-β (74,105,106). IL-2 enhances this activity but is not an absolute requirement. Furthermore, after a cell has switched to IgA, antibody secretion can be enhanced with IL-5 and IL-6 (107,108). It has been challenging to assess the in vivo importance of these cytokines to IgA switching because mice that have these genes deleted show chronic inflammation, which makes it difficult to analyze the model.

IgA Function During Infection

The role of IgA in protecting against infections of intestinal origin is a paramount question. Previous findings supporting the importance of IgA to the mucosal immune system include the large body of investigations involved in developing the oral vaccine against polio and studies of rotavirus interaction with the mucosal immune system. Studies revealing that sIgA was produced in much greater quantities after mucosal rather than parenteral vaccination with the live attenuated polio vaccine were an important step in realizing the significance of this immune molecule in conferring immunity to enteric pathogens. Rotavirus has also been extensively studied as an inducer of intestinal IgA response. Mouse models have been developed for rotavirus infection that have given investigators great insight into the mechanisms involved in viral clearance and immunity. Published research has clearly demonstrated the correlation between rotavirus clearance, subsequent protection against reinfection, and mucosal IgA production. One important finding to come out of the rotavirus studies was that a significant degree of IgA protection can occur without T-cell help. Specifically, this was observed in nude mice and T-cell receptor knockout animals. In these situations, viral clearance correlated with T-cell–independent IgA production, which again reinforces the importance of B1 cells described earlier.

Although these findings show significant evidence of the immunologic importance of IgA in protecting against enteric pathogens, other investigations into the effects of IgA deficiency demonstrate perplexing, and sometimes conflicting, outcomes. On the one hand, individuals with selective IgA deficiency suffer from frequent GI infections and develop nodular follicular hyperplasia, which is thought to result from the local immune response to local antigens. On the other hand, redundancies in the mucosal immune system, such as the overproduction of IgG and IgM that respond when IgA is deficient, make it difficult to fully assess the global role of IgA in human immunity. Animal models only add to this confusion, as IgA, J-chain, or pIgR-negative mice, seem to show no signs of ill health.

The end result of direct encounters between bacterial products and cells of the classic adaptive immune system as presented by A. Lanzavecchia is the generation of memory B cells. Immunoglobulin M memory cells differ from B1 and PP marginal zone B cells in that they have a mutated Ig, the hallmark of an adaptive B-cell response. These cells proliferate and secrete IgM in response to the microbial pathogen-associated molecular pattern CpG, in conjunction with bystander help in the form of IL-2 or IL-15. By contrast, naive B cells require surface Ig engagement. Thus, microbial products could even be essential for the maintenance of long-term B-cell memory by continuous polyclonal activation. This is necessary because even long-lived plasma cells have a half-life of months and thus cannot explain how human B-cell memory is maintained for many years.

Even the TCR can function as an exotic type of PRR for superantigens. Following whole genome sequencing of several strains of *Staphylococcus aureus*, it became apparent that the number of superantigens and superantigen-like sequences had been strongly underestimated. Superantigens bridge the gap between MHC class II molecules and subsets of TCRs and thereby act as extremely potent T-cell mitogens. However, this cannot be their only function. One particularly large and diverse family, the *set*-cluster, is present in every *S. aureus* strain analyzed. Crystallization analysis revealed that the general superantigen structure is well conserved in Set proteins; however, they do not stimulate T cells and are thus, by definition, not superantigens (109). Elucidation of the function of this new group of potential virulence factors will be a task for the future.

Multiple lines of information have shown that the classic B- and T-cell response represents the final, and certainly most effective, means of countering microbial attack. However, other cell types, ranging from epithelial cells to the B cells and T cells of the intermediate immune system, situated at the mucosal surfaces and at central checkpoints such as the spleen or GI tract, have a more important role in pathogen defense than was previously thought. Rapid advances are expected in our better understanding of this system, which will surely have a major impact on the strategies developed to counter pathogens.

Common Infections of the Gastrointestinal Tract and Immune Responses to Them

Gastric

The gastric mucosa separates the underlying tissue from the antigenic universe of the stomach lumen; the extreme pH of this space, as low as 1, is important in aiding digestion of food, activating enzymes, and altering the ionic state of iron to a form that is readily absorbed. HCl also renders the gastric epithelium free from bacterial colonization or infection, with the exception of one important human pathogen: *Helicobacter pylori*. This bacterium has developed means to survive the harsh environment of the stomach, actively move through the mucosa layer, attach to the epithelium, evade immune responses, and achieve persistent colonization.

Epidemiology

Approximately 50% of the world's population is infected with *H. pylori*. The route of transmission is believed to be fecal-oral, acquired most commonly in early childhood through family transmission. Infection rates are directly correlated with socioeconomic status; developing nations, with poor water quality and reduced sanitation, contain the highest rates of infection, approaching 100% in certain areas of South America.

Pathogenesis

Every person infected with *H. pylori* displays inflammation of the gastric epithelium (chronic superficial gastritis); a portion of those individuals develop severe disease, either ulcerative disease or gastric cancer. Over 80% of gastroduodenal ulcers are caused by *H. pylori*. Furthermore, the extremely high correlation of *H. pylori* with MALT lymphoma and gastric carcinoma has led to its classification as a class I carcinogen, on a par with asbestos and cigarette smoke.

H. pylori has a genome of 1.65 base pair (bp) and codes for about 15,000 proteins. This organism has a high degree of genetic diversity, evidenced by the clinical isolation of many different strains, several of which have been fully sequenced. Indeed, *H. pylori* was the first bacteria to have genomes sequenced and compared from two different strains. An important discovery from this work was the existence of a pathogenicity island in many of the strain's genomes. The first gene sequenced from this 29-gene cluster was *cagA* (cytotoxin-associated gene A) and was used to name this cag pathogenicity island (cag PAI). Several of these genes in the cag PAI encode a predicted type IV secretion system. CagA is a 120-kd protein that is inserted into the host cell, is phosphorylated, and binds to SHP-2 phosphatase. Cag$^+$ *H. pylori* strains are known as type I strains and cag$^-$ strains are designated type II. Although the correlation between *H. pylori* strain and disease manifestation is complicated, it has been clearly shown that cag$^+$ *H. pylori* strains are dramatically more capable of inducing proinflammatory epithelial cell responses, of which IL-8 release appears to be central (Fig. 3.4).

Another significant *H. pylori* pathogenesis factor is vacuolating cytotoxin VacA. This bacterial gene product is not a part of the cag PAI, and is expressed by the majority of strains. VacA inserts itself into the epithelial-cell membrane and forms a hexameric anion-selective, voltage-dependent channel that creates large vacuoles in host epithelial cells. VacA also affects the mitochondrial membrane, where it causes release of cytochrome *c* and induces apoptosis.

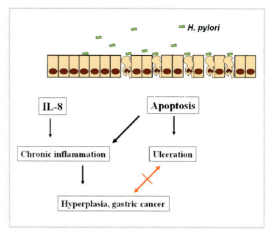

Figure 3.4. Two important epithelial responses to *H. pylori* infection are interleukin-8 (IL-8) release and apoptosis. These phenomena have been observed both in vivo and in vitro. Interestingly, however, clinical data suggest that these outcomes are part of divergent pathways, as development of gastric cancer is much less prevalent in patients with gastroduodenal ulcers.

As the single bacteria infecting the human gastric surface, *H. pylori* has evolved a mechanism by which it shields itself from the acidic milieu long enough to colonize the gastric mucosa; the major protein produced by this gram-negative spirochete is urease. Urease is created in large quantities by *H. pylori*, representing up to 15% of total cellular protein production. The autolysis of a portion of a *H. pylori* population results in adjacent viable bacteria coating themselves with this free enzyme. As bacteria move through the extreme pH of the lumen onto the moderately acidic pH of the mucosa layer, urease catalyzes the conversion of endogenous urea into carbon dioxide and ammonia. The ammonia buffers the HCl to near neutrality, allowing *H. pylori*'s survival until it can burrow into the less severe environment of the mucosa.

H. pylori: *Epithelial Cell Interactions*

Infection with this gram-negative, flagellated spirochete is extracellular; therefore, colonization is dependent on, and pathogenesis is induced by, *H. pylori* binding and interaction with surface receptors on the gastric epithelium. *H. pylori* binds tightly to epithelial cells by utilizing several bacterial surface proteins. The best characterized adhesin, BabA, is a 78-kd outer-membrane protein (Hop) that binds to the fuco-

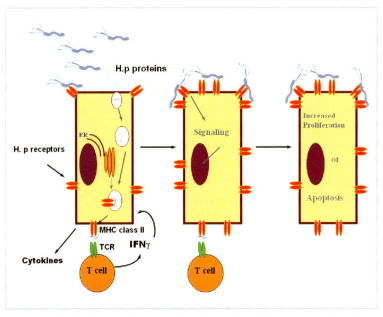

Figure 3.5. *H. pylori* receptors must be apically expressed on the gastric epithelium to allow binding of this noninvasive bacteria. Once binding occurs, proinflammatory signal transduction events are initiated. Gastric epithelial cells are equipped with the molecular tools needed to process and present antigens in the major histocompatability complex (MHC) class II pathway, including the expression of MHC class II. Because MHC class II is expressed on both apical and basolateral surfaces of the gastric epithelium, it is available to bind *H. pylori* and then present its antigens to lamina propria T cells.

sylated Lewis B blood-group antigen. Several other members of the Hop protein family also mediate adhesion to epithelial cells. However, in vitro and in vivo epithelial responses to *H. pylori* binding demonstrate that host signaling receptors are required. One such *H. pylori*–binding epithelial surface protein, capable of transducing signals, is MHC class II molecules. The gastric epithelium is the only mucosal surface to constitutively express this heterodimeric protein complex. Furthermore, the inflammation associated with *H. pylori* infection upregulates MHC class II expression throughout the stomach, which offers one answer to the important question of how this bacteria benefits by inducing an inflammatory immune response in the host. Studies indicate that *H. pylori* urease is the adhesin for MHC class II binding (110) (Fig. 3.5).

Local Immune Response to H. pylori

Because *H. pylori* is an extracellular infection, the local immune response to infection is initiated by gastric epithelial cells. The release of IL-8 in response to *H. pylori* binding recruits and activates neutrophils, which in turn release reac-

tive oxidative species (ROS) that propagate the inflammatory response. Reactive oxidative species represent an important source of host cell injury during *H. pylori* infection, as demonstrated by the reduction of *H. pylori*–induced apoptotic cell death in the presence of antioxidant compounds. This initial inflammatory insult leads to the accumulation of Th1 cytokines IFN-γ, TNF-α, and IL-1β. In many cases, the epithelial barrier is compromised, which is demonstrated clinically by the high percentage of ulcer patients infected with *H. pylori* (~80%) and the ability to heal many bacteria-induced ulcerative disease cases with antibiotic therapy. A breach in the mucosal epithelial layer permits *H. pylori* and its breakdown products access to the underlying tissue, including APCs, LP T cells, and myofibroblasts. Significant numbers of local T cells in infected individuals are *H. pylori* specific, and seroconversion to bacterial antigens is virtually guaranteed.

Looking closer at the immune response to *H. pylori*, the induction of a distinct innate and adaptive response is evident. Although these defensive measures mirror those of classic peripheral immunity, there are distinctive aspects to both the innate and adaptive immune response within the gastric mucosa that reflect the specialized nature of *H. pylori* infection.

Innate Response

As with any classic innate immune response, the two most important factors in nonspecific immunity to *H. pylori* are an early defense against bacterial proliferation and a signal of infection to the adaptive immune effectors.

However, because *H. pylori* exists, at least during the early stages of infection, exclusively within the luminal cavity, there is minimal contact between the bacteria and the agents of an innate response. Thus, much work has been done to understand the epithelium's role in both innate and adaptive immunity. Clearly, gastric epithelial production of IL-8 with the subsequent recruitment of neutrophils is among the first of the innate immune responses. The requirement of *H. pylori*–epithelial cell binding for IL-8 production is well documented (111,112). An obvious candidate for such an innate response-associated receptor is the TLR family.

There is ongoing research to understand the importance of TLRs in *H. pylori* infection, but little has been established at this point. Current dogma suggests that professional APCs respond to bacterial products through their TLRs by producing proinflammatory mediators such as TNF-α, IL-1β, and IL-8. Furthermore, these cytokines are found in increased levels during *H. pylori* infection. Although *H. pylori* contact with APCs is possible during later stages of infection, an important question is the role of these innate immunity-signaling receptors during early infection. Current studies focus on the role of TLRs 2, 4, 5, and 9 in *H. pylori* infection. Published studies have shown that the gastric epithelium expresses TLR4, TLR5, and TLR9 (113). Their expression is found on both the apical and basolateral surfaces. Interestingly, the expression of TLR5 and TLR9 changed to the basolateral surface during infection. A different study suggested that whereas gastric epithelial cells express TLR4, this TLR4 does not signal (114).

Adaptive Response

Although the innate immune response to *H. pylori* is insufficient to clear the infection, the inflammatory response that it elicits contributes to the breaching of the epithelial barrier. The presence of *H. pylori* and its products has been documented in the LP underlying the epithelium. This may be explained by the ability of *H. pylori* to alter the composition and function of the apical-junctional complex and the eventual disruption of the epithelial barrier function (115). Here, bacterial antigens are taken up and processed by APCs, initiating an adaptive immune response.

Evidence of both cellular and humoral specific immune responses to *H. pylori* is abundant. Both CD4 and CD8 T cells are induced locally during *H. pylori* infection. Furthermore, CD4⁺ T cells specific for *H. pylori* have been isolated from the antrum of infected patients. Further investigation into the CD4 T-cell response to *H. pylori* revealed that gastric T cells isolated from infected animals and humans produce IFN-γ and TNF-α, but not IL-4, suggesting a Th1 polarization (116,117). However, despite the apparent dominance of Th1-type CD4 T cells, the antibody response to *H. pylori* is significant; virtually all those infected with *H. pylori* seroconvert in response to multiple *H. pylori* antigens, and these bacteria-specific antibodies can be found both in the peripheral blood and locally in the stomach. Both IgG and IgA antibodies are found, with specificity to urease, flagellin, LPS, as well as other membrane proteins. In fact, up to 10% of all mononuclear cells isolated from infected individuals produce IgA.

Despite a large body of research clearly demonstrating a local, specific immune response to *H. pylori*, the role of this immune response in shaping *H. pylori*–mediated disease states is poorly understood. Suspecting that these divergent clinical outcomes are mediated by differing T-cell responses, investigators have recently found that T cells isolated from infected patients with differing disease states have different antigen specificities; for example, *CagA* seems to be immunodominant in cases of peptic ulcer (118). Moreover, patients who have bacteria-induced ulceration are seemingly protected from MALT lymphoma and adenocarcinoma. Furthermore, *H. pylori*–infected individuals in countries where helminth infection is high demonstrate a much lower rate of ulcer formation, suggesting a parasite-induced increase in Th2 response, which could balance an otherwise long-lasting and self-destructive cycle of Th1-induced inflammation and ulceration. However, recent studies suggest that a shift toward Th2 polarization promotes the growth of neoplastic cells in the context of *H. pylori* infection, although these studies have not been well duplicated to date.

Despite many unanswered questions, it is clear that the mucosal immune response to *H. pylori* is inadequate to clear the infection, yet pivotal in influencing the course of bacteria-mediated disease. The inability of *H. pylori*–infected patients to clear the bacteria has led to the suspicion that *H. pylori* may interfere with the host's immune response. Antigen processing and presentation are pivotal events in the development of an adaptive immune response. The vacuolating cytotoxin of *H. pylori* impairs these processes (119). An obvious consideration in trying to eradicate this gastric mucosal pathogen is that it is essential to elucidate the mechanisms implicated in skewing the host's T-cell response and how it avoids mucosal immunity.

Intestinal

Pathogenic bacterial colonization of the large and small intestine results in millions of human deaths each year. Various microorganisms, mostly gram-negative pathogenic bacteria, have evolved methods to evade host defenses, and in some cases, to use these defense mechanisms to their own advantage. Coevolution of the human host with these pathogenic bacteria results in burdens for both the host and the invading bacteria; all potentially harmful enterics must overcome physical, chemical, and immune defense barriers to achieve colonization or infection. Conversely, the mucosal immune system of the host must be established in the form of redundant systems of defense to counter the complex invasion strategies of various bacteria.

A properly functional intestinal mucosal immune system is commissioned at birth. Although sterile in utero, the colonization of the intestinal mucosa with commensal bacteria begins within hours of delivery, with a mature flora of over 400 species developing after 3 to 4 weeks. Establishment of these commensal bacteria is important as an innate defense against pathogenic species, and failure of proper colonization, or a dysregulation of this commensal population, is suspected as a primary cause of many intestinal diseases, including ulcerative colitis and Crohn's disease. Therefore, in addi-

tion to providing competition to pathogenic bacteria, the communal flora plays a crucial role in shaping the local immune system that will be charged with defending the host from pathogenic species.

Thus, the response of the intestinal mucosal immune system to invading bacteria takes place within a dynamic environment of both self and nonself; within the discrete immune tissues of the large and small intestine, a diverse arsenal of both innate and adaptive immune weapons is used to prevent the establishment and propagation of harmful bacteria.

Epidemiology

Pathogenic gram-negative enteric bacteria kill more than 3 million people each year. The vast majority of these illnesses are diarrheal/dysenteric in nature. Although several organisms can cause dysentery, indicated by bloody diarrhea, *Shigella* is the most important. *Shigella dysenteriae* type 1 (Sd1) is the most virulent of the four serogroups of *Shigella*. It is the only cause of epidemic dysentery. *Shigella* is a major problem in the developing world, with over 150 million cases and 1 million deaths per year. *Salmonella*, although not as life threatening as *Shigella*, has major health and economic impacts. Every year, the Centers for Disease Control and Prevention (CDC) receives reports of 40,000 cases of salmonellosis in the United States. The agency estimates that 1.4 million people in this country are infected, however, and that 1000 people die each year of salmonellosis. Symptoms are most severe in the elderly, infants, and people with chronic conditions. Currently, there are four recognized classes of enterovirulent *E. coli* (collectively referred to as the EEC group). Like *Salmonella*, the various strains of virulent enteric *E. coli* can cause both a watery and a bloody diarrhea. Enterohemorrhagic *E. coli* (EHEC) O157:H7 is the serotype most commonly associated with hemorrhagic colitis. *Yersinia enterocolitica* is the causative agent of enteric yersiniosis, which results in enterocolitis.

Pathogenesis

Enteric pathogenic organisms utilize a relatively conserved set of molecular tools to evade and suppress the host mucosal immune system, resulting in a vast number of pathologies. The similarities between certain pathogenic bacterial species, combined with advances in genomic investigation, have led to the discovery of the common ancestry of pathogens such as *E. coli* and *S. enteritica*. Indeed, similarities between species at the genotypic and phenotypic levels are striking and provide clues to the relationship and shared strategies of these organisms. Conversely, the genome of a pathogenic organism must retain a certain degree of plasticity to persist in infection, counter immune strategies of the host, and take advantage of their fast reproductive cycles. Decades of investigation of pathogenic bacteria has revealed that some of the most virulent components of their genomes are also the most mobile and transmissible. Examples of these genomic pathogenesis vehicles include plasmids, bacteriophages, and pathogenicity islands. Pathogenicity islands are discrete loci within the genome of certain pathogenic bacteria that are absent in their nonvirulent parental strains. These genomic additions have a high proportion of insertion sequences and encode for virulence factors, including bacteriophage receptors (120).

One such virulence factor carried by many pathogenicity islands is the family of proteins encoding macromolecular secretion systems. Gram-negative bacteria must transport molecules into and out of a dual-membrane system, including pathogenic toxins important in infection. This selective pressure resulted in the evolution of at least five secretion systems, ranging for simple to complex.

Type I Secretion System

This secretion system consists of a complex of three secretory proteins: an inner membrane transport adenosine triphosphatase (ATPase) (termed ABC protein for ATP-binding cassette), which provides the energy for protein secretion; an outer membrane protein, which is exported via the sec pathway; and a membrane fusion protein, which is anchored in the inner membrane and spans the periplasmic space.

Type II Secretion System

The type II system consists of 12 to 14 proteins that transport fully folded proteins from the periplasmic space through the outer membrane (121). *Vibrio cholerae* utilizes a type II secretion system to export its primary virulence factor, cholera toxin. This toxin, through the stimulation of chloride secretion by enterocytes, is responsible for the massive fluid secretion into the lumen, and subsequent diarrheal disease of the host.

Type III Secretion System

This secretion system is found in pathogenic enteric bacteria such as *Salmonella*, *Shigella*, and enteropathogenic *E. coli*. Five classes (virotypes) of *E. coli* that cause diarrheal diseases are now recognized: enterotoxigenic *E. coli* (ETEC), enteroinvasive *E. coli* (EIEC), enterohemorrhagic *E. coli* (EHEC), enteropathogenic *E. coli* (EPEC), and enteroaggregative *E. coli* (EAggEC). Each class falls within a serologic subgroup and causes distinct pathogeneses. This secretory system is encoded by at least 20 genes and exports toxins through both bacterial membranes (122–124). Furthermore, a syringe-like macromolecular projection injects bacterial proteins directly into host cells. Unlike the less-complex type II system used by *V. cholerae* (a noninvasive pathogen), *Salmonella*, *Shigella*, and EPEC use the type III secretion system to gain entry into nonphagocytic cells. Although EPEC uses the type III secretion system to enter the cell, *Yersinia* use this secretion system to avoid uptake by phagocytic cells. This is done by the injection of toxins known as *Yersinia* outer proteins (Yops) into the host cell, which block cytoskeletal changes required for bacterial uptake (125). Yops as well as SipB and IpaB produced by *Salmonella* and *Shigella*, respectively, induce apoptosis of phagocytic cells (126,127).

Type IV Secretion System

Unlike the type III secretion system, which is built from the basic components of the flagellar machinery, the type IV secretion system utilizes bacterial conjugation proteins. To date, three types of substrates have been found to be injected into host cells by this secretion system: DNA conjugation intermediates, pertussis toxin (PT), and monomeric proteins such as the *H. pylori* CagA protein (128).

Type V Secretion System

The type V secretion system sends proteins across the outer membrane via a transmembrane pore formed by a self-encoded β barrel assembly. This secretion system represents the largest family of protein-translocating outer membrane porins in gram-negative bacteria (129).

Epithelial Cell Interactions with Intestinal Pathogens

Shigella

Shigella initially breaches the mucosal epithelium at M cells of the colonic follicle-associated epithelium (FAE) (130). From here, the bacteria have access to the basolateral surface of epithelial cells, which is the most permissive site for cellular invasion. Prior to and after moving inside epithelial cells, however, epithelial cells respond to this pathogen by producing significant amounts of IL-8, leading to the recruitment of neutrophils, which disrupt the tight junctions of the epithelium, and allow *Shigella* direct access to epithelial subsurfaces (131). The induction of this inflammatory cytokine in epithelial cells has been traced to two factors, in addition to the actual bacterial penetration into the epithelial cell. The first factor is LPS contact with apical epithelial surfaces during early infection. Second, *Shigella* elicits massive amounts of IL-1b from macrophages as well as inducing macrophage apoptosis (132). The massive IL-1b response of infected macrophages is a crucial aspect of the inflammatory process leading to clinical disease. Thus, the bacteria are able to use the epithelial host cell response as a means to increase their cellular infection rate via inducing significant epithelial cytokine release as well as physical alteration of the mucosal barrier.

Salmonella

Salmonella, like *Shigella*, subverts the host epithelial cytoskeletal proteins to its own advantage. Upon contact with epithelial cells, *Salmo-nella* causes the enterocyte microvilli to degenerate (51). This is followed by epithelial membrane ruffling localized at the point of bacterial adhesion. Membrane ruffling is accompanied by profuse macropinocytosis, which allows the bacteria entry into the cell. Once inside the host epithelium, *Salmonella* resides within membrane-bound vesicles, and cell morphology returns to its normal state.

The mechanics of *Salmonella* invasion of host epithelial cells is dependent on at least two virulence factor-encoding gene clusters of the bacteria. The first is a pathogenicity island named SPI1 (*Salmonella* pathogenicity island 1), which encodes a type III secretion system that is used to inject virulence proteins into the epithelium. The second is a bacteriophage genome that encodes the virulence factors SopE1 and SopE2, which are injected by the secretion system and act to dramatically increase actin nucleation, a process important in facilitating bacterial entry into the cell induction of ruffling and macropinocytosis. Remarkably, at the same time these membrane-altering proteins are injected into the epithelial cell, a virulence protein found on SPI1 called SptP is also delivered to the host cytoplasm. SptP is antagonistic to SopE and turns off the G proteins activated by SopE, thus facilitating the return of the epithelial cell membrane to its original form (133).

Enteropathogenic E. coli

Although *Shigella* and *Salmonella* secrete proteins through the plasma membrane of their host cells that mediate cell entry directly, EPEC induces microvilli destruction and the formation of pedestals, known as attaching and effacing (A/E). This process requires that a protein called the translocated intimin receptor (Tir) be delivered into the host cell by the type III secretion system. This secretion system and all the proteins necessary for establishing A/E are located on a pathogenicity island termed the locus of enterocyte effacement (LEE). Tir associates intimately with the bacterial outer membrane adhesion, intimin, thus allowing the attachment of EPEC to its host. Consequently, EPEC does not have to search for a eukaryotic receptor; it simply carries its own (134).

Local Immune Response to Intestinal Pathogens

Innate Response

As previously mentioned, the ability of the host to utilize preexisting defenses or mount de-novo barricades against virulent enteric bacteria is critical in escaping serious disease. The innate immune system of the intestine is composed of structural, chemical, macromolecular, and cellular barricades to bacterial invasion and proliferation.

Just as in the stomach, the critical component of the structural portion of the innate immune system of the intestinal tract is the epithelial lining. While providing a physical barrier to pathogenic enterics, the epithelium is also responsible for signaling the initial alarm to the rest of the local immune system that a potential pathogen is present. Most often, this is done in the form of cytokine and chemokine release. The most intensively studied of these molecules, in relation to modulation of the innate response to enteric bacteria, are IL-8, MCP-1, TNF-α, GM-CSF, and IL-6. The release of these molecular sentinels is often initiated by the transmembrane pattern recognition receptors discussed earlier. The TLRs are cornerstones of the innate immune system due to their ability to recognize and react to various patterns unique to pathogens before they have a chance to invade cells. Interestingly, recent investigations have provided results indicating that intestinal epithelial cells contain internal pattern recognition receptors to complement the extracellular TLR receptors. Two such cytosolic receptors are NOD1 and NOD2, which are members of a family of proteins known as mammalian nucleotide binding site (NBS) leucine-rich repeat (LRR) proteins. NOD1 expression is ubiquitous in humans. This protein recognizes the terminal two amino acids found in gram-negative bacteria-derived peptidoglycan. NOD2, found mainly in myeloid cells, is similar to NOD1 and recognizes a different portion of peptidoglycan that is found in gram-positive as well as gram-negative bacteria. The functional results of NOD1,2 binding with their bacterial-product ligands include NF-κB activation and apoptosis potentiation. Furthermore, very recent findings show that NOD2 is involved in the killing of intracellular *Salmonella* in intestinal epithelial cells. The full consequence of NOD1,2-mediated signal transduction is still being investigated, along with the coordination of this response with TLR-induced signaling (135).

A key element of the host innate immune response to these intestinal pathogens is the phagocytic action of both resident and newly recruited cells in the tissue underlying the epithelium. These cells include macrophages, dendritic cells, and neutrophils, and they serve to kill bacteria outright, limit their replication, and release cytokine and chemokine mediators that recruit and activate additional immune cells. Once inside phagosomes, bacteria such as *Salmonella* are susceptible to chemical defenses produced within host cells. Reduced nicotinamide adenine dinucleotide phosphate (NADPH) oxidase, a multicomponent enzyme that catalyzes the reduction of molecular oxygen, and inducible nitric oxide synthase (iNOS) combine to provide both immediate and long-term bacterial killing. Furthermore, NO-derived species, in addition to synergizing with oxyradicals to kill *Salmonella*, exhibit prolonged oxidase-independent bacteriostatic effects (136).

Adaptive Response

The ability of the mucosal innate immune system to halt or slow the spread of infectious enteric microorganisms as a first line of defense is critical to maintaining host health. However, an equally important role for these front-line cells of the MALT is to alert and inform the adaptive immune system of the type of infection and its location. Pathogen-induced production of cytokines and chemokines by enterocytes such as epithelial cells and local macrophages initiates a cascade of activity critical to the adaptive immune response, including dendritic cell maturation, Th1/Th2 polarization, and effector cell chemotaxis. These events, combined with the humoral component of the adaptive immune system, profoundly affect the long-term fate of the invading pathogen and provide the foundation for immunologic memory should the host be challenged in the future.

Clearance of a severe *Salmonella* infection, and the subsequently acquired resistance (immunologic memory), is a classic example of the importance of a combinatorial adaptive immune response to a mucosal pathogen.

Specifically, it is well documented that clearance of this pathogen requires CD28–dependent activation of T cells as well as a specific antibody response (136). *Shigella*, on the other hand, induces an adaptive immune response that is heavily weighted toward humoral effectors. Serum IgG and intestinal secreted IgA targeting LPS and IpaA-D characterize the most important immune response to this pathogen. Although in vivo priming of CD4+ T cells by *Shigella* antigens has been reported, it is doubtful that these cells play a significant role in bacterial clearance because *Shigella* exist almost entirely either within the cytoplasm (beyond the lysosomes of the class II MHC pathway) or extracellularly. Conversely, the MHC class I–mediated CD8$^+$ T-cell response was initially assumed to be an important defense against *Shigella*. However, no significant cytotoxic T-cell priming against this bacteria has been demonstrated. Similarly, the immune response to *Vibrio* relies heavily on antibodies. Specifically, protective immunity is based on secretory IgA, which is successful against *Vibrio* because this pathogen is noninvasive, and not subject to the phagocytic or cytotoxic effectors of the local immune system.

Overt Immune Responses of the Gastrointestinal Tract

The mucosal immune system is a first line of defense against foreign antigens, including microbial and dietary antigens. Under normal circumstances, the mucosal immune system employs tightly regulated dynamic intramucosal communication consisting of unique sites for the induction of an appropriate immunologic homeostasis between the host and mucosal environments. The common mucosal immune system (CMIS), which interacts between inductive (e.g., PP) and effector (e.g., intestinal LP) tissues for the induction of the IgA response, is well characterized. Recent data provide strong evidence for the presence of a CMIS-independent IgA induction pathway. Two distinct subsets of mucosal IgA-committed B cells, termed B1 and B2, which were described earlier, are associated with CMIS-independent and CMIS-dependent cascades, respectively. In some cases, the breakdown of this tightly regulated mucosal immune system leads to pathologic responses to different gut environmental antigens. As a result, disorders such as allergic gastroenteropathy and inflammatory bowel disease (IBD), can be induced in the GI tract.

Disease Mechanisms of Food Allergy

Adverse reactions to foods are caused by an inappropriate immune response to the ingestion of food antigens. Food allergy has to be differentiated from food intolerance, an adverse reaction to food not linked to an abnormal immune response. In contrast, true food allergy is an immunologically mediated process occurring only in susceptible individuals in response to specific food antigens. The gastrointestinal tract plays an important role in protecting the host against the development of allergic reactions. Two main mechanisms that seem to be important are limitation of the absorption of foreign antigens across the digestive epithelium and control of the systemic immune response to these antigens. Each part of the digestive tract plays a role in this process, and the intestine is considered a central part. However, the role of the gastric barrier has also been recognized. This barrier has a physical double-layer structure, comprising the epithelial cells covered by an adherent mucus layer. The integrity of the barrier is assured by the continuity of the epithelial cell layer maintained by the intercellular tight junctions (137,138), by the integrity of epithelial cell membranes, and by the thickness and composition of the mucus layer. In normal conditions, the gastric barrier constitutes an almost total barrier to the retrodiffusion of ions (H$^+$ ions into the gastric wall and Na$^+$ ions into the lumen). Physical abrasion as well as chemical or bacterial agents may damage the barrier at any time, and may lead to increased passage of various molecules (antigens) across this barrier.

The gastric epithelium also constitutes an important barrier against the penetration of bacterial, viral, and food antigens into the small intestine. Studies have shown that achlorhydria may be associated with an increased proliferation of gram-positive bacteria in the intestine, with an increased incidence of GI infections and with hypersensitivity reactions to macromolecular antigens (139). On the other hand, animals fed with bicarbonate mixed with proteins had increased intestinal transport of macromolecules, suggesting that protein hydrolysis may affect the antigenic properties of proteins or the amount of protein antigens being absorbed.

Multiple lines of evidence suggest that both the intestine and the stomach are potential targets for allergic sensitization. The intestine, with its largely developed immune (lymphocytic cells) system, is classically considered a central organ of food sensitivity reactions. It has also been shown, however, that the gastric epithelium, like the small intestine epithelium, is able to absorb small amounts of macromolecules and that this antigen absorption may induce the IgE-mediated sensitivity reactions to these antigens (140).

The role of genetic background in allergic diseases, including food allergy, is strongly supported by clinical observations and epidemiologic studies showing that heredity plays a significant role in the development of sensitization to food antigens, and that a higher frequency of these diseases is observed in twins (141,142) and in children of parents with allergic diseases than in the control population. Other studies on twins found that quantifiable traits associated with allergic diseases, such as total serum IgE levels and skin test results, show intrapair correlation coefficients twofold higher for monozygotic than for dizygotic twins. As a minimum, two independently segregating disease-susceptibility genes are thought to come together with environmental factors to result in allergic inflammation in a particular tissue. The genetic studies have implicated multiple regions in the human and mouse genomes that are currently being evaluated for harboring putative atopy genes (143). For instance, susceptibility to peanut allergy could be determined by the HLA class II genetic polymorphism. It is not known why a specific antigen leads to an abnormal immune response, but environmental factors, such as bacterial and viral stimuli, probably also play a role. These factors could modify the intestinal permeability to food antigens or activate the co-stimulatory molecules at the surface of local APCs, thus favoring the development of an immune response instead of the normal suppressive response, which is the basis of oral tolerance.

Pathogens and Food Allergy

Among environmental factors that modulate oral tolerance, the bacterial intestinal microflora is an important one. Contradictory data exist regarding the effect of intestinal microorganisms on the immune response to luminal antigens. Although the commensal microflora is necessary for full induction and maintenance of oral tolerance (144), including the IgE production system (145), a strong immune response to an orally administered antigen is obtained when the antigen encounters the GALT together with microorganisms capable of stimulating antigen presentation (144). This has been recently explained by the fact that dendritic cells of the intestinal mucosa play an important role in inducing oral tolerance (146) and by the fact that the regulation of intestinal responses to soluble antigens through dendritic cell presentation depends on the presence or absence of inflammatory signals. It also is significant that nonpathogenic enteric bacteria, interacting directly with human epithelial cells grown in vitro, have been shown to attenuate the synthesis of proinflammatory effector molecules such as NF-κB elicited by diverse proinflammatory stimuli including pathogenic bacteria. The mechanism of an inhibitory effect consists of the blockade of inhibitory κ-β-α degradation, preventing subsequent nuclear translocation of the active NF-κB dimer and then the transcription of genes coding for inflammatory cytokines.

Inflammatory Bowel Disease

The mucosal immune system faces the delicate task of coexistence with a copious commensal intestinal bacterial flora (10^{12} bacteria per gram of feces in the colon, and roughly 10^3 different species, with anaerobes predominating). Yet a protective immune response to invasive enteric pathogens is also mandatory. Any commensal organism has the potential to become a pathogen in the appropriate circumstances, and the magnitude of this balancing act is illustrated by the similarity between proteins of the harmless commensal *E. coli* and its pathogenic derivatives (or the *Shigella* genus). The essential differences between innocent and harmful bacteria reside in toxin production and qualities of adherence to, or penetration of, the intestinal epithelial cell layer.

Observations from multiple studies support the notion that IBDs result from an activation of immune and inflammatory responses initiated by a stimulation of the luminal flora or their products. Genetically determined variations in key mucosal functions, including cell activation by prototypic bacterial molecular patterns, lead to differential susceptibility to the development of these disorders, probably reflecting an interrelated activation of the innate and adaptive immune responses. The persistence and amplification of inflammation is likely to reflect the continuing presence of the driving stimulus and the complex, self-reinforcing activation of select T-helper subtypes and macrophages and other APCs, mediated by several cytokines. These cytokines include IL-2, -12, and -18 as well as IFN and macrophage migration inhibitory factors. The production of other broadly proinflammatory cytokines, most notably TNF and IL-1 and -6, enhances related inflammatory processes that eventually lead to many of the clinical manifestations of IBD. The overall severity of the inflammatory process reflects a balance between leukocyte recruitment and downregulatory mucosal repair processes.

Over the past couple of years the study of genetically manipulated rodents has contributed enormously to the understanding of the circumstances that predispose to intestinal inflammation. Ablating the function of a large number of different immunologic genes, including IL-2, IL-10, or α T-cell receptor, or inserting HLA-B27, each independently renders the animal liable to develop spontaneous intestinal inflammation that may usually be attenuated or avoided by breeding and keeping the animals in very clean (SPF) or germ-free facilities. Although genetic loci linked to human IBD have been described, the hunt for the genes themselves continues, so many of the animal genetic abnormalities may be somewhat artificial. Nevertheless, they do provide support, in well-defined conditions, for the concept that upsetting the delicate balance among the mucosal immune system, the epithelial cell layer, and the commensal bacterial flora results in chronic intestinal inflammation.

Duchmann et al. (147) have examined the reactivity of T-cell clones, derived from IBD intestinal mucosa, against commensal bacteria. It is clear that the mucosa of active Crohn's disease contains an increased proportion of activated T cells, and T-cell cloning has generally proved a powerful immunologic technique, as it provides a culture of T cells with a single receptor with specificity for short peptide epitopes (9 to 15 amino acids long). From such clones the major antigenic determinants for helper and cytotoxic T cells in viral and bacterial infections have been elucidated. These groups have previously presented data that T cells isolated from the intestinal mucosa of control subjects proliferate in vitro in response to relatively crude fractions of bacteria isolated from the intestinal (heterologous) flora of a different individual, but not from their own flora. Interestingly, patients with Crohn's disease have shown intestinal T cells capable of responding to their own (autologous) flora. Thus Crohn's disease could be interpreted as a failure of mucosal tolerance to the indigenous flora, an idea that is in keeping with the data from the animal models and with the clinical effectiveness of fecal stream diversion. Despite the diversity of the human intestinal microflora, there is considerable homology between proteins of related species and common carriage of many species by different individuals, so the differences that bring about responses to the heterologous flora in normal subjects are still unclear.

It has been difficult to get T cells from the intestinal mucosa to proliferate well in response to antigens, so to produce clones of identical cells, the stimulation process had to be nonspecific phytohemagglutinin (PHA) followed by expansion on irradiated allogenic feeder cells (the classic way). Although the idea was to obtain representative clones, the responses to bacterial sonicates may not reflect the antigen specificities of the initial T cells. With this cocktail, there were three main possible T-cell responses. First, there was considerable cross-reactivity in the response of CD4+ clones to anaerobic (Bifidobacterium and Bacteroides) and aerobic enterobacteria. Second, there were cells from patients with IBD responding to crude preparations of the autologous flora (T-cell clones). Third, the authors analyzed which bacterial species within a heterologous mixed isolate could stimulate a T-cell clone from a patient with ulcerative colitis and showed that aerobic enterobacteria were mainly responsible, and curiously some colonies of a bacterial species (e.g., E. coli) might stimulate this clone whereas others would not.

The data suggest that there seems to be cross-reactivity in the proliferative responses of T-cell clones from patients with IBD between different bacterial species. Also, the responses to the heterologous flora involve many common aerobic species. The question that needs to be addressed is in which moment "tolerance" to the autologous flora develops or collapses at the T-cell level. The beauty of T-cell clones is that specificities to individual protein molecules (or other structural bacterial components) can be determined, if bacterial proteins are first purified, and this could sort out the molecular basis of cross-reactivity. Unfortunately, each person is likely to be different because of the diversity in the human population of MHC class II, which present antigenic peptides to $CD4^+$ T cells. Therefore, in addition to the presence or absence of LP $CD4^+$ T cells that respond to commensal bacterial determinants, there are many other levels of regulation, including unresponsive T cells and those that produce downregulatory cytokines but do not proliferate. The relative contributions of these mechanisms in health and their defects in IBD are still under study.

Conclusion

We have reviewed the current state of knowledge on the mucosal immune system of the human GI tract. From the barrier defenses of the epithelial mucosa and its pattern recognition receptors to the molecular defenses of antimicrobial peptides, the local innate immune system is a critical component of the mucosal immune system of the gut. The pathogen-specific immunity provided by local lymphocytes completes the immune response to enteric pathogens and often provides extended immunity via memory cells. Also discussed were commensal bacteria that play a critical role in the maturation of the mucosal immune system, as well as providing an innate defense via competition with pathogenic strains. The importance of these probiotics is realized by observing not only what they keep out, but also the deleterious effects of an inappropriate response to them. We have reviewed some of the overt actions that can occur in the GI tract when the delicate balance between immunity and tolerance to these microorganisms is disrupted.

Indeed, the interactions between the human mucosal immune system and pathogenic enteric bacteria represent an extremely dynamic model of coevolution. The complex and multilayered features of the human GI immune system are constantly challenged by the relatively limited tools of any given enteric pathogen. However, bacteria are able to compensate for a small genome that encodes a discrete set of molecular weapons by adapting to and compensating for the defenses of the host. This is accomplished by virtue of bacteria's fast life cycle and multiple genetic tools enabling the horizontal and vertical transfer of virulence factors. Moreover, not only do some virulent bacteria evade eukaryotic immune cells and their defensive chemical products, they often use these immune cells as the primary staging area for their pathogenesis. Indeed, sometimes the clearest insight into host defense strategy is gained by elucidating bacterial offensive strategy. What then prevents such adaptable microorganisms from overwhelming their hosts much more regularly than they do? The answer most certainly lies in the redundancy of the human immune system. The innate and acquired immune effectors found in the GI mucosa combine to provide a multitiered system of defense that is profoundly organized by way of structure and function. Certainly, the testament to the success of the mucosal immune system lies in the staggering plasticity of bacterial pathogenic strategy required to overcome it.

Acknowledgments

The authors wish to acknowledge the support from different sources. David A. Bland was a recipient of a fellowship from the McLaughlin Fellowship Fund and was also supported by the Immunology and Mucosal Defense Training Program (T32 AI007626). This work was also supported by National Institutes of Health grant DK050669 (V.E.R.) and by Texas Gulf Coast Digestive Diseases Center grant DK056338.

References

1. Youakim A, Ahdieh M. Interferon-gamma decreases barrier function in T84 cells by reducing ZO-1 levels and disrupting apical actin. Am J Physiol 1999;276(5 pt 1):G1279–G1288.
2. Kucharzik T, Walsh SV, Chen J, Parkos CA, Nusrat A. Neutrophil transmigration in inflammatory bowel disease is associated with differential expression of epithelial intercellular junction proteins. Am J Pathol 2001;159(6):2001–2009.
3. Dignass A, Lynch-Devaney K, Kindon H, Thim L, Podolsky DK. Trefoil peptides promote epithelial migration through a transforming growth factor beta-independent pathway. J Clin Invest 1994;94(1):376–383.
4. Dignass AU, Podolsky DK. Cytokine modulation of intestinal epithelial cell restitution: central role of transforming growth factor beta. Gastroenterology 1993;105(5):1323–1332.

5. Eckmann L, Kagnoff MF, Fierer J. Epithelial cells secrete the chemokine interleukin-8 in response to bacterial entry. Infect Immun 1993;61(11):4569–4574.

6. Yang SK, Eckmann L, Panja A, Kagnoff MF. Differential and regulated expression of C-X-C, C-C, and C-chemokines by human colon epithelial cells. Gastroenterology 1997;113(4):1214–1223.

7. Reinecker HC, Loh EY, Ringler DJ, Mehta A, Rombeau JL, MacDermott RP. Monocyte-chemoattractant protein 1 gene expression in intestinal epithelial cells and inflammatory bowel disease mucosa. Gastroenterology 1995;108(1):40–50.

8. Izadpanah A, Dwinell MB, Eckmann L, Varki NM, Kagnoff MF. Regulated MIP-3alpha/CCL20 production by human intestinal epithelium: mechanism for modulating mucosal immunity. Am J Physiol Gastrointest Liver Physiol 2001;280(4):G710–G719.

9. Dwinell MB, Lugering N, Eckmann L, Kagnoff MF. Regulated production of interferon-inducible T-cell chemoattractants by human intestinal epithelial cells. Gastroenterology 2001;120(1):49–59.

10. Shibahara T, Wilcox JN, Couse T, Madara JL. Characterization of epithelial chemoattractants for human intestinal intraepithelial lymphocytes. Gastroenterology 2001;120(1):60–70.

11. Kunkel EJ, Campbell JJ, Haraldsen G, et al. Lymphocyte CC chemokine receptor 9 and epithelial thymus-expressed chemokine (TECK) expression distinguish the small intestinal immune compartment: epithelial expression of tissue-specific chemokines as an organizing principle in regional immunity. J Exp Med 2000; 192(5):761–768.

12. Wurbel MA, Philippe JM, Nguyen C, et al. The chemokine TECK is expressed by thymic and intestinal epithelial cells and attracts double- and single-positive thymocytes expressing the TECK receptor CCR9. Eur J Immunol 2000;30(1):262–271.

13. Pan J, Kunkel EJ, Gosslar U, et al. A novel chemokine ligand for CCR10 and CCR3 expressed by epithelial cells in mucosal tissues. J Immunol 2000;165(6): 2943–2949.

14. Wang W, Soto H, Oldham ER, et al. Identification of a novel chemokine (CCL28), which binds CCR10 (GPR2). J Biol Chem 2000;275(29):22313–22323.

15. Jung HC, Eckmann L, Yang SK, et al. A distinct array of proinflammatory cytokines is expressed in human colon epithelial cells in response to bacterial invasion. J Clin Invest 1995;95(1):55–65.

16. Reinecker HC, MacDermott RP, Mirau S, Dignass A, Podolsky DK. Intestinal epithelial cells both express and respond to interleukin 15. Gastroenterology 1996;111(6):1706–1713.

17. Maaser C, Schoeppner S, Kucharzik T, et al. Colonic epithelial cells induce endothelial cell expression of ICAM-1 and VCAM-1 by a NF-kappaB-dependent mechanism. Clin Exp Immunol 2001;124(2):208–213.

18. Maaser C, Eckmann L, Paesold G, Kim HS, Kagnoff MF. Ubiquitous production of macrophage migration inhibitory factor by human gastric and intestinal epithelium. Gastroenterology 2002;122(3):667–680.

19. Jordan NJ, Kolios G, Abbot SE, et al. Expression of functional CXCR4 chemokine receptors on human colonic epithelial cells. J Clin Invest 1999;104(8): 1061–1069.

20. Colgan SP, Hershberg RM, Furuta GT, Blumberg RS. Ligation of intestinal epithelial CD1d induces bio-active IL-10: critical role of the cytoplasmic tail in autocrine signaling. Proc Natl Acad Sci USA 1999;96(24):13938–13943.

21. Barrera C, Ye G, Espejo R, et al. Expression of cathepsins B, L, S, and D by gastric epithelial cells implicates them as antigen presenting cells in local immune responses. Hum Immunol 2001;62(10):1081–1091.

22. Li Y, Yio XY, Mayer L. Human intestinal epithelial cell-induced CD8+ T cell activation is mediated through CD8 and the activation of CD8-associated p56lck. J Exp Med 1995;182(4):1079–1088.

23. Toy LS, Yio XY, Lin A, Honig S, Mayer L. Defective expression of gp180, a novel CD8 ligand on intestinal epithelial cells, in inflammatory bowel disease. J Clin Invest 1997;100(8):2062–2071.

24. Nakazawa A, Dotan I, Brimnes J, et al. The expression and function of costimulatory molecules B7H and B7-H1 on colonic epithelial cells. Gastroenterology 2004;126(5):1347–1357.

25. Dwinell MB, Eckmann L, Leopard JD, Varki NM, Kagnoff MF. Chemokine receptor expression by human intestinal epithelial cells. Gastroenterology 1999;117(2):359–367.

26. Brumell JH, Steele-Mortimer O, Finlay BB. Bacterial invasion: force feeding by Salmonella. Curr Biol 1999; 9(8):R277–R280.

27. Galan JE. Salmonella interactions with host cells: type III secretion at work. Annu Rev Cell Dev Biol 2001;17:53–86.

28. Kaisho T, Akira S. Critical roles of Toll-like receptors in host defense. Crit Rev Immunol 2000;20(5):393–405.

29. Kelly CP, O'Keane JC, Orellana J, et al. Human colon cancer cells express ICAM-1 in vivo and support LFA-1-dependent lymphocyte adhesion in vitro. Am J Physiol 1992;263(6 pt 1):G864–G870.

30. Cario E, Rosenberg IM, Brandwein SL, Beck PL, Reinecker HC, Podolsky DK. Lipopolysaccharide activates distinct signaling pathways in intestinal epithelial cell lines expressing Toll-like receptors. J Immunol 2000;164(2):966–972.

31. Gewirtz AT, Navas TA, Lyons S, Godowski PJ, Madara JL. Cutting edge: bacterial flagellin activates basolaterally expressed TLR5 to induce epithelial proinflammatory gene expression. J Immunol 2001; 167(4):1882–1885.

32. Dogan A, Wang ZD, Spencer J. E-cadherin expression in intestinal epithelium. J Clin Pathol 1995;48(2): 143–146.

33. Yio XY, Mayer L. Characterization of a 180-kDa intestinal epithelial cell membrane glycoprotein, gp180. A candidate molecule mediating t cell-epithelial cell interactions. J Biol Chem 1997;272(19):12786–12792.

34. Huang GT, Eckmann L, Savidge TC, Kagnoff MF. Infection of human intestinal epithelial cells with invasive bacteria upregulates apical intercellular adhesion molecule-1 (ICAM-1) expression and neutrophil adhesion. J Clin Invest 1996;98(2):572–583.

35. Parkos CA, Colgan SP, Diamond MS, et al. Expression and polarization of intercellular adhesion molecule-1 on human intestinal epithelia: consequences for CD11b/CD18-mediated interactions with neutrophils. Mol Med 1996;2(4):489–505.

36. Yamanaka T, Straumfors A, Morton H, Fausa O, Brandtzaeg P, Farstad I. M cell pockets of human

Peyer's patches are specialized extensions of germinal centers. Eur J Immunol 2001;31(1):107–117.

37. Meijer LK, Schesser K, Wolf-Watz H, Sassone-Corsi P, Pettersson S. The bacterial protein YopJ abrogates multiple signal transduction pathways that converge on the transcription factor CREB. Cell Microbiol 2000;2(3):231–238.

38. Schesser K, Spiik AK, Dukuzumuremyi JM, Neurath MF, Pettersson S, Wolf-Watz H. The yopJ locus is required for Yersinia-mediated inhibition of NF-kappaB activation and cytokine expression: YopJ contains a eukaryotic SH2-like domain that is essential for its repressive activity. Mol Microbiol 1998;28(6): 1067–1079.

39. Lehrer RI, Ganz T. Antimicrobial peptides in mammalian and insect host defence. Curr Opin Immunol 1999;11(1):23–27.

40. Agerberth B, Charo J, Werr J, et al. The human antimicrobial and chemotactic peptides LL-37 and alpha-defensins are expressed by specific lymphocyte and monocyte populations. Blood 2000;96(9):3086–3093.

41. Ouellette AJ, Bevins CL. Paneth cell defensins and innate immunity of the small bowel. Inflamm Bowel Dis 2001;7(1):43–50.

42. Lehrer RI, Ganz T. Defensins of vertebrate animals. Curr Opin Immunol 2002;14(1):96–102.

43. Durr M, Peschel A. Chemokines meet defensins: the merging concepts of chemoattractants and antimicrobial peptides in host defense. Infect Immun 2002; 70(12):6515–6517.

44. Masson PL, Heremans JF. Lactoferrin in milk from different species. Comp Biochem Physiol B 1971;39(1): 119–129.

45. Vorland LH. Lactoferrin: a multifunctional glycoprotein. APMIS 1999;107(11):971–981.

46. Ellison RT, III, Giehl TJ, LaForce FM. Damage of the outer membrane of enteric gram-negative bacteria by lactoferrin and transferrin. Infect Immun 1988;56(11): 2774–2781.

47. Tomita M, Bellamy W, Takase M, Yamauchi K, Wakabayashi H, Kawase K. Potent antibacterial peptides generated by pepsin digestion of bovine lactoferrin. J Dairy Sci 1991;74(12):4137–4142.

48. Olson KA, Verselis SJ, Fett JW. Angiogenin is regulated in vivo as an acute phase protein. Biochem Biophys Res Commun 1998;242(3):480–483.

49. Strydom DJ. The angiogenins. Cell Mol Life Sci 1998; 54(8):811–824.

50. Hooper LV, Stappenbeck TS, Hong CV, Gordon JI. Angiogenins: a new class of microbicidal proteins involved in innate immunity. Nat Immunol 2003; 4(3):269–273.

51. Takeuchi A. Electron microscope studies of experimental Salmonella infection. I. Penetration into the intestinal epithelium by Salmonella typhimurium. Am J Pathol 1967;50(1):109–136.

52. Alexopoulou L, Holt AC, Medzhitov R, Flavell RA. Recognition of double-stranded RNA and activation of NF-kappaB by Toll-like receptor 3. Nature 2001;413(6857):732–738.

53. Means TK, Hayashi F, Smith KD, Aderem A, Luster AD. The Toll-like receptor 5 stimulus bacterial flagellin induces maturation and chemokine production in human dendritic cells. J Immunol 2003;170(10): 5165–5175.

54. Zhou X, Giron JA, Torres AG, et al. Flagellin of enteropathogenic Escherichia coli stimulates interleukin-8 production in T84 cells. Infect Immun 2003;71(4):2120–2129.

55. Into T, Kiura K, Yasuda M, et al. Stimulation of human Toll-like receptor(TLR) 2 and TLR6 with membrane lipoproteins of Mycoplasma fermentans induces apoptotic cell death after NF-kappa B activation. Cell Microbiol 2004;6(2):187–199.

56. Jurk M, Heil F, Vollmer J, et al. Human TLR7 or TLR8 independently confer responsiveness to the antiviral compound R-848. Nat Immunol 2002;3(6):499.

57. Hemmi H, Takeuchi O, Kawai T, et al. A Toll-like receptor recognizes bacterial DNA. Nature 2000;408(6813): 740–745.

58. Goodyear CS, Narita M, Silverman GJ. In vivo VL-targeted activation-induced apoptotic supraclonal deletion by a microbial B cell toxin. J Immunol 2004; 172(5):2870–2877.

59. Regnault A, Cumano A, Vassalli P, Guy-Grand D, Kourilsky P. Oligoclonal repertoire of the CD8 alpha alpha and the CD8 alpha beta TCR-alpha/beta murine intestinal intraepithelial T lymphocytes: evidence for the random emergence of T cells. J Exp Med 1994; 180(4):1345–1358.

60. Rocha B. Characterization of V beta-bearing cells in athymic(nu/nu) mice suggests an extrathymic pathway for T cell differentiation. Eur J Immunol 1990;20(4):919–925.

61. Saito H, Kanamori Y, Takemori T, et al. Generation of intestinal T cells from progenitors residing in gut cryptopatches. Science 1998;280(5361):275–278.

62. Farstad IN, Halstensen TS, Lien B, et al. Distribution of beta 7 integrins in human intestinal mucosa and organized gut-associated lymphoid tissue. Immunology 1996;89(2):227–237.

63. Jones BD, Ghori N, Falkow S. Salmonella typhimurium initiates murine infection by penetrating and destroying the specialized epithelial M cells of the Peyer's patches. J Exp Med 1994;180(1):15–23.

64. Kawanishi H, Saltzman L, Strober W. Mechanisms regulating IgA class-specific immunoglobulin production in murine gut-associated lymphoid tissues. II. Terminal differentiation of postswitch sIgA-bearing Peyer's patch B cells. J Exp Med 1983;158(3):649–669.

65. Kawanishi H, Saltzman LE, Strober W. Mechanisms regulating IgA class-specific immunoglobulin production in murine gut-associated lymphoid tissues. I. T cells derived from Peyer's patches that switch sIgM B cells to sIgA B cells in vitro. J Exp Med 1983; 157(2):433–450.

66. Fukaura H, Kent SC, Pietrusewicz MJ, Khoury SJ, Weiner HL, Hafler DA. Induction of circulating myelin basic protein and proteolipid protein-specific transforming growth factor-beta1-secreting Th3 T cells by oral administration of myelin in multiple sclerosis patients. J Clin Invest 1996;98(1):70–77.

67. London SD, Rubin DH, Cebra JJ. Gut mucosal immunization with reovirus serotype 1/L stimulates virus-specific cytotoxic T cell precursors as well as IgA memory cells in Peyer's patches. J Exp Med 1987; 165(3):830–847.

68. Farstad IN, Halstensen TS, Fausa O, Brandtzaeg P. Heterogeneity of M-cell-associated B and T cells in human Peyer's patches. Immunology 1994;83(3):457–464.

69. Iwasaki A, Kelsall BL. Localization of distinct Peyer's patch dendritic cell subsets and their recruitment by chemokines macrophage inflammatory protein(MIP)-3alpha, MIP-3beta, and secondary lymphoid organ chemokine. J Exp Med 2000;191(8):1381–1394.

70. Iwasaki A, Kelsall BL. Unique functions of CD11b+, CD8 alpha+, and double-negative Peyer's patch dendritic cells. J Immunol 2001;166(8):4884–4890.

71. Schieferdecker HL, Ullrich R, Hirseland H, Zeitz M. T cell differentiation antigens on lymphocytes in the human intestinal lamina propria. J Immunol 1992;149(8):2816–2822.

72. De Maria R, Boirivant M, Cifone MG, et al. Functional expression of Fas and Fas ligand on human gut lamina propria T lymphocytes. A potential role for the acidic sphingomyelinase pathway in normal immunoregulation. J Clin Invest 1996;97(2):316–322.

73. Bowman EP, Kuklin NA, Youngman KR, et al. The intestinal chemokine thymus-expressed chemokine (CCL25) attracts IgA antibody-secreting cells. J Exp Med 2002;195(2):269–275.

74. Coffman RL, Lebman DA, Shrader B. Transforming growth factor beta specifically enhances IgA production by lipopolysaccharide-stimulated murine B lymphocytes. J Exp Med 1989;170(3):1039–1044.

75. Crago SS, Kutteh WH, Moro I, et al. Distribution of IgA1-, IgA2-, and J chain-containing cells in human tissues. J Immunol 1984;132(1):16–18.

76. Faure F, Jitsukawa S, Triebel F, Hercend T. Characterization of human peripheral lymphocytes expressing the CD3-gamma/delta complex with anti-receptor monoclonal antibodies. J Immunol 1988;141(10):3357–3360.

77. Cepek KL, Shaw SK, Parker CM, et al. Adhesion between epithelial cells and T lymphocytes mediated by E-cadherin and the alpha E beta 7 integrin. Nature 1994;372(6502):190–193.

78. Brandtzaeg P, Farstad IN, Helgeland L. Phenotypes of T cells in the gut. Chem Immunol 1998;71:1–26.

79. Rocha B, Vassalli P, Guy-Grand D. The V beta repertoire of mouse gut homodimeric alpha CD8+ intraepithelial T cell receptor alpha/beta + lymphocytes reveals a major extrathymic pathway of T cell differentiation. J Exp Med 1991;173(2):483–486.

80. Williams N. T cells on the mucosal frontline. Science 1998;280(5361):198–200.

81. Groh V, Steinle A, Bauer S, Spies T. Recognition of stress-induced MHC molecules by intestinal epithelial gammadelta T cells. Science 1998;279(5357):1737–1740.

82. Wu J, Groh V, Spies T. T cell antigen receptor engagement and specificity in the recognition of stress-inducible MHC class I-related chains by human epithelial gamma delta T cells. J Immunol 2002;169(3):1236–1240.

83. Gapin L, Cheroutre H, Kronenberg M. Cutting edge: TCR alpha beta+ CD8 alpha alpha+ T cells are found in intestinal intraepithelial lymphocytes of mice that lack classical MHC class I molecules. J Immunol 1999;163(8):4100–4104.

84. Franco MA, Greenberg HB. Role of B cells and cytotoxic T cells in clearance of and immunity to rotavirus infection in mice. J Virol 1995;69(12):7800–7806.

85. Brandtzaeg P, Farstad IN, Haraldsen G. Regional specialization in the mucosal immune system: primed cells do not always home along the same track. Immunol Today 1999;20(6):267–277.

86. van Egmond M, Damen CA, van Spriel AB, Vidarsson G, van Garderen E, van de Winkel JG. IgA and the IgA Fc receptor. Trends Immunol 2001;22(4):205–211.

87. Conley ME, Delacroix DL. Intravascular and mucosal immunoglobulin A: two separate but related systems of immune defense? Ann Intern Med 1987;106(6):892–899.

88. Macpherson AJ, Hunziker L, McCoy K, Lamarre A. IgA responses in the intestinal mucosa against pathogenic and non-pathogenic microorganisms. Microbes Infect 2001;3(12):1021–1035.

89. Weisbart RH, Kacena A, Schuh A, Golde DW. GM-CSF induces human neutrophil IgA-mediated phagocytosis by an IgA Fc receptor activation mechanism. Nature 1988;332(6165):647–648.

90. Monteiro RC, Kubagawa H, Cooper MD. Cellular distribution, regulation, and biochemical nature of an Fc alpha receptor in humans. J Exp Med 1990;171(3):597–613.

91. Deo YM, Sundarapandiyan K, Keler T, Wallace PK, Graziano RF. Bispecific molecules directed to the Fc receptor for IgA (Fc alpha RI, CD89) and tumor antigens efficiently promote cell-mediated cytotoxicity of tumor targets in whole blood. J Immunol 1998;160(4):1677–1686.

92. Patry C, Herbelin A, Lehuen A, Bach JF, Monteiro RC. Fc alpha receptors mediate release of tumour necrosis factor-alpha and interleukin-6 by human monocytes following receptor aggregation. Immunology 1995;86(1):1–5.

93. van Egmond M, van Garderen E, van Spriel AB, et al. FcalphaRI-positive liver Kupffer cells: reappraisal of the function of immunoglobulin A in immunity. Nat Med 2000;6(6):680–685.

94. Kaetzel CS, Robinson JK, Lamm ME. Epithelial transcytosis of monomeric IgA and IgG cross-linked through antigen to polymeric IgA. A role for monomeric antibodies in the mucosal immune system. J Immunol 1994;152(1):72–76.

95. Kaetzel CS, Robinson JK, Chintalacharuvu KR, Vaerman JP, Lamm ME. The polymeric immunoglobulin receptor (secretory component) mediates transport of immune complexes across epithelial cells: a local defense function for IgA. Proc Natl Acad Sci U S A 1991;88(19):8796–8800.

96. Orlans E, Peppard J, Reynolds J, Hall J. Rapid active transport of immunoglobulin A from blood to bile. J Exp Med 1978;147(2):588–592.

97. Jackson GD, Lemaitre-Coelho I, Vaerman JP, Bazin H, Beckers A. Rapid disappearance from serum of intravenously injected rat myeloma IgA and its secretion into bile. Eur J Immunol 1978;8(2):123–126.

98. Kramer DR, Cebra JJ. Early appearance of "natural" mucosal IgA responses and germinal centers in suckling mice developing in the absence of maternal antibodies. J Immunol 1995;154(5):2051–2062.

99. Mackie RI, Sghir A, Gaskins HR. Developmental microbial ecology of the neonatal gastrointestinal tract. Am J Clin Nutr 1999;69(5):1035S–1045S.

100. Kang HS, Chin RK, Wang Y, et al. Signaling via LTbetaR on the lamina propria stromal cells of the gut is required for IgA production. Nat Immunol 2002;3(6):576–582.

101. Macpherson AJ, Gatto D, Sainsbury E, Harriman GR, Hengartner H, Zinkernagel RM. A primitive T cell-independent mechanism of intestinal mucosal IgA responses to commensal bacteria. Science 2000; 288(5474):2222–2226.

102. Fagarasan S, Honjo T. Intestinal IgA synthesis: regulation of front-line body defences. Nat Rev Immunol 2003;3(1):63–72.

103. McWilliams M, Phillips-Quagliata JM, Lamm ME. Mesenteric lymph node B lymphoblasts which home to the small intestine are precommitted to IgA synthesis. J Exp Med 1977;145(4):866–875.

104. Newberry RD, McDonough JS, McDonald KG, Lorenz RG. Postgestational lymphotoxin/lymphotoxin beta receptor interactions are essential for the presence of intestinal B lymphocytes. J Immunol 2002;168(10): 4988–4997.

105. Kunimoto DY, Harriman GR, Strober W. Regulation of IgA differentiation in CH12LX B cells by lymphokines. IL-4 induces membrane IgM-positive CH12LX cells to express membrane IgA and IL-5 induces membrane IgA-positive CH12LX cells to secrete IgA. J Immunol 1988;141(3):713–720.

106. Defrance T, Vanbervliet B, Briere F, Durand I, Rousset F, Banchereau J. Interleukin 10 and transforming growth factor beta cooperate to induce anti-CD40-activated naive human B cells to secrete immunoglobulin A. J Exp Med 1992;175(3):671–682.

107. Beagley KW, Eldridge JH, Kiyono H, et al. Recombinant murine IL-5 induces high rate IgA synthesis in cycling IgA-positive Peyer's patch B cells. J Immunol 1988;141(6):2035–2042.

108. Kunimoto DY, Nordan RP, Strober W. IL-6 is a potent cofactor of IL-1 in IgM synthesis and of IL-5 in IgA synthesis. J Immunol 1989;143(7):2230–2235.

109. Arcus VL, Langley R, Proft T, Fraser JD, Baker EN. The Three-dimensional structure of a superantigen-like protein, SET3, from a pathogenicity island of the Staphylococcus aureus genome. J Biol Chem 2002; 277(35):32274–32281.

110. Fan X, Gunasena H, Cheng Z, et al. Helicobacter pylori urease binds to class II MHC on gastric epithelial cells and induces their apoptosis. J Immunol 2000;165(4): 1918–1924.

111. Crabtree JE, Farmery SM, Lindley IJ, Figura N, Peichl P, Tompkins DS. CagA/cytotoxic strains of Helicobacter pylori and interleukin-8 in gastric epithelial cell lines. J Clin Pathol 1994;47(10):945–950.

112. Crowe SE, Alvarez L, Dytoc M, et al. Expression of interleukin 8 and CD54 by human gastric epithelium after Helicobacter pylori infection in vitro. Gastroenterology 1995;108(1):65–74.

113. Schmausser B, Andrulis M, Endrich S, et al. Expression and subcellular distribution of toll-like receptors TLR4, TLR5 and TLR9 on the gastric epithelium in Helicobacter pylori infection. Clin Exp Immunol 2004;136(3):521–526.

114. Backhed F, Rokbi B, Torstensson E, et al. Gastric mucosal recognition of Helicobacter pylori is independent of Toll-like receptor 4. J Infect Dis 2003;187(5):829–836.

115. Amieva MR, Vogelmann R, Covacci A, Tompkins LS, Nelson WJ, Falkow S. Disruption of the epithelial apical-junctional complex by Helicobacter pylori CagA. Science 2003;300(5624):1430–1434.

116. Bamford KB, Fan X, Crowe SE, et al. Lymphocytes in the human gastric mucosa during Helicobacter pylori have a T helper cell 1 phenotype. Gastroenterology 1998;114(3):482–492.

117. Karttunen R, Karttunen T, Ekre HP, MacDonald TT. Interferon gamma and interleukin 4 secreting cells in the gastric antrum in Helicobacter pylori positive and negative gastritis. Gut 1995;36(3):341–345.

118. D'Elios MM, Manghetti M, Almerigogna F, et al. Different cytokine profile and antigen-specificity repertoire in Helicobacter pylori-specific T cell clones from the antrum of chronic gastritis patients with or without peptic ulcer. Eur J Immunol 1997;27(7): 1751–1755.

119. Molinari M, Salio M, Galli C, et al. Selective inhibition of Ii-dependent antigen presentation by Helicobacter pylori toxin VacA. J Exp Med 1998;187(1):135–140.

120. Perna NT, Mayhew GF, Posfai G, et al. Molecular evolution of a pathogenicity island from enterohemorrhagic Escherichia coli O157:H7. Infect Immun 1998;66(8):3810–3817.

121. Russel M. Macromolecular assembly and secretion across the bacterial cell envelope: type II protein secretion systems. J Mol Biol 1998;279(3):485–499.

122. Kubori T, Matsushima Y, Nakamura D, et al. Supramolecular structure of the Salmonella typhimurium type III protein secretion system. Science 1998;280(5363): 602–605.

123. Blocker A, Gounon P, Larquet E, et al. The tripartite type III secreton of Shigella flexneri inserts IpaB and IpaC into host membranes. J Cell Biol 1999;147(3): 683–693.

124. Knutton S, Rosenshine I, Pallen MJ, et al. A novel EspA-associated surface organelle of enteropathogenic Escherichia coli involved in protein translocation into epithelial cells. EMBO J 1998;17(8):2166–2176.

125. Donnenberg MS. Pathogenic strategies of enteric bacteria. Nature 2000;406(6797):768–774.

126. Hersh D, Monack DM, Smith MR, Ghori N, Falkow S, Zychlinsky A. The Salmonella invasin SipB induces macrophage apoptosis by binding to caspase-1. Proc Natl Acad Sci U S A 1999;96(5):2396–2401.

127. Hilbi H, Moss JE, Hersh D, et al. Shigella-induced apoptosis is dependent on caspase-1 which binds to IpaB. J Biol Chem 1998;273(49):32895–32900.

128. Christie PJ, Vogel JP. Bacterial type IV secretion: conjugation systems adapted to deliver effector molecules to host cells. Trends Microbiol 2000;8(8):354–360.

129. Yen MR, Peabody CR, Partovi SM, Zhai Y, Tseng YH, Saier MH. Protein-translocating outer membrane porins of gram-negative bacteria. Biochim Biophys Acta 2002;1562(1–2):6–31.

130. Sansonetti PJ, Arondel J, Cantey JR, Prevost MC, Huerre M. Infection of rabbit Peyer's patches by Shigella flexneri: effect of adhesive or invasive bacterial phenotypes on follicle-associated epithelium. Infect Immun 1996;64(7):2752–2764.

131. Perdomo JJ, Gounon P, Sansonetti PJ. Polymorphonuclear leukocyte transmigration promotes invasion of colonic epithelial monolayer by Shigella flexneri. J Clin Invest 1994;93(2):633–643.

132. Zychlinsky A, Fitting C, Cavaillon JM, Sansonetti PJ. Interleukin 1 is released by murine macrophages during apoptosis induced by Shigella flexneri. J Clin Invest 1994;94(3):1328–1332.

133. Stebbins CE, Galan JE. Modulation of host signaling by a bacterial mimic: structure of the Salmonella effector SptP bound to Rac1. Mol Cell 2000;6(6):1449–1460.

134. Celli J, Deng W, Finlay BB. Enteropathogenic Escherichia coli (EPEC) attachment to epithelial cells: exploiting the host cell cytoskeleton from the outside. Cell Microbiol 2000;2(1):1–9.

135. Hisamatsu T, Suzuki M, Reinecker HC, Nadeau WJ, McCormick BA, Podolsky DK. CARD15/NOD2 functions as an antibacterial factor in human intestinal epithelial cells. Gastroenterology 2003;124(4):993–1000.

136. Mastroeni P. Immunity to systemic Salmonella infections. Curr Mol Med 2002;2(4):393–406.

137. Karczewski J, Groot J. Molecular physiology and pathophysiology of tight junctions III. Tight junction regulation by intracellular messengers: differences in response within and between epithelia. Am J Physiol Gastrointest Liver Physiol 2000;279(4):G660–G665.

138. Gasbarrini G, Montalto M. Structure and function of tight junctions. Role in intestinal barrier. Ital J Gastroenterol Hepatol 1999;31(6):481–488.

139. Sarker SA, Gyr K. Non-immunological defence mechanisms of the gut. Gut 1992;33(7):987–993.

140. Sampson HA. Food allergy: immunology of the GI mucosa towards classification and understanding of GI hypersensitivities. Pediatr Allergy Immunol 2001; 12(Suppl 14):7–9.

141. Bardella MT, Fredella C, Prampolini L, Marino R, Conte D, Giunta AM. Gluten sensitivity in monozygous twins: a long-term follow-up of five pairs. Am J Gastroenterol 2000;95(6):1503–1505.

142. Sicherer SH, Furlong TJ, Maes HH, Desnick RJ, Sampson HA, Gelb BD. Genetics of peanut allergy: a twin study. J Allergy Clin Immunol 2000;106(1 pt 1): 53–56.

143. Pietrzyk JJ. [Genetic background of atopy]. Pediatr Pol 1996;71(1):7–10.

144. Par A. Gastrointestinal tract as a part of immune defence. Acta Physiol Hung 2000;87(4):291–304.

145. Sudo N, Sawamura S, Tanaka K, Aiba Y, Kubo C, Koga Y. The requirement of intestinal bacterial flora for the development of an IgE production system fully susceptible to oral tolerance induction. J Immunol 1997;159(4):1739–1745.

146. Mowat AM, Donachie AM, Parker LA, et al. The role of dendritic cells in regulating mucosal immunity and tolerance. Novartis Found Symp 2003;252:291–302.

147. Duchmann R, Neurath MF, Meyer zum Buschenfelde KH. Responses to self and non-self intestinal microflora in health and inflammatory bowel disease. Res Immunol 1997;148(8–9):589–594.

4

Virology of the Gastrointestinal Tract

Richard L. Ward, Xi Jiang, Tibor Farkas, and Dorsey M. Bass

Viruses of the Gastrointestinal Tract

The gastrointestinal (GI) tract is one of the most common portals of entry for pathogens, and viruses are frequently spread by fecal–oral routes. Historically, poliovirus was the first enteric virus known to cause a disease responsible for extensive morbidity and mortality. Vaccines against poliovirus were also the first to be developed against an enteric viral infection and included both a parenterally administered inactivated vaccine and an orally administered live attenuated vaccine. Poliovirus vaccines have been extremely effective, and no cases of wild-type poliovirus have been reported in the Western Hemisphere since 1989. Although poliovirus enters the host via the GI tract, poliovirus vaccines protect against systemic paralytic disease by inducing serum neutralizing antibodies that prevent extraintestinal virus spread. Serum neutralizing antibodies may play little role in protective immunity against other enteric virus infections whose morbidity and mortality is associated only with intestinal replication. Thus, development of effective vaccines against these agents presents new challenges based on the need for mucosal rather than systemic immunity.

Viruses have been recognized as likely etiologic agents of severe gastroenteritis since at least the 1940s, when GI disease was successfully transmitted to volunteers challenged with bacteria-free stool filtrates derived from gastroenteritis outbreak specimens (1–5). In the 1960s, virus particles with wheel-like appearances, later recognized as characteristic of rotaviruses, were visualized using electron microscopy in intestinal contents of both mice and calves with gastroenteritis (6,7) and in rectal swabs obtained from a healthy monkey (8). However, it was not until the discovery of Norwalk virus in the stools of children with epidemic viral gastroenteritis in 1972 (9) and rotavirus in the vomitus of children with severe gastroenteritis in 1973 (10) that specific viral agents were associated with human GI disease. Although other viruses have been associated with GI diseases, the three enteric virus groups established today as the most common causes of severe GI illnesses are rotaviruses, caliciviruses, and astroviruses. Therefore, important features of the structure, replication cycles, pathogenesis, and immunity of each group are highlighted in this chapter. The only obvious prophylactic method that can be reliably used to prevent illness due to these viruses, as with poliovirus over 40 years ago, is vaccination, but no licensed vaccines are in use for any of these GI viruses. However, large phase 3 clinical trials have recently been conducted with two candidate rotavirus vaccines, and it is anticipated that vaccines against this agent may soon be available for routine immunization of infants throughout the world.

Rotaviruses

Rotaviruses are the single most important cause of severe infantile gastroenteritis worldwide. In the United States alone, these viruses cause over

50,000 hospitalizations in young children annually and approximately 40 deaths (11). It is further estimated that rotavirus diarrhea results in 600,000 visits to a physician's office or emergency room each year. Direct medical costs associated with rotavirus disease in the United States have been estimated to be at least $500 million annually, with an additional $1 billion in nonmedical expenses (12). On a world scale, rotaviruses are estimated to be responsible for more than 600,000 deaths each year (U.D. Parashar, 2004, personal communication). For these reasons, rotaviruses have received a high priority as a target for vaccine development.

Rotavirus transmission occurs by the fecal-oral route, which provides a highly efficient mechanism for universal exposure. Approximately 90% of children in both developed and developing countries experience a rotavirus infection by 3 years of age (13). The symptoms associated with rotavirus disease typically are diarrhea and vomiting, accompanied by fever, nausea, anorexia, cramping, and malaise that can be mild and of short duration or produce severe dehydration. Severe disease occurs primarily in young children, most commonly between 6 and 24 months of age. Rotavirus infection normally provides short-term protection and immunity against subsequent severe illnesses but does not provide lifelong immunity; furthermore, there are numerous reports of sequential illnesses. Neonates also can experience rotavirus infections, and these occur endemically in some settings but typically are asymptomatic (14–17). These neonatal infections have been reported to reduce the morbidity associated with a subsequent rotavirus infection (15,16). Rotavirus illnesses also occur in adults and the elderly but, as with other sequential rotavirus infections, the symptoms are generally mild. However, recent studies conducted in Japan suggest that rotaviruses can cause hospitalization in adults (18) and significant morbidity in adolescents (19).

Because of the frequency of rotavirus infections and the reduced severity of illness typically associated with sequential infections, a realistic goal for a rotavirus vaccine may be to protect against severe disease. Several vaccine candidates have been developed and evaluated in infants with promising results. Incorporation of an effective rotavirus vaccine into the infant immunization schedule in developed countries could reduce hospitalizations due to diarrhea in young children by 40% to 60% and total diar-rheal deaths by 10% to 20% (20). Until an effective vaccine is available, control of rotavirus disease is limited to nonspecific methods, primarily rehydration therapy.

History of Rotaviruses

Viruses with morphologic features later associated with rotaviruses were first observed in 1963 in intestinal tissues and rectal swab specimens from mice and monkeys by electron microscopy (6,8). These agents, called epizootic diarrhea of infant mice virus and simian agent 11, respectively, were described as 70-nm particles that had a wheel-like appearance. Hence, they later were designated as "rota" viruses from the Latin for wheel (13). In 1969, Mebus and colleagues (7) demonstrated the presence of these particles in stools of calves with diarrhea, thus associating these viruses with a diarrheal disease in cattle. The correlation between these viruses and human diarrheal disease was first reported in 1973 by Bishop and colleagues (10) who used electron microscopy to examine biopsy specimens of duodenal mucosa from children with acute gastroenteritis. In this seminal study, the investigators identified viral particles with the characteristic wheel-like appearance found earlier in feces of animals. Within a short time, these and other investigators confirmed the association between the presence of rotavirus in feces and acute gastroenteritis. Today these human viruses along with their animal rotavirus counterparts have been classified as members of the *Rotavirus* genus within the *Reoviridae* family.

Properties of the Rotavirus Particle

A computer-generated image of the rotavirus particle obtained by cryoelectron microscopy (cryo-EM) (Fig. 4.1) showed that it is about 100 nm in diameter and has a capsid composed of three concentric protein layers (21,22). The outer layer contains the VP7 glycoprotein (780 molecules/virion) and 60 dimers of the VP4 protein, the latter of which forms spikelike projections that extend through and 11 to 12 nm beyond the VP7 layer (21–24). The VP4 protein is anchored to the intermediate layer of the particle composed of 780 molecules of the VP6 protein. The innermost layer contains 120 molecules of the VP2 protein that interact with 12

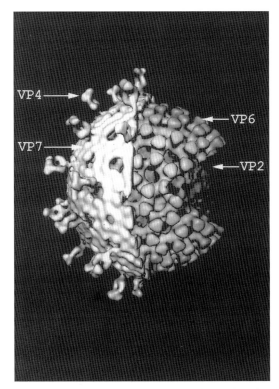

Figure 4.1. Computer-generated image of the triple-layered rotavirus particle obtained by cryoelectron microscopy. The cutaway diagram shows the outer capsid composed of VP4 spikes and a VP7 shell, an intermediate VP6 layer, and an inner VP2 layer surrounding the core containing the 11 double-stranded RNA segments and VP1 and VP3 proteins. (*Source*: Courtesy of B.V.V. Prasad, Baylor College of Medicine, Houston, Texas.)

molecules each of the viral transcriptase (VP1) and guanylyltransferase (VP3) along with the 11 segments of the double-stranded RNA genome. These genome segments encode the six structural proteins and six nonstructural proteins designated NSP1 to NSP6 (Table 4.1). Each segment except segment 11, which is bicistronic (25), encodes one known rotavirus protein whose functions has been investigated but are still not fully understood. The genome segments have sizes ranging from approximately 660 to 3300 base pair (bp), and their encoded proteins have molecular weights of approximately 12 to 125 kd.

Classification of Rotaviruses

Rotavirus Groups

In addition to their distinctive morphologies, rotaviruses were found to share a group antigen (13) that was later determined to be the highly conserved VP6 protein that comprises the intermediate capsid layer. In 1980, particles that were indistinguishable morphologically from established rotavirus strains but lacked the common group antigen were discovered in pigs (26,27). This subsequently led to the identification of rotaviruses belonging to six additional groups (B to G) based on a common group antigen, with the original rotavirus strains classified as group A. Only groups A to C have been associated with

Table 4.1. Sizes of rotavirus gene segments and properties of encoded proteins

RNA segment	No. of base pairs	Encoded protein	Molecular weight of protein ($\times 10^{-4}$)	Properties of protein
1	3300	VP1	12.5	Inner core protein, RNA binding, RNA transcriptase
2	2700	VP2	10.2	Inner capsid protein, RNA binding
3	2600	VP3	9.8	Inner core protein, guanylyl transferase
4	2360	VP4	8.7	Outer capsid protein, HA, NP, receptor binding, fusogenic protein
5	1600	NSP1	5.9	Nonstructural protein, RNA binding, contains zinc fingers, host range determinant(?)
6	1360	VP6	4.5	Intermediate capsid protein, group and subgroup antigen
7	1100	NSP3	3.5	Nonstructural protein, RNA binding, translational control
8	1060	NSP2	3.7	Nonstructural protein, RNA and NSP5 binding, NTPase
9	1060	VP7	3.7	Outer capsid glycoprotein, NP
10	750	NSP4	2.0	Nonstructural glycoprotein, transmembrane protein, enterotoxin
11	660	NSP5	2.2	Nonstructural protein, phosphorylated, O-glycosylated, interacts with NSP2, virosomes
		NSP6	1.2	Nonstructural protein, interacts with NSP5

HA, hemagglutinin; NP, neutralization protein; NTPase, nucleotide triphosphate hydrolase.

human diseases, the vast majority of which have been caused by group A strains. However, group B rotaviruses have been associated with large outbreaks in China, particularly in adults (28,29), and numerous smaller outbreaks of disease due to group C rotaviruses have been reported throughout the world but particularly in Japan (30), which suggests that these non–group A strains could become major pathogens in the future. This suggestion is supported by numerous seroepidemiology studies showing high prevalence of group C rotavirus antibody in different countries.

Electropherotypes and Genotypes of Rotavirus

A variety of classification schemes have been used to characterize rotaviruses for epidemiologic purposes. Each scheme, however, is intertwined with a unique property of viruses with segmented genomes, that is, the ability to form reassortants. During the rotavirus replication cycle, newly formed plus strand viral RNAs that are destined to be packaged within viral particles can freely associate prior to incorporation into replication intermediates in the first stages of virus assembly (31). From these genomic precursors are selected the appropriate number and combination of segments for assembly of progeny viruses. Co-infection of cells with more than one virus permits reassortment of plus strand RNAs from both parents. If co-infection is between different strains of virus, reassortment of messenger RNAs (mRNAs) results in progeny that are genetic mosaics of the co-infecting strains. These new strains, or reassortants, are identified by their specific array of genome segments, usually through their electrophoretic mobilities during polyacrylamide gel electrophoresis (i.e., electropherotypes). The properties of the new virus strains depend on which segments are inherited from which parent and the functional behavior of each particular combination of segments and their protein products.

Rotavirus reassortants form readily in cell culture and in co-infected experimental animals, which at least partially is responsible for the variety of rotavirus strains found in nature. Reassortant formation between rotavirus strains, however, is not a universal phenomenon. For example, there is no evidence that reassor-

tants form between strains belonging to different rotavirus groups (32). Even within group A rotaviruses, there are severe limitations within strain combinations that are capable of forming stable reassortants, limitations that appear to be related directly to the degree of genetic variation between strains (33). One outcome of restricted reassortant formation between rotavirus strains is the concept of genetic families or genogroups (34). A genogroup is composed of rotavirus strains whose gene segments form interstrain RNA-RNA hybrids of sufficient stability to migrate as defined bands during polyacrylamide gel electrophoresis (34). Thus, members of a genogroup share a high degree of genetic relatedness and have significantly less genetic homology with members of other genogroups. Because rotavirus genogroups appear to be species-specific (34,35), interspecies transmission of rotaviruses should be detectable by genogroup analyses. Almost all human rotaviruses belong to either the Wa or DS-1 genogroup (36,37), a designation developed from these prototype strains. The concept of genogroup has been used extensively to determine the origin of rotaviruses causing human infections and disease, particularly to detect viruses or reassortants with gene segments of animal origin.

Serotypes of Rotavirus

Both outer capsid proteins of rotavirus, VP4 and VP7, contain neutralization epitopes, and, thereby, both are involved in serotype determination. Originally, serotyping was based solely on differences in the VP7 protein because animals that are hyperimmunized with rotaviruses develop most neutralizing antibody to this protein. Cross-neutralization studies conducted with these hyperimmune sera readily separated the strains into VP7 serotypes (38,39). When it was found later that VP4 could, in some cases, be the dominant neutralization protein (40–43), a dual serotyping scheme was required. Although VP7 serotypes could be determined readily by cross-neutralization studies, this was more difficult for VP4. Therefore, two numeric systems were devised to classify the VP4 protein in rotavirus strains. One is based on comparative nucleic hybridization and sequence analyses (genotypes), and this designation is provided within brackets. The second is based on neutralization (serotypes) using antisera against baculovirus-expressed VP4 proteins (44) or reassortants with specific VP4 genes (45).

Rotavirus classification based on VP4 and VP7 is designated P and G types to describe the protease sensitivity and glycosylated structure of these two proteins, respectively (46). Until very recently, 15 G serotypes and 22 P genotypes had been identified (47,48). However, in 2003, Liprandi et al. (49) described a porcine group A rotavirus (strain A34) that belongs to a potential new P genotype based on sequence analysis of part (amino acids 13 to 250) of the VP8 region of the VP4 gene of this virus and have tentatively classified it as P(23). Furthermore, based on a full sequence comparison of its VP4 gene, a strain of macaque rotavirus named TUCH was identified as a new P genotype (50). It will be classified as either P[23] or P[24] based on a full sequence analysis of strain A34.

Human rotaviruses belonging to 10 G serotypes have been isolated, but until very recently the vast majority have been identified as G1, G2, G3, or G4 (13). The severity of illness among viruses belonging to these four serotypes has varied little if at all (51–53). Likewise, 10 P genotypes have been found in humans, but almost all illnesses have been associated with P genotypes 4, 6, and 8 (13). However, other G and P types have been the most frequently isolated in some settings, particularly G9 strains, which have been found worldwide, sometimes representing a large fraction of the isolates.

Linkages Between Rotavirus Genome Segments

If the G and P types of rotaviruses found in humans could associate freely during reassortant formation, it is anticipated that the combinations of types for these proteins would be generated randomly. However, this is clearly not the case. For example, G1[P8] and G2[P4] rotaviruses similar to the prototype Wa and DS-1 strains, respectively, frequently are isolated but belong to two distinct genogroups of human rotaviruses (34). Therefore, they rarely should form stable reassortants, an assumption that has been substantiated through analyses of numerous rotavirus strains. Other associations between gene segments have also been found. The VP6 protein or group antigen can be divided into four subgroups (I, II, I/II, and non-I/II), based on antigenic differences within this

protein (54,55). Almost all G2 and G8 human rotaviruses belong to subgroup I, whereas G1, G3, G4, and G9 human rotaviruses belong almost solely to subgroup II. G3 also is a common serotype in animal strains, but in contrast with results found with G3 human strains, almost all G3 animal rotaviruses belong to subgroup I. In addition, subgroup I human, but not animal, strains have been found to have a characteristic "short" electropherotype associated with an inversion in the migration order of segments 10 and 11 (56). Thus, distinct genetic linkages have been found by serotype, genotype, subtype, and electropherotype analysis as well as by genogroup determination.

Replication of Rotavirus

Cell Attachment

Tissue tropism for rotaviruses in vivo is very specific, and these viruses typically infect only enterocytes on the tips of the intestinal villi. Although they bind to many cell types in vitro, rotaviruses efficiently infect only cell lines derived from the kidney and intestine. This implies that both pre- and postbinding selection steps regulate rotavirus replication. Cell binding requirements differ between rotavirus strains. Some strains, particularly those obtained from animals, need sialic acid (SA) for their initial attachment, but this association is not essential because variants that no longer require SA have been isolated (57,58). Furthermore, many animal rotaviruses and most human strains do not require this receptor. For these strains, several cell surface proteins have been implicated as the initial attachment molecules including GM1 and GM3 gangliosides and the integrin $\alpha2\beta1$ (59–62). The VP5* region (amino acids 308–310) of the VP4 protein of some rotaviruses contains the $\alpha2\beta1$ ligand sequence DGE (60,63), thus suggesting that the initial interaction between virus and cell is sometimes with this portion of the VP4 molecule. In contrast, amino acids 155 and 188 to 190 of the VP8* region of VP4 have been reported to play an essential role in the SA-binding activity of this protein (64). After contact with primary receptors, rotaviruses subsequently bind to one or more secondary receptors prior to entry into the cell (Fig. 4.2). These include, but are undoubtedly not limited to, heat shock protein hsc 70 (65) and the integrins

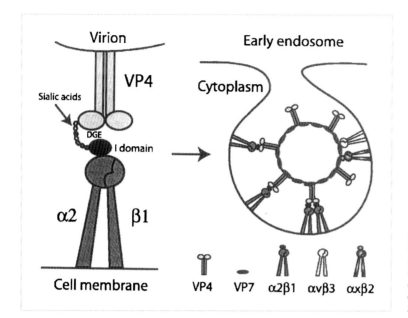

Figure 4.2. Model for involvement of integrins in both attachment of rotavirus to cellular membranes and uptake of rotavirus particles into the cell within endosomes. [*Source*: Graham et al. (60), with permission of the American Society for Microbiology.]

$\alpha X\beta 2$, $\alpha 4\beta 1$, and $\alpha V\beta 3$ (60,63,66). Binding to the latter integrins appears to be through interactions with the VP7 rather than the VP4 protein (60,65). The VP7 protein has also been found to modulate some VP4-mediated phenotypes, including receptor binding (67,68).

Sphingolipid- and cholesterol-enriched lipid microdomains or rafts have been proposed to exist in cell membranes as a result of differential affinity associations with these lipids (69,70). Rafts that are resistant to treatment with certain detergents have been found to both selectively bind rotavirus and be enriched with rotavirus receptors such as ganglioside GM1, integrin subunits $\alpha 2$ and $\beta 3$, and hsc 70 (71). This finding suggests that these detergent-resistant microdomains may provide a platform to facilitate efficient interaction of rotavirus receptors with the virus particle.

Cell Entry and Transcriptional Activation

Activation of rotavirus by trypsin cleavage of the VP4 protein into VP5* and VP8* subunits is required for bound virus to yield a productive infection. Two mechanisms have been proposed for the internalization of these bound viral particles. Early studies based on electron microscopy suggested that entry occurred by endocytosis, and particles in endosomes were rapidly transported to lysosomes for uncoating and transcriptional activation (72,73), a mechanism typically used by other RNA viruses. Later it was reported that this may be the route used by non–trypsin-activated rotaviruses, which disappear from the cell surface with a half-life of 30 to 50 minutes and result in an abortive infection (74,75). In contrast, the activated viruses appeared to enter cells by direct penetration with a half-life of 3 to 5 minutes. Further evidence against rotavirus activation in lysosomes was provided by finding that lysosomotropic agents had little effect virus replication (74–77).

Recently new evidence has suggested that bound rotavirus is assimilated within endosomes as first suggested (78,79) (Fig. 4.2). However, instead of being transported to lysosomes for uncoating, the outer capsid proteins of the virus may become detached due to low Ca^{2+} concentrations within the endosomal compartments. From this point, it is suggested that the endosomal membranes become permeabilized and that transcriptionally active, double-layered rotavirus particles are released into the cytoplasm (78). During the entry process, the cellular membrane becomes temporarily permeabilized, possibly through its association with the VP5* subunit protein of the bound virus (80). This leads to the release of radioactive chromium from the cell and internalization of toxins such α-sarcin (75,81,82). Although other mechanisms of entry and transcriptional activation can be proposed based on reported data, this series of steps provides a reasonable melding of accumulated observations.

Transcription and Replication

The immediate outcome of the release of transcriptionally active particles into the cytoplasm is the synthesis of 11 viral plus-strand RNAs that are extruded from the virus cores through channels in the VP2 and VP6 protein layers at the vertices of the viral particles (83,84). Every particle contains 12 molecules of VP1 and VP3, and one molecule of each protein is thought to be associated with every genome segment, which is transcribed by VP1 and capped through the enzymatic activities of VP3. These initial transcripts must serve as mRNAs for the production of viral proteins. Translational efficiency is dependent on the production of the nonstructural rotavirus protein NSP3 that efficiently binds to both the 3' termini of the viral mRNAs and the eukaryotic initiation factor eIF4G (85). This cellular protein acts as a scaffold that brings together other translational initiation proteins and promotes the circularization of the mRNAs for efficient translation. Because of its high avidity for eIF4G, NSP3 blocks the circularization of cellular mRNAs and thereby inhibits cellular protein synthesis (86,87).

Once produced during the early stages of the rotavirus replication cycle, two other nonstructural proteins, NSP2 and NSP5, form cytoplasmic electrodense particulate structures called virosomes within which viral RNA packaging and replication as well as assembly of double-layered particles is thought to occur (88–90). Particle assembly may be initiated by the formation of complexes within the viroplasm that contain plus-strand RNAs from the 11 genome segments along with VP1 and VP3 and RNA-binding nonstructural proteins NSP2, NSP5, and NSP6 (91,92). Although the mechanism is unknown, the virus faithfully assembles and packages one of each of the plus-strand RNAs within individual precursor viral complexes. These complexes eventually lose their nonstructural proteins, evolve into double-layered viral particles with the sequential addition of VP2 and VP6, and convert their single-stranded RNAs into double-stranded genome segments. Because reassortment of viral genome segments between two co-infecting rotaviruses is so efficient, and

because plus-strand RNAs appear to be excluded from the viroplasm after their release into the cytosol (84), transcriptionally active, double-layered particles derived from co-infecting rotaviruses appear to be incorporated within a shared viroplasm where production and reassortment of plus-strand RNAs destined to be packaged into viral particles occurs. This explains why development of a reverse genetics system for rotavirus has failed (84).

Virus Maturation and Release

Five of the rotavirus proteins synthesized within the cytosol of the infected cell appear to have no role in assembly of double-layered particles and, therefore, do not appear to be transported into the viroplasm. One is NSP3, whose role during translation of viral mRNAs has already been described. Another is the nonessential NSP1 protein (93) that has been reported to interact with interferon regulatory factor 3, thereby inhibiting activation of IRF-3 and diminishing the cellular interferon response (94). The other three (i.e., VP4, VP7, and NSP4) are proteins that play specific and essential roles during the final steps of virus maturation. Both VP7 and NSP4 are glycoproteins that are synthesized on ribosomes associated within the endoplasmic reticulum (ER) and are inserted, via signal peptide sequences, into the ER membrane. In contrast, the VP4 protein has been reported to localize to the space between the periphery of the viroplasm and the outside of the ER (95,96). It appears that during the final steps of viral maturation, fully assembled double-layered rotavirus particles within the viroplasm become associated with the outer membrane of the ER. It is possible that this is the site of addition of VP4. Very recently, however, it has been suggested that addition of VP4 may be a final step in the maturation process (97). Regardless of when VP4 is added, it has been reported that immature viral particles bud through the ER in association with VP7 and NSP4 and temporarily acquire an envelope that is removed as the particles move to the interior of the ER (79). Whether VP4 is contained in these particles or added at a later step remains to be determined.

The mechanism by which fully mature triple-layered particles are released from the infected cell has not been fully explained. It has been suggested either that rotavirus infections stimulate increased intracellular Ca^{2+} concentrations that result in accidental cell death (oncosis) (98) or that these infections trigger programmed cell death (apoptosis) (99). Jourdan et al. (100) have proposed that, prior to cell death, fully mature rotavirus particles are transported from the ER to the apical surface of the cell in "nonconventional vesicles." More recently it has been suggested that mature rotavirus particles associate with lipid rafts within the cell and are transported to the cell surface (101). Because VP4 has been found associated with these rafts, it has been suggested that this is the site of final maturation of the virus particles with the addition of VP4 (97). The structure of these rafts is similar to those contained within the cell membrane and reported to be enriched for rotavirus receptors (71), thus suggesting that both viral entry and exit may involve similar cell membrane/virus interactions. Because most double- and triple-layered rotavirus particles remain associated with cellular debris after cell lysis, it is likely that they are bound to these cell structures while within the cell as suggested by Musalem and Espejo (102).

Rotavirus Pathogenesis

Disease Manifestations of Rotavirus Infection in Humans

Rotavirus is the most common etiologic agent of dehydrating diarrhea in children. The primary site of rotavirus infection is the mature enterocytes on the tips of the intestinal villi. The incubation period following consumption of rotavirus is about 2 to 4 days before the abrupt onset of vomiting and diarrhea. Disease usually is self-limited, lasting 4 to 8 days. When hospitalization is required, the stay is usually short, with an average of 4 days and a range of 2 to 14 days (103). Other clinical findings associated with intestinal rotavirus infections include fever, abdominal distress, and mild dehydration. Several reports associate respiratory symptoms, such as cough, pharyngitis, otitis media, and pneumonia, with rotavirus infections, but the relationship of these symptoms to rotavirus is unclear, and the ability to isolate rotavirus from respiratory secretions has been variable. Other clinical manifestations associated either etiologically or incidentally with rotavirus infection include encephalitis and meningitis, Kawasaki syndrome, sudden infant death syndrome, hepatic abscess, pancreatitis, neonatal necrotizing enterocolitis (104), and diabetes (105,106). The relationship between intussusception and natural rotavirus infection is of great interest because of the reported link between the tetravalent RRV (Rotashield™, Wyeth, Philadelphia, PA) vaccine and intussusception that resulted in removal of this licensed vaccine from the U.S. market less than 1 year after its introduction in 1998 (107). Several studies have investigated the possible infectious etiology of intussusception and most conclude that natural rotavirus infections are not a major cause (108,109). Although not proven, it is suggested that the rare association of intussusception with Rotashield vaccination is due to unique properties of RRV not expected to be found in other rotavirus strains.

Changes in Intestinal Villi After Rotavirus Infection

After fecal–oral transmission of rotavirus, infection is initiated in the small intestine and typically leads to a series of histologic and physiologic changes. Visual pathology due to rotavirus infection is almost solely limited to this site and has been primarily studied in animal models. The extent of intestinal histologic changes following rotavirus infection varies greatly between animals from little or none (e.g., adult mice) to extensive (e.g., neonatal calves and piglets). Studies in calves and piglets revealed that rotavirus infection caused the villus epithelium to change from columnar to cuboidal, which resulted in shortening and stunting of the villi (110,111). The cells at the villus tips became denuded (Fig. 4.3), whereas in the underlying lamina propria, the numbers of reticulum-like cells increased and mononuclear cell infiltration was observed. The infection started at the proximal end of the small intestine and advanced distally. The most pronounced changes usually, but not always (112), were associated with the proximal small intestine.

The pathology of rotavirus infection in mice has been examined in many studies, several of which were conducted with heterologous rotavirus strains that require orders of magnitude more virus than needed to elicit infection with murine strains because of their restricted replication in mice. The histologic changes induced by these heterologous strains, however, are similar to those reported after murine

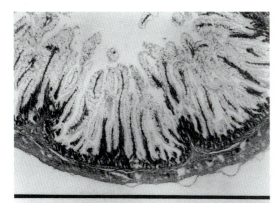

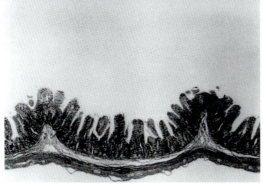

Figure 4.3. Top: Normal histologic appearance of ileum from an 8-day-old gnotobiotic pig. Normal mature vacuolate absorptive cells cover the villi. Hematoxylin and eosin stain. Bottom: Ileum from an 8-day-old gnotobiotic pig after oral inoculation with virulent human rotavirus (Wa strain). Severe villous atrophy and early crypt hyperplasia are evident. Hematoxylin and eosin stain. (Source: Courtesy of L.A. Ward, Ohio Agricultural Research and Development Center, Ohio State University, Wooster, OH.)

rotavirus infection. Mice are susceptible to rotavirus diarrhea during their first 2 weeks of life and recently a series of strikingly clear results on murine rotavirus infection in neonatal mice were reported (113). These included histologic changes (Fig. 4.4), kinetics of rotavirus replication (Fig. 4.5), shortening of intestinal villi (Fig. 4.6), induction of apoptosis, and alterations in cell migration kinetics (Fig. 4.7). A few studies have examined the pathologic changes in the intestines of humans, and the results appeared to be similar to those found in calves and piglets (114,115).

Mechanisms of Diarrhea

Although rotaviruses cause severe diarrhea in numerous species, including humans, the mech-anisms responsible have not been determined and may be due to multiple factors. An early study in piglets indicated that net Na^+ and Cl^- fluxes were not different between control and infected animals, but glucose-mediated sodium adsorption was diminished by rotavirus infection (116). Based on this and other physiologic changes, the authors concluded that retarded differentiation of uninfected enterocytes that migrated at an accelerated rate from the crypts after the virus had invaded villus cells was responsible for adsorptive abnormalities. Another study with piglets led to the conclusion that destruction of the villus tip cells causes carbohydrate maladsorption and osmotic diarrhea (117). In mice it has been reported that carbohydrate maladsorption did not occur as in piglets, and, therefore, crypt cell secretions may be the cause of fluid loss (118). Additional studies in animals and humans concerning changes in the adsorption of macromolecules across the intestinal surface after rotavirus infection have revealed no general pattern. Uptake of some molecules, such as horseradish peroxidase and 2-rhamnose, is increased; uptake of other molecules, such as lactulose and D-xylose, is decreased. Therefore, the relationship between the absorptive properties of intestinal mucosa induced by rotavirus infection and development of diarrhea remains unclear.

Diarrhea also has been induced in infant mice and rats by intraperitoneal inoculation with the rotavirus NSP4 protein as well as with a 22-amino acid peptide derived from this protein (119,120). It was observed that this protein and its peptide caused an increase in Ca^{2+} concentration in insect cells when added exogenously (121). In subsequent experiments, it was found that NSP4 and its peptide can increase the levels of intracellular Ca^{2+} (122) by activating a calcium-dependent signal transduction pathway that mobilizes transport of this ion from the endoplasmic reticulum (121,123). Further reports suggest that NSP4 possesses membrane destabilization activity (124,125) that may result from increased intracellular Ca^{2+} concentrations resulting in cytoskeleton disorganization and cell death (98,126–128). Thus, binding NSP4 to intestinal epithelium after its release from infected cells may contribute to altered ion transport and diarrhea. Another possible target for secreted NSP4 is the enteric nervous system, which lies under the villus epithelium. It has been reported that rotavirus infection can acti-

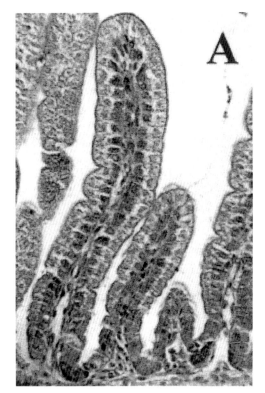
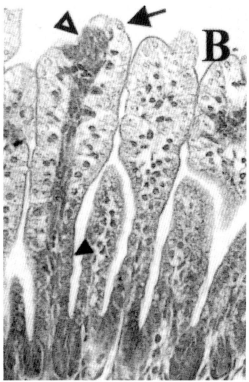

Figure 4.4. Histopathological lesions in the mouse small intestine at 1 day after murine rotavirus infection. A: In control animals, enterocytes are polarized and the nuclei are localized at their base. B: In infected mice, enterocytes have numerous vacuoles, the villus tips are swollen (arrow), the bases are constricted, and the nuclei are irregularly positioned within the cell (solid arrowhead). In many villi, lesions are present at the tips (open arrowhead). [Source: Boshuizen et al. (113), with permission of the American Society for Microbiology.]

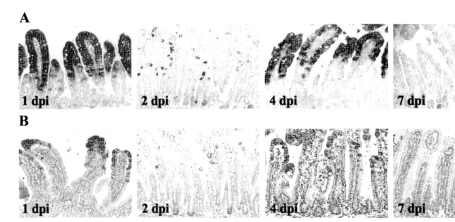

Figure 4.5. Kinetics of rotavirus replication in the mouse small intestine. Levels of NSP4 messenger RNA (mRNA) (A) and protein (B) expression in the jejunum at different days after murine rotavirus infection (dpi) determined by in situ hybridization and immunohistochemistry, respectively. [Source: Boshuizen et al. (113), with permission of the American Society for Microbiology.]

vate this system in mice, and drugs that block nerve activity attenuate rotavirus induced fluid secretion in vitro and attenuate diarrhea in vivo (129). Whether the activity of NSP4 is a major contributor to diarrhea occurring after rotavirus infection remains to be determined. Some additional studies in mice support a role for NSP4 as a cause of diarrhea (130,131), whereas others indicate that mutations in NSP4 are not responsible for attenuation of rotavirus in either mice or humans (132–134), thus questioning its importance as a cause of diarrhea in nature.

The molecular basis for the pathogenicity of rotaviruses, defined by their abilities to induce diarrhea, has not been established. Offit and coworkers (135) reported that the virulence of reassortants generated between heterologous rotaviruses and tested in a mouse model correlated with the presence of the VP4 protein from the more virulent virus. Neither rotavirus strain used in that study (a simian and a bovine strain) replicated efficiently in mice, which suggested that the observation may have limited applicability. A later study with murine/simian rotavirus reassortants revealed no association between the VP4 protein and virulence (136). In that study, the strongest association between virulence and a gene product was with NSP1, a nonstructural protein. Associations between

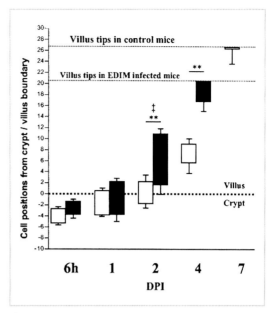

Figure 4.7. Cell migration in the mouse small intestine (ileum) after mouse rotavirus infection (dpi). The positions of the foremost and least-progressed labeled cells are expressed as the number of cell positions from the crypt-villus boundary. Control animals (open bars); infected animals (filled bars). From 2 to 7 dpi, labeled cells in infected animals migrated significantly higher up the villi than in respective control animals (**$p < .01$). The number of cell positions between the foremost and least-advanced cells was also increased at 2 dpi (‡ $p < .05$). EDIM, epizootic diarrhea of infant mice. [Source: Boshuizen et al. (113), with permission of the American Society for Microbiology.]

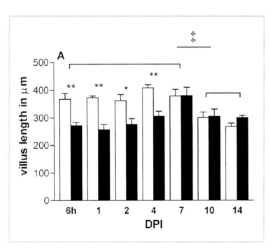

Figure 4.6. Villus length in the jejunum of control mice (open bars) and murine rotavirus-infected mice (solid bars) on the days after infection (dpi). Data are expressed as mean villus height from three to five animals and standard error of the mean (SEM) (error bars). *$p < .05$; **$p < .01$ (Student's t-test). Controls at 6 and 7 dpi were compared to controls at 10 and 14 dpi and analyzed by analysis of variance followed by an unpaired t-test (‡ $p < .05$). [Source: Boshuizen et al. (113), with permission of the American Society for Microbiology.]

virulence and specific gene segments also were examined in piglets. Virulence variants that appeared to differ only in their VP4 genes were isolated from the feces of an infected pig (137). In another study with reassortants between a virulent porcine virus and a human strain attenuated for piglets, it was found that the porcine rotavirus genes encoding VP3, VP4, VP7, and NSP4 all were required for virulence in piglets (138). Whether either of these observations has general applicability or pertains only to a limited combination of rotavirus strains because of specific interactions between their proteins remains to be determined.

Rotavirus Epidemiology

Age-Dependent Susceptibility to Rotavirus Disease

Although rotavirus infection of previously uninfected animals or humans is not age-restricted, there are strict age restrictions in all species associated with rotavirus disease. Mice are susceptible to rotavirus diarrhea for only their first 2 weeks of life (139). Similarly, piglets and calves are most susceptible to rotavirus diarrhea during their first days or weeks of life (140,141). Even nonhuman primates appear to be susceptible to rotavirus diarrhea for only a few days after birth (50). In contrast, severe human rotavirus disease in humans is most common between 6 and 24 months of age (Fig. 4.8), but milder rotavirus illnesses occur throughout our lifetimes. Determination of possible causes for these age restrictions, particularly in humans, has been the subject of intense investigations. Non-immunologic, age-dependent changes occur within the intestine that could account for the reduced severity with increasing age, including an observed decrease in virus-specific receptors on enterocytes between suckling and adult mice (142). A similar suggestion has been made for calves (143). This may also partially explain why young children are more susceptible to rotavirus illnesses than older children or adults.

One contributory factor that could help explain the resistance of human neonates to rotavirus disease is decreased concentrations of proteases needed to cleave the VP4 protein in intestinal secretions of newborns relative to older infants (144). However, the favored explanation for the resistance of human infants to severe rotavirus disease during their first months of life is the presence of circulating transplacental maternal antibody. It has been reported that the onset of rotavirus disease in infants coincided with the decline of maternal antibody titers to low concentrations (145). Furthermore, excellent correlations have been observed between responsiveness to live rotavirus vaccines and transplacental neutralizing antibody titers to the vaccine strains (R.L. Ward et al., unpublished results). Mechanisms by which transplacental antibody might protect against intestinal infection are unclear. Passive transfer of neutralizing antibody to the intestine of both humans and animals is associated with protection, but circulating rotavirus immunoglobulin G (IgG) appears to confer little, if any, protection in animals. Possibly, maternal IgG in humans is taken into the intestine where it neutralizes rotaviruses prior to infection.

The reduced severity of rotavirus disease in older children and adults probably is due primarily to immune responses stimulated by previous rotavirus infections. Protection against rotavirus infection and disease in both previously infected children and adults has been correlated with titers of circulating and intestinal rotavirus antibodies (146–152). It remains to be determined whether these antibodies are responsible for protection.

Cross-Species Rotavirus Transmission

Rotaviruses have an extremely wide host range, but natural cross-species infections may be rare, particularly those between animals and humans. However, a number of human isolates appear to be animal strains or animal–human rotavirus reassortants, as determined by genogroup and sequence analyses. The importance of these strains in human disease may be limited. It has been suggested, however, that once adapted to replication in humans, such strains may become important human pathogens (34). Experimental studies in animals have shown that intestinal replication of rotaviruses in heterologous species is generally limited, and if shedding of progeny viruses is detectable, it often occurs only when animals are inoculated with very high doses of the heterologous viruses.

The genetic basis for host range restriction is unknown and may involve the collective properties of several genes. When reassortants between a murine and a simian rotavirus were used in a mouse model, however, a significant linkage to host range restriction was associated with gene 5 encoding NSP1 (136). Other studies also report

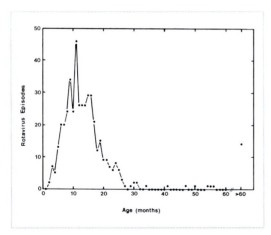

Figure 4.8. Age-related incidence of clinically significant rotavirus episodes in the Matlab region of Bangladesh for residents from 1985–86. [*Source*: Ward et al. (190).]

nonrandom selection of gene 5 in progeny after co-infection of cells in culture (153) and in mice (154), thus suggesting a possible growth advantage associated with this gene. The *NSP1* gene is the most variable of the 11 rotavirus genes (80) and shows a high amount of sequence divergence between rotaviruses of different species (46), thus supporting its possible role in host restriction. Because NSP1 appears to be a nonessential protein for virus replication (93), and the only suggested function for this protein is to interfere with interferon responses (94), it is possible that this function helps maintain species specificity. It should be noted, however, in a study where the *NSP1* gene from a bovine rotavirus that produces an abortive infection in pigs was substituted in a porcine rotavirus that replicates productively in pigs, the new reassortant still demonstrated productive replication in piglets (155). Thus, *NSP1* is not the only determinant of host range.

Rotavirus Seasonality and Sources of Epidemic Strains

As with other respiratory and enteric viruses, distinct seasonality is associated with rotavirus disease. This particularly is evident in temperate climates, where rotaviruses probably are responsible for the large increase in diarrheal deaths found during the winter season. The seasonality of rotavirus disease is less apparent in tropical climates but still is more prevalent in the drier, cooler months (156). The cause for the seasonality of rotavirus disease is a topic of considerable interest but remains unknown. Because rotavirus illnesses decrease to almost undetectable levels during the off-season, the virus must be retained in a less active state during the majority of each year. It is unlikely that human rotavirus is retained in animal reservoirs between seasons because of their low interspecies transmissibility. Therefore, the virus may continue to replicate at low levels in humans until conditions are favorable for the annual epidemic. The occasional rotavirus illnesses that occur in the off-season support the suggestion that humans are a reservoir. It also is possible that the virus survives in the environment that provides continuous exposure throughout the year but results in sustained rotavirus illnesses only during seasonal epidemics. Rotaviruses are shed in extremely high concentrations (157),

retain their infectivities for many months at ambient temperatures (158,159), and are readily detectable on environmental surfaces (160). Therefore, the environment could be a reservoir for human rotavirus and a possible source for the initiation of seasonal epidemics.

To provide clues regarding the origin of rotavirus strains responsible for epidemics, many extensive studies have been performed to characterize the circulating viruses, primarily using electropherotypes and serotypes. From these, it has been determined that rotavirus strains in a specific locale can vary little over sequential seasons or change dramatically, even within a single season. Furthermore, multiple strains often are present within a region at any period during an epidemic. Because gene reassortment can be extensive after rotavirus co-infection, it is difficult to identify the source of new strains within a defined geographic area. They could be derived from outside sources, they could be obtained from local reservoirs, or they could arise by gene reassortment of circulating strains. Clearly, if the source of virus responsible for initiating annual rotavirus epidemics could be identified, much would be learned about the epidemiology of rotavirus.

Immunity to Rotavirus

The immunologic effectors that prevent rotavirus disease have been partially identified, particularly through studies with animal models, but in humans remain poorly understood. Because rotaviruses replicate in intestinal enterocytes, resulting in the associated GI symptoms, it is generally assumed that effector mechanisms must be active at the intestinal mucosa. The most obvious immunologic effector is secretory IgA. Following infection of mice with a high dose of heterologous rotavirus, a large fraction of all IgA cells in the lamina propria of the intestine can be rotavirus-specific (161). Furthermore, protection against rotavirus infection in orally immunized mice correlates with levels of intestinal (stool) and serum rotavirus IgA but not serum rotavirus IgG (162,163). In humans, titers of serum rotavirus IgG and IgA as well as intestinal rotavirus IgA correlate with protection following natural infection. However, the titer of any isotype of rotavirus-specific antibody could not be consistently correlated with protection after either natural infection or vaccination.

Thus, the possibility remains that rotavirus antibody is merely an indicator of protection and not the actual effector.

The most obvious mechanism of protection by antibody is by virus neutralization. Passive protection has been definitively linked with the consumption of neutralization antibody in both animal and human studies. Evidence that active immunity induced by oral inoculation with live rotavirus or natural rotavirus infection is due to neutralizing antibody is varied (164–166). For example, initial vaccine trials with both bovine and simian rotaviruses suggested that protection developed in the absence of neutralizing antibody to the circulating human rotavirus strains. Protection, however, was inconsistent, and subsequent vaccine trials with a rhesus rotavirus (RRV) strain suggested that protection may be serotype-specific (167,168). These results led to the development of bovine and simian rotavirus vaccine strains containing genes for human rotavirus neutralization proteins, which have been or are currently being evaluated in infants (11). Even in these trials, the relationship between serum neutralizing antibody titers and protection was inconsistent, and protection was much greater than the serotype-specific neutralizing antibody responses to the circulating human rotavirus strains (169).

Most data from animal studies indicate that classic neutralization is not the only mechanism of protection. The most immunogenic protein is VP6, which does not appear to stimulate neutralizing antibody responses. Evidence, however, suggests that IgA antibodies directed at VP6 are protective by as yet incompletely understood mechanisms (170,171). Vaccination with either virus-like particles (VLPs) that lack the outer capsid proteins and thus, do not induce neutralizing antibody, or a chimeric VP6 protein can also elicit protective immunity against infection in adult mice (172,173). Passive protection against murine rotavirus disease in neonatal mice has also been produced by adoptive transfer of CD8[+] T cells from spleens of mice previously infected (orally) with either homologous or heterologous rotavirus strains (174). Similarly, CD8[+] splenic or intraepithelial lymphocytes from rotavirus-infected mice can eliminate chronic rotavirus shedding in severe combined immunodeficiency (SCID) mice (175). Thus, at least passive protection against rotavirus disease and resolution of rotavirus shedding can be promulgated with cytotoxic T cells.

An adult mouse model of rotavirus infection has been particularly useful in examining the mechanisms of active immunity against rotavirus in mice (176). Because adult mice become infected with rotavirus but do not develop disease, this model uses protection against infection as its end point. According to this model, protection against live oral murine rotavirus infection is not correlated with either serum or intestinal neutralizing antibody titers against the challenge virus (166). However, it is correlated with total serum and stool rotavirus IgA titers (162,163,177) as well as high titers of rotavirus-specific IgA at the intestinal mucosa surface (178). Subsequently, the use of B-cell–deficient mice that cannot produce antibody has shown that long-term protection against rotavirus infection after a previous rotavirus infection depends at least partially on antibody (179,180). Even after parenteral immunization, migration of antigen-presenting cells from the peripheral lymphoid tissues to the gut-associated lymphoid tissues may contribute to mucosal IgA responses and protection (181). Although protection in this model is typically associated with rotavirus IgA, genetically modified mice that cannot produce IgA are also protected after live virus immunization, presumably due to increased titers of rotavirus IgG (182). Studies have also demonstrated the importance of integrin-mediated B-cell homing to the intestine for their antirotaviral effectiveness (183).

Resolution of rotavirus shedding and protection against subsequent rotavirus infection of mice has also been associated with rotavirus-specific CD8[+] T cells. Depletion of CD8[+] cells in B-cell–deficient mice prior to oral inoculation with live murine rotavirus prevents resolution of the initial infection (179,180). Thus, cytotoxic T cells appear to be critical for the initial resolution of virus shedding when antibody is not present. In fully immunocompetent mice, however, CD8[+] cell depletion merely delays the resolution of shedding, which occurs with the appearance of antibody. More recently it was shown that intranasal or oral inoculation of mice with a chimeric VP6 protein, or even a 14-amino-acid peptide of VP6, along with an effective adjuvant consistently elicited more than 95% reductions in rotavirus shedding after challenge (173,184). CD4[+] T cells were subsequently found to be the only lymphocytes required to elicit this protection (185). Therefore, B, CD8[+], and CD4[+] T

Table 4.2. Mechanisms of resolution and protection identified in the adult mouse model

Mouse strain	Immunization	Outcome	Reference
BALB/c (normal)	Oral, several live homologous and heterologous rotaviruses	Protection correlates with serum rotavirus IgA	163
BALB/c	Oral, live RRV, EDIM	Protection correlates with intestinal rotavirus IgA	162,178
J$_H$D (B-cell-efficient)	Oral, live murine rotaviruses	Resolution dependent on CD8 T cells; protection primarily dependent on antibody	179,180
BALB/c	Intramuscular with live murine rotavirus	Intestinal IgA production after parenteral immunization	181
BALB/c	Intranasal immunization with VLPs or VP6 plus adjuvant	Almost total protection against rotavirus shedding	172,173
IgA −/−	Oral, live murine rotavirus	Intestinal IgG associated with protection	182
β7 −/−	Adoptive transfer of immune B or CD8$^+$ T cells into chronically shedding Rag-2–deficient mice	B but not CD8$^+$ T cells require α4β7 homing receptor	183
J$_H$D	Intranasal immunization with VP6	CD4 T cells are only lymphocytes required for protection	185

EDIM, epidemic disease of infant mice; IgA, immunoglobulin A; RRV, rhesus rotavirus.

cells have all been identified as effectors of protection against rotavirus shedding in mice, and the relative importance of each appears to be dependent on the immunogen and the method of immunization. A summary of the major findings on immune mechanisms in this adult mouse model are listed in Table 4.2.

Control and Prevention of Rotavirus

Nonspecific supportive measures such as oral or intravenous rehydration have been the only methods available to overcome rotavirus illness. Therefore, vaccines are being developed to prevent these illnesses. Based on the belief that protection from rotavirus is best achieved by inducing local intestinal immune responses and the finding that natural rotavirus infections induce at least partial protection against subsequent rotavirus disease, vaccine efforts have been primarily directed at the development of live-attenuated, orally deliverable rotavirus vaccines (11). Most of these efforts have concentrated on the use of animal rotavirus strains that are naturally attenuated for humans and stimulate largely heterotypic immune responses [e.g., RIT 4237 (bovine), WC3 (bovine), RRV (simian)]. More recently, human rotavirus genes have been introduced into these animal strains by creating reassortant viruses to increase their serotypic relatedness to human rotaviruses [e.g., Rotashield, RotaTeq™ (Merck, West Point, PA)]. Rotashield, a tetravalent reassortant vaccine

based on the simian RRV strain, is the only rotavirus vaccine to be licensed in the U.S. but was removed from the market in 1999 by its manufacturer due to association with a small number of cases of intussusception (107). The RotaTeq vaccine contains five reassortant rotaviruses on the bovine WC3 strain background. This vaccine candidate was recently evaluated for safety in more than 35,000 infants during phase 3 trials and is being prepared for licensure in the U.S. (P. Heaton, personal communication, 2004).

Human rotaviruses have also been developed as vaccine candidates. Most are neonatal strains that may be naturally attenuated. However, the most extensively evaluated human rotavirus vaccine candidate is strain 89-12, a G1[P8] obtained from the stool of a symptomatic child and attenuated by multiple cell culture passages (186). This strain has been modified by Glaxo-SmithKline (Rixensart, Belgium) and the new candidate vaccine is called Rotarix™. It has also recently been evaluated in more than 35,000 infants in multiple countries during phase 3 trials (B. DeVos, personal communication, 2004) and in July 2004 was licensed in Mexico. The critical studies leading to the development of these candidate vaccines are summarized in Table 4.3.

Subunit and DNA vaccines, various expression vectors, synthetic peptides, and VLPs produced from baculovirus-expressed rotavirus capsid proteins are also being considered as alternative vaccine candidates. The VLPs are nonreplicating particles that are safe, highly immunogenic, and capable of inducing protective immunity

Table 4.3. Selected published studies important in the development of the Rotashield™, RotaTeq™, and Rotarix™ vaccine candidates (104)

Vaccine	Country	Number of subjects	Number of doses	Percentage[a] protection (overall/severe disease)
RIT 4237	Finland	178	1	50/58
	Finland	328	2	58/82
	Rwanda	245	3	0/0
	Gambia	185	3	0/37
	Peru	391	3	40/75
WC3	U.S. (Philadelphia, PA)	104	1	43/89
	U.S. (Cincinnati, OH)	206	1	17/41
	Central African Republic	472	2	0/36
RRV	U.S. (Rochester, NY)	176	1	0/0
	Venezuela	247	1	68/100
	Finland	200	1	38/67
	Venezuela	320	1	64/90
	U.S. (Rochester, NY)	223	1	66/N.D.[b]
	U.S. (Indian Reservation)	321	1	0/N.D.
RRV reassortants				
RRV G1	Finland	359	1	67/N.D.
RRV G2				66/N.D.
RRV G1	U.S. (Rochester, NY)	223	1	77/N.D.
RRV G1	U.S.	898	3	69/73
RRV TV				64/82
RRV G1	U.S.	1187	3	54/69
RRV TV				49/80
RRV TV	Finland	2273	3	66/91
RRV TV	Venezuela	2207	3	48/88
WC3 reassortants				
WC3 G1	U.S. (Rochester, NY)	325	3	64/87
WC3 TV	U.S.	417	3	73/73
89–12	U.S.	215	2	89/100

[a] Measured in the first year after vaccination.
[b] Not determined.

(172,187–189). Intranasal or oral inoculation of a chimeric VP6 protein along with a mucosal adjuvant has also been shown to provide excellent protection against rotavirus shedding in the adult mouse model (173).

Based on the finding that sequential illnesses with even the same serotypes of rotaviruses are not uncommon, it is difficult to envision how any live virus vaccine delivered orally can, by itself, stimulate complete and lasting protection against all rotavirus illnesses. Therefore, a reasonable goal for present vaccine candidates is to eliminate severe rotavirus disease in children during their most vulnerable period, between 6 months and 2 years of age. To do this, the vaccine must be delivered at an early age, a time when maternal components such as transplacental antibody and possible innate resistance factors may limit immune responses to it, and when the immune system is immature. To overcome possible age-dependent inhibitory factors and stimulate more durable immune responses,

parenteral rotavirus vaccines are receiving serious consideration. Studies on animals suggest that this route of immunization may provide excellent protection either alone or in combination with oral immunization. Novel and less invasive means of vaccine delivery by the parenteral route are being investigated, and these may enhance the feasibility of this approach and help overcome a general resistance toward the development of additional parenteral childhood vaccines. An important goal that remains is to establish a clear correlate of protection. If this could be achieved for children, it would greatly simplify future evaluations of new rotavirus vaccine candidates.

Caliciviruses

Caliciviridae include a group of morphologically similar, but genetically and antigenically diversified viruses classified into four genera

(191). The *Norovirus* (NV) and *Sapovirus* (SV) genera, previously also called "Norwalk-like viruses" and "Sapporo-like viruses," respectively, based on the names of the prototype Norwalk and Sapporo viruses, mainly cause acute gastroenteritis in humans and, therefore, are referred to as human caliciviruses (HuCVs). Members of the other two genera, *Vesivirus* and *Lagovirus*, are not found in humans but cause a variety of diseases including respiratory infection, abortion, hemorrhagic diseases, and gastroenteritis in animals. Morphologically, SVs have a typical CV morphology composed of a rigid surface structure for the virions similar to that of many animal CVs, whereas NVs have atypical CV morphology composed of a smooth surface structure of the virions. Based on this physical feature, NVs were also previously called "small round structured viruses" (SRSVs). Genetically, NVs and SVs are distantly related and SVs are genetically closer to animal CVs than to NVs and have a genomic organization similar to that of the rabbit hemorrhagic disease virus (192).

The HuCVs are transmitted by the fecal–oral pathway and cause acute gastroenteritis that typically self-resolves in 2 to 3 days. These viruses may only replicate in the GI tract. Because HuCV infection does not typically induce a strong immune response after infection, individuals can be infected with the same strains later, a feature making CV gastroenteritis one of the most frequent human diseases. The wide genetic diversity of both NVs and SVs could be another reason for the frequency of the disease, because individuals infected with one strain appear to be susceptible to other antigenically distinct strains. The NVs have a particular importance in causing epidemics of acute gastroenteritis because these viruses are readily transmitted by contaminated water and food, which often results in large outbreaks. Such outbreaks can occur in closed or semiclosed settings such as schools, child care centers, restaurants, hospitals, nursing homes for the elderly, cruise ships, and military facilities. Because these outbreaks usually have a high attack rate, affect all age groups, and can cause public panic, NVs have been listed as category B agents in the National Institutes of Health (NIH)/Centers for Disease Control and Prevention (CDC) biodefense program. The SVs occasionally also cause outbreaks of acute gastroenteritis in adults but mainly infect young children. Due to the limited diagnostic methods available for SVs, the distribution and importance of this genus remain to be defined. This chapter include findings obtained on SVs but the main focus is on NVs.

Brief History of Caliciviruses

Studies on NVs were initiated a half century ago when researchers performed investigations with human volunteers to seek a possible viral etiology for acute gastroenteritis. These studies repeatedly demonstrated that the oral administration of filterable fecal materials from patients with acute gastroenteritis could transmit the diseases to healthy volunteers, suggesting viral pathogens as the causative agents. In the early 1970s, by using immune electron microscopy (IEM), Kapikian et al. (9) observed the first viral pathogen, the Norwalk virus, in stool specimens from patients involved in a large outbreak of acute gastroenteritis that occurred in an elementary school in Norwalk, Ohio, in 1968 (193). This discovery opened the door for characterization of viral pathogens associated with diarrheal diseases. Soon after the discovery of the Norwalk virus, a number of Norwalk-like viruses with similar morphologies and associated with acute gastroenteritis were discovered, including strains such as the Hawaii virus, Taunton virus, Montgomery virus, and Snow Mountain virus (194–197). During this same period, Sapporo virus, the first HuCV that revealed typical CV morphology and caused acute gastroenteritis in young children, was also described (198). Over the period following the discovery of the Norwalk virus, HuCV research entered its first period concerned with understanding basic features of clinical manifestation, immunology, and epidemiology of these viruses. However, the inability to grow HuCVs in cell culture and lack of an animal model soon became major disabilities for rapid advancement.

The cloning and sequencing of the prototype Norwalk virus (199) and, subsequently, many other Norwalk-like viruses in the 1990s opened a new page of molecular virology in HuCV research. The determination of the gene sequences of many HuCVs (199,200) allowed the genetic classification of NVs into the Caliciviridae family. Sequence information also allowed the development of reverse-transcriptase polymerase chain reaction (RT-PCR) assays for the diagnosis of HuCVs (201,202). Application of these new assays in surveillance of acute gastroenteritis in different populations and countries by many laboratories resulted in a rapid accumulation of information on the genetic variation, prevalence, and distribution of HuCVs as a cause of acute gastroenteritis. Advanced molecular techniques also allowed studies of the viral genomic RNA, functional proteins, and viral genomic replication. The successful expression of the NV capsid proteins in baculovirus (203) and other expression systems (204,205) and the fact that these proteins spontaneously form VLPs (203) provided valuable approaches for studying the virus host interaction, including immune responses (206,207) and virus-receptor

recognition (208,209). These recombinant VLPs also provided valuable reagents for diagnostic tests (210–216), vaccine development (205,217–219), and determination of the atomic structure of NVs (220,221).

Properties of Calicivirus Particles

Both NVs and SVs are small (ca. 38 nm), round viruses as determined by electron microscopy (Fig. 4.9). Originally the Norwalk virus particles were described as 27 nm in diameter. Following the baculovirus expression of the Norwalk viral capsid protein, a more accurate measurement of the VLPs was performed by cryo-EM and atomic

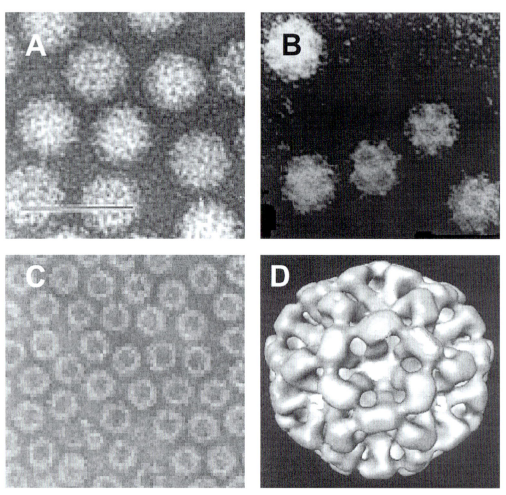

Figure 4.9. Electron micrographs of *Norovirus* (A), *Sapovirus* (B), and baculovirus-expressed recombinant virus-like particles of *Norovirus* (C), and its computer-generated three dimensional structure obtained by cryo-electron microscopy (D). Bar = 50 nm. [*Source*: Part D from Prasad et al. (308). Reprinted with permission from Springer Science+Business Media.]

structure analysis where the Norwalk VLPs were found to have an average diameter of 38 nm (222). The appearances of the NV and SV virions are distinctly different. The SVs have typical CV morphology with the "Star of David" appearance that is similar to many animal CVs. The surface structure of the NVs is smooth and normally does not reveal the "Star of David" appearance.

The structure of the NV capsid has been largely elucidated by cryo-EM (Fig. 4.9) and crystallography using recombinant capsid proteins of the prototype Norwalk virus expressed in baculovirus (220–222). The baculovirus-expressed Norwalk virus capsid proteins self-assemble into VLPs that are morphologically and antigenically similar to authentic virions. The Norwalk capsid is composed of 180 molecules of a single major capsid protein, called viral protein 1 (VP1), encoded in open reading frame (ORF) 2. Recent reports showed that the protein encoded by ORF3 (VP2) is also present in virions, possibly associated with the viral genome. Analysis of the recombinant Norwalk VLPs by cryo-EM and computer image processing has revealed a distinct architecture with T a = 3 icosahedral symmetry. The capsid is made of 90 dimers of the capsid protein that is composed of a shell (S) domain and an arch-like protruding (P) domain. These arches are arranged so that there are large hollows at the icosahedral fivefold and threefold positions.

Atomic structure resolution of the recombinant Norwalk VLPs confirmed the results of the cryo-EM study (220,222). The N-terminal 225 residues constitute the S domain and fold into a classic eight-stranded antiparallel β-sandwich. The S domain is responsible for forming the icosahedral shell. Expression of the S domain resulted in the formation of smooth particles smaller than the intact capsid particles. The rest of the protein constitutes the P domain that is connected to the S domain through a short flexible hinge and forms the arch-like protrusions. The P domain consists of two subdomains: P1 and P2. The P2 subdomain is located on the surface of the capsid, exhibits a larger sequence variation among NVs than the P1 domain, and has recently been shown to play a critical role in pathogen/host interaction. Sequence homology comparison followed by site-directed mutagenesis demonstrated that a pocket located in the P2 domain is responsible for receptor binding (223,224).

Classification of Caliciviruses

Genetic Classification

The classification of Norwalk virus as a CV was suggested by the morphology and structure of the viral capsid before molecular cloning of the Norwalk viral genome. Biochemical studies of the viral proteins isolated from stool samples of infected volunteers revealed the presence of a single virion-associated protein of 60 kd. This characteristic was consistent with data obtained on prototype animal CVs that have a single structural protein with molecular weights ranging from 60 to 70 kd (225). More direct evidence that NVs are CVs was obtained from the analysis of the first Norwalk viral complementary DNA (cDNA) showing sequence similarity with the feline CVs (199). Subsequent description of the full-length Norwalk genome (200) and procurement of partial and full-length genomic sequences of many NVs confirmed these viruses were Caliciviridae. Like animal CVs, NVs contain a single-stranded, positive-sense, poly A-tailed RNA genome of about 7.7 kilobase (kb). The viral genome of Norwalk virus contains three major ORFs. ORF1 encodes a polyprotein that is cleaved into multiple nonstructural proteins, ORF2 encodes the capsid protein, and ORF3 encodes a minor structural protein (200).

Following the accumulation of sequence data for NVs, the prototype Sapporo virus (226) and many Sapporo-like viruses, such as the Manchester, Houston 90, Houston 86, Parkville, and London 92 viruses, have been cloned. Sequence analysis showed that SVs also contain a typical CV genome, but phylogenetically are more closely related to animal CVs than to NVs. The SV genome contains only two major ORFs (192). The first ORF encodes both the nonstructural and the capsid proteins, in which the capsid protein is fused to the C-terminus of the nonstructural polyprotein. The classification of HuCVs within the CV family is supported by the identification of a number of animal CVs that are genetically closely related to HuCVs and cause diarrhea in domestic animals (227–230). Evidence that CV gastroenteritis is a zoonotic disease is lacking.

The genetic classification and nomenclature used for HuCVs is inconsistent even though the International Taxonomy Committee for Viruses (ITCV) issued a guideline in 2000 (191). Most sequence data are from the RNA polymerase region, although the sequences of the entire capsid gene are required to establish a new cluster or genogroup. Katayama et al. (231) showed that the N-terminal shell (S) domain of the capsid is sufficient for correct genotyping, but the recent discovery of NV receptors and the "binding pocket" on the capsid suggested that the protruding (P) domain should not be excluded (224). Ando et al. (232) first proposed numbering the genetic clusters or genotypes for NVs, a suggestion followed by Schuffenecker et al. (233) for SVs. Others, however, prefer to cluster representative viruses by the names of their prototypes. Both systems have been used, but inconsistency remains, such as a different number or "prototype" being designated to the same cluster, which is difficult to follow even by the experts. With the continuous discovery of new unique strains and naturally occurring recombinants, a consistent and accepted nomenclature is greatly needed.

Antigenic Classification

Both the NV and the SV genera contain multiple members. Before cloning the Norwalk virus in 1990, individual members were named after their location of discovery as previously noted. Phylogenic analyses of the viral RNA showed a wide genetic variation in both genera. According to the level of sequence identities, currently known NVs and SVs are divided into genogroups and genetic clusters. The NV genus contains at least 20 genetic clusters within three genogroups (Fig. 4.10) (232), although Alphatron and similar strains are often regarded as genogroup IV and the recently discovered murine NV as a fifth genogroup (234). The SV genus was previously thought to be less diverse than the NV genus, but recent data shows it contains at least nine clusters within five genogroups (Fig. 4.10).

The antigenic relationships among different HuCVs remain to be fully described because HuCVs are antigenically diverse and currently available assays do not cover all types. Even before the cloning of NVs made more extensive tests available, several "serotypes" had been described among the few prototypes based on cross-challenge studies in volunteers (197). After the development of recombinant enzyme immunoassays (EIAs), the antigenic relation-

ships among several strains of HuCVs were studied (235). In general, the antigenic types correlate with the genetic types (236,237). These results remain preliminary because additional recombinant capsid proteins and antibodies raised against them are needed. In addition, the assays used in the antigenic typing are based on antigen recognition, not neutralization; future studies to determine the neutralization types based on a cell culture system or animal model are needed. Finally, human NVs have been found to have a wide spectrum of host specificities according to the histo-blood types of humans as will be discussed in detail later.

Replication of Caliciviruses

In Vitro Cultivation of Caliciviruses

Despite the efforts of many investigators, HuCVs remain refractory to growth in cell culture. One early hypothesis was that the restriction of HuCV replication in vitro may be due to the lack of proper receptor(s) on the cultured cells for viral attachment and entry. This hypothesis has been challenged by recent data. White et al. (238) used recombinant VLPs as probes for studying the host/pathogen interaction and demonstrated specific binding of Norwalk VLPs to 13 cell lines of different origins. Differentiated Caco 2 cells, a human colon carcinoma cell line, bound significantly more VLPs than other cell lines and about 7% of the VLPs were internalized within the cells. Later Marionneau et al. (209) showed that this binding is mediated by the H type 1 and/or H types 3/4 histo-blood group antigens (HBGAs) on the surface of Caco 2 and gastroduodenal epithelial cells. However, the Caco 2 cells, even after differentiation, failed to support NV replication, suggesting a blockade of the viral replication in steps after attachment and penetration.

A recent report by Duizer et al. (239) summarized the efforts undertaken by two laboratories in attempting to cultivate NVs in cell culture. Although the outcome was negative, it should be mentioned because of the diverse approaches used. The studies were based on a hypothesis that successful replication of NVs in vitro depends on the ability to mimic the exact stage of differentiation of the intestinal epithelial cells and on the best match of the luminal microenvironment in a cell culture system. These

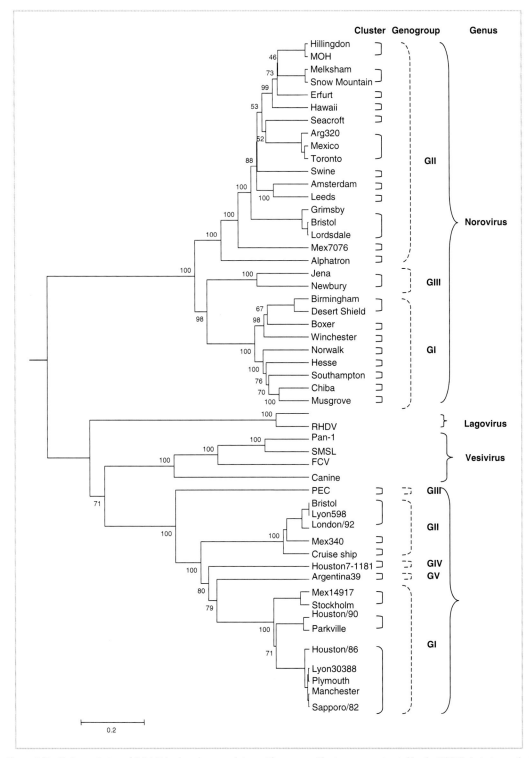

Figure 4.10. Phylogenetic tree of Caliciviridae based on complete capsid sequences. The tree was constructed by the UPGMA clustering method (MEGA v2.1). Bootstrap values are indicated as percent of 125 replicates. Genera, genogroups, and genetic clusters are indicated.

studies utilized more than 20 human and primate cell lines and evaluated different incubation conditions, including the addition of supplements to the culture medium, different treatments of stool specimens for preparation of the virus inoculum, conditions for maintenance of the cell monolayers, methods of inoculation of the cells, and different virus strains. Serial blind passages were performed and each passage was monitored for CPE (cytopathic effect) and for newly synthesized viral products by immunologic and RT-PCR assays. Although positive signals were detected in certain passages, no reproducible NV-induced CPE was observed, and all RT-PCR positive cultures became negative after continued passage.

A more promising system reported recently was to analyze NV replication in mammalian cells by expressing the native forms of Norwalk virus RNA devoid of extraneous nucleotide sequences derived from the expression vector (Asanaka et al., presentation at the Second International Calicivirus Conference, Dijon, France, 2004). When the viral subgenomic RNA was expressed, empty virus particles (VP1 and VP2) were recovered. When the full-length viral genomic RNA was expressed, nonstructural proteins, the capsid protein (VP1), as well as the subgenomic RNA were detected. These results indicate that replication of the genomic RNA occurred and that the subgenomic RNA generated from the genomic RNA was translated into VP1. Coexpressing the genomic and subgenomic RNA in the system resulted in production of virus particles containing the genomic RNA with a density in CsCl gradient similar to that of the authentic virions purified from stool.

The adaptation and cultivation of a porcine CV (PEC) is the only example of in vitro cultivation of an enteric CV (240). PEC/Cowden genetically belongs to the SV genus and causes gastroenteritis in domestic pigs. The virus was initially adapted to grow in primary porcine kidney cells and then in a continuous porcine kidney cell line (LLC-PK) in the presence of intestinal contents (ICs) from uninfected gnotobiotic pigs as a medium supplement. The multiple passages of PEC in LLC-PK cells resulted in adaptation and attenuation because it no longer causes diarrhea in gnotobiotic pigs (241,242). Sequence comparison of wild-type and tissue culture adapted PEC/Cowden revealed 2 amino acid changes in the RNA polymerase and 1 distant and 3 clustered amino acid changes in

the capsid protein. The clustered mutations occurred in the hypervariable region and led to a localized higher hydrophilicity. The hypervariable region of PEC corresponds to the protruding (P) domain of Norwalk virus capsid protein, which is believed to be responsible for antigen recognition and receptor binding. Thus, these amino acid changes in the capsid hypervariable region may be associated with the cell culture adaptation and attenuation. Even after tissue culture adaptation, propagation of PEC/Cowden still requires the supplementation of the culture medium with IC. Different intestinal enzymes of porcine origin such as trypsin, pancreatin, alkaline phosphatase, enterokinase, elastase, and lipase did not support PEC replication. The recent study by Chang et al. (243) showed that bile acids were the active factor in the IC that is essential for PEC replication in tissue culture. A mechanism involving the protein kinase A cell-signaling pathway and a possible downregulation of innate immunity has been proposed to be required for PEC replication. It is hoped that similar adaptation mechanisms will also apply to NVs.

Recently tissue culture adaptation of the murine norovirus 1 (MNV-1) in cultured dendritic cells and macrophages has been reported (244). Murine norovirus 1 growth was inhibited by the interferon αβ receptor and STAT-1 (signal transducer and activater of transcription), and was associated with extensive rearrangements of intracellular membranes. Serial passage and plaque purification resulted in strains with attenuated virulence. Although MNV-1 is not a typical enteric calicivirus and originally was found in immunodeficient mice, these findings should aid strategies for HuCV propagation.

In Vivo Replication and Pathogenesis of Caliciviruses

The clinical manifestation of NV-associated illness has been primarily described in outbreaks and volunteer studies of acute gastroenteritis (197,245–247). The incubation time of NV gastroenteritis is 12 to 24 hours and the illness is generally mild and self-limiting, with symptoms lasting for 24 to 48 hours. The main clinical features include the sudden onset of vomiting, diarrhea, abdominal cramps, nausea, malaise, and occasionally fever. The diarrheal stools are often watery, without mucus, blood, or leukocytes. Vomiting occurs more frequently than diarrhea in children, whereas diarrhea is more frequent in adults. Whether disease is strain specific remains unknown. The viruses can be shed in stools for several days. Recent studies with more sensitive molecular diagnostic assays showed that significant numbers of volunteers had subclinical infection and the virus is shed in stools for longer periods (7 to 21 days) than previously recognized. These findings are important in outbreak control and prevention, particularly when food handlers are the source of outbreak.

Longitudinal studies performed in the Netherlands and Finland showed that clinical features of gastroenteritis are similar between NVs and SVs. The SVs were more frequently detected in infants and toddlers than in school-aged children, whereas NVs were found in children of all ages (248). The major symptoms were similar, but vomiting was more common for NV gastroenteritis (249,250).

The pathogenesis of NV-associated illness also has been described based on studies of volunteers challenged with Norwalk virus and other NVs. The NVs are believed to replicate in the proximal small intestines. Histologic changes were seen in jejunal biopsies of infected volunteers. Symptomatic illness was correlated with a broadening and blunting of the intestinal villi, crypt cell hyperplasia, cytoplasmic vacuolization, and infiltration of polymorphonuclear and mononuclear cells into the lamina propria, but the mucosa itself remained intact. The illness also is accompanied by a small intestinal brush border enzymatic activity decrease and mild nutrient malabsorption. Gastric secretion of HCl, pepsin, and intrinsic factor was associated with these histologic changes, and gastric emptying was delayed. The reduced gastric motility is believed to be responsible for the nausea and vomiting associated with NV gastroenteritis.

One useful animal model of CV-associated gastroenteritis is the porcine enteric CV that induces diarrheal disease in gnotobiotic pigs (241,242,251). Although the overall histopathology and pathophysiology of the animals following infection with the wild-type PEC/Cowden were similar to those observed for HuCVs (241,242,251), this model has provided additional information about the sites of virus replication and the stages of viremia following PEC infection. PEC-infected enterocytes were detected by immunofluorescent staining of mucosal impression smears of the duodenum and jejunum. This result is consistent with the pathologic changes that PEC mainly infects villous epithelial cells of the proximal small intestine and induces lesions in the duodenum and jejunum of infected gnotobiotic pigs. Examination of the colon and extraintestinal tissue or organs did not reveal significant signs of virus replication. Thus, the upper portion of the small intestine is the major site of PEC replication.

One interesting finding is that, in addition to intact virions, a truncated capsid protein was found in stool specimens of patients infected with NVs. The concentration of this protein must be high because it is easily detected by Western blot analysis. This truncated protein has been mapped to the C-terminus containing the entire P domain of the capsid protein (252). A trypsin cleavage site was identified in the hinge/P domain junction that is responsible for the generation of the P protein in vitro as well as in vivo (252). Similar truncated proteins also have been found in insect cell cultures expressing the Norwalk viral capsid protein, although an upstream chymotrypsin digestion site was responsible for the cleavage (203). Because the P domain is located on the surface of the capsid and contains the highest genetic variation, it is believed that the P domain plays an important role in virus replication and antigenicity. Recent data show that in fact the P domain is responsible for the recognition of histo-blood group antigens (HBGAs) and at least is important for virus attachment and penetration (223,224). However, these functions do not explain why only the P protein without the S domain is found in stools. Whether additional functions of this truncated protein exist, for example, in virus replication and pathogenesis, remains unknown.

Host Range of Noroviruses

The hypothesis that a genetic factor is involved in NV host-specificity was suggested in the early 1970s following some unique observations of NV infection and immunity in outbreaks and volunteer studies. Volunteers who had a high level of antibody against Norwalk virus were more susceptible to Norwalk virus challenge than volunteers who did not have the antibody. Some individuals even without detectable antibodies were never infected following challenge with Norwalk virus. Some studies showed that short-term immunity to Norwalk virus (6 to 14 weeks) exists (197), but individuals can be reinfected by NVs following rechallenge after 27 to 42 months (246). Finally, infection of NVs tended to be clustered in families during large outbreaks. These observations suggested that a genetic factor in addition to the acquired immunity of the host must play a role in the susceptibility or resistance to NV infection.

The linkage of human HBGAs with NV infection was first suggested by studies on the prototype Norwalk virus after the report that rabbit hemorrhagic disease virus (RHDV), an animal CV, binds to antigens of the ABH-histo-blood group family (253). The first study showed that the Norwalk virus recognizes human HBGAs in the intestinal tissues and saliva of secretors (expressing H antigen) but not of nonsecretors (209). Using oligosaccharide conjugates containing human HBGA epitopes and monoclonal antibodies specific to these oligosaccharide epitopes, it was found that the fucosyl residue on the human HBGAs added by the 1,2-fucosyl-transferase (FUT-2) is responsible for the binding. The specificity of this binding was further confirmed by blocking with human milk from a secretor woman and by binding of Norwalk VLPs to Chinese hamster ovary (CHO) cells transfected with a fucosyl-transferase gene. Using hemagglutination assays and oligosaccharides as HBGA-specific reagents, Hutson et al. (254) also showed a specific interaction occurs between recombinant Norwalk VLPs and human HBGAs.

The linkage of HBGAs with NV infection in clinical settings also was suggested. In a retrospective study of volunteers challenged with Norwalk virus, the type O individuals had significantly higher relative risk of infection than individuals with other blood types, and type B individuals possessed the lowest risk (255). The same observation was also found in an outbreak possibly caused by a NV (256). Direct evidence that Norwalk virus recognizes the H antigens (secretor gene product) as receptors for infection was obtained in a subsequent volunteer study performed by Lindesmith et al. (257). Of 77 volunteers challenged with Norwalk virus, 22 were nonsecretors and the remaining 55 were secretors according to blood typing on saliva samples. Of the 22 nonsecretors, none was infected following challenge with Norwalk virus and none of their saliva bound Norwalk VLPs. Furthermore, saliva of the type B individuals did not bind or bound weakly to Norwalk VLPs and volunteers with this blood type had the lowest risk of infection after Norwalk virus challenge. Among the 55 secretors, 34 (62%) were infected based on clinical symptoms, detection of virus in stools, or antibody responses. The dynamics of secretory IgA responses measured in the saliva samples also were different between infected and uninfected individuals, indicating that acquired immunity also played a role in NV infection.

Following the initial description of the binding pattern of Norwalk virus, research has been rapidly expanded to other NVs by different laboratories. Using the same saliva binding assays and a panel of recombinant NV capsid antigens, Huang et al. (208) showed that different NVs recognize different HBGAs, and at least four receptor-binding patterns of NVs exist based on the ABO, secretor, and Lewis blood types of the saliva donors (Fig. 4.11). The prototype Norwalk virus represents one of the four binding patterns that recognizes the types A and O, but not type B, secretors. The other three binding patterns are binders of A, B, and O secretors (VA387), A and B secretors (MOH), and Lewis positive secretors and nonsecretors (VA207). According to the biosynthesis pathways of human HBGAs, the binding targets of each of the four binding patterns have been deduced. Thus, this study for the first time raised the possibility of the existence of host-range variability within NVs. Because almost all known antigens in the three human HBGA families are involved in NV binding, it is likely that all humans are susceptible to NV infection. However, because no NV has been identified that binds to all HBGAs, it is predicted that the ability of a NV strain to infect all humans is unlikely.

The different binding patterns of NVs described above suggested possible differences in host range for different individuals that may help explain the epidemiology of NVs, although direct evidence is lacking that connects receptor binding data with infection for three of the four binding patterns described above. For example, the broad spectrum of receptor binding of VA387, a Lordsdale-like strain that recognizes the types A, B, and O secretors, which represent approximately 80% of the general populations, explains why viruses of this cluster are so common throughout the world. The predicted host-specificity also explains the consistent finding that some volunteers who do not have

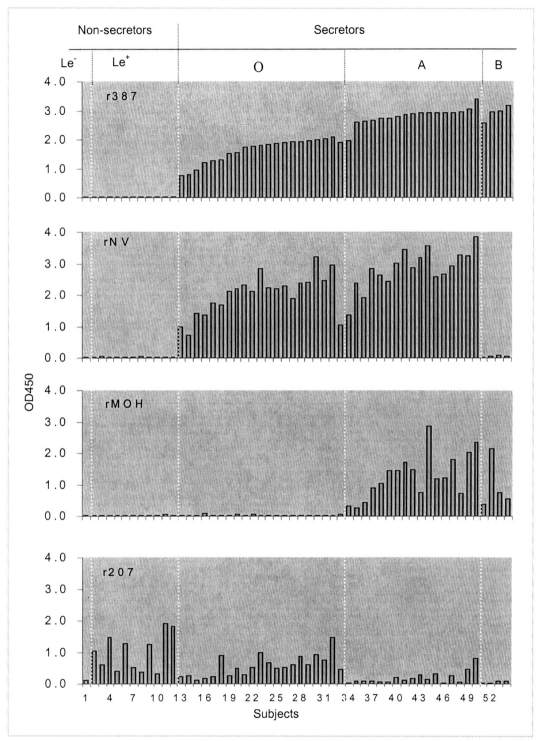

Figure 4.11. Binding of recombinant *Norovirus* (NV) capsid antigens to saliva samples from 51 volunteers of European and three of non-European descent (type B). Saliva samples were tested at a dilution of 1 : 5000. The histo-blood group types of the individuals are shown at the top, and subject numbers at the bottom. The 54 subjects were grouped by their histo-blood types and the magnitudes of saliva binding within each group were sorted by optical density (OD) readings from the lowest to the highest to strain VA387.

preexisting antibody against NV do not become infected following challenge. It also explains why some individuals who have a high level of antibody against NVs are more susceptible than individuals who do not have the antibody (246,258), a finding discussed below (see Immunity to Noroviruses).

These findings suggest that NV receptors may have a significant impact on other fields of NV research. For example, an understanding of the structure of the viral capsid and its interaction with cellular receptors may permit the rational design of antiviral drugs for NVs. Understanding the host specificity may overcome longstanding impediments for studying these viruses, such as their inability to replicate in cell culture or unreliable use in an animal model. In addition, studying the interaction between NVs and HBGA receptors may add insights to our understanding of the coevolution of microbial pathogens and their human host. Finally, studies on NV/receptor interactions may serve as a model for other pathogens that also rely on HBGA as their receptors. In conclusion, the study of NV receptors and host specificity is a new and promising field with the anticipation of major developments within the near future.

Age Restrictions for Caliciviruses

There is no clear age restriction to NV infection, although both young children and the elderly are more susceptible to severe disease than individuals between these age groups. In the case of young children, immaturity of the immune system may be responsible. For the elderly, poor health and possible deterioration of immunity that accompanies old age could play roles. However, many reports on outbreaks in the elderly are from nursing homes in which the environmental factors, such as crowding in the nursing care facilities and poor hygiene, could contribute to the high prevalence of illness.

Sapovirus-associated infection and acute gastroenteritis clearly are more common in young children than in adults. One possible explanation is that natural SV infections result in longer protection than infections with NVs. Therefore, adults are possibly protected by an acquired immunity obtained from childhood exposures to circulating strains, although direct evidence of such long-term immunity is lacking. A study

conducted in 1985 demonstrated that the presence of serum antibody correlated with resistance to SV gastroenteritis because only 3% of the infants with preexisting antibodies developed illness during an outbreak compared to 75% of infants without preexisting antibodies (259). Differences in host antibody prevalence has been observed between NVs and SVs in a study that showed a significantly greater number of infants (90%) possessed maternal antibodies against a Lordsdale-like NV than against a London 92–like SV (23%) in the first week of life (Farkas, T., unpublished results). The clinical explanation and significance of this observation remains unknown.

Epizoology of Caliciviruses

There is no direct evidence of cross-species transmission of CVs. The recent finding of human NVs recognizing human HBGAs suggested that cross-species transmission of CVs is unlikely because it is known that the major HBGA epitopes of humans are different from those of other mammals. Using recombinant Norwalk VLPs as a probe, Hutson et al. (254) showed that Norwalk virus hemagglutinated only human and chimpanzee red blood cells. Using the same recombinant VLP assay, the swine CV (SW918) did not bind to a panel of salivas from 52 human donors representing all known human HBGA types (Farkas, T., unpublished results).

The discovery of many NVs and SVs associated with acute gastroenteritis in domestic animals indicates they may be a reservoir for human disease agents. Although these animal CVs represent independent lineages (genogroups or genetic clusters) from human strains, and worldwide surveillance reports indicate they do not readily cross the species barrier to infect humans, the high variability and mutability of the viral single-stranded RNA genome and possible low requirements for adaptation indicate that a breakdown in the species barrier may have occurred in the past and could happen in the future.

Epidemiology of Caliciviruses

Noroviruses have been found to be the most important cause of nonbacterial acute gastroenteritis across all ages in both developing and developed countries. Noroviruses commonly cause outbreaks of acute gastroenteritis in closed or semiclosed communities and in a variety of institutions, such as schools, restaurants, hospitals, and nursing homes, but are particularly common on large cruise ships and battleships. The transmission of NVs and SVs is believed to be mainly through the fecal–oral route although different pathways may also be utilized (Fig. 4.12). Several authors have suggested transmission of HuCVs by aerosols created from vomitus of infected individuals (260–262).

Outbreaks resulting from contamination of community or family water systems have been documented, but waterborne outbreaks due to contamination of municipal water systems are rare. Food-borne outbreaks resulting from consumption of contaminated food such as uncooked shellfish and precooked food, such as salad, ham, and sandwiches, are common. Both NV and SV infections occur year-round, but winter seasonal predominance has been suggested. Washing of hands remains the most effective personal intervention to stop person-to-person transmission. To help the global control of NVs, several reporting systems have been created including CaliciNet (CALICINET@CDC.GOV) by the CDC and the Food-Borne Viruses in Europe Network (WWW.EUFOODBORNEVIRUSES.CO.UK) by the European Union.

The importance of NVs as a cause of acute gastroenteritis in children was first demonstrated by serosurveillance following the development of new recombinant enzyme immunoassays in the early 1990s (263–268). Studies in many countries showed that children acquire antibody against NVs when very young and the antibody prevalence continues to increase into adulthood. Antibody prevalence has been higher in developing countries than in developed countries. The importance of NVs as a cause of illness in children also has been demonstrated by detection of the viruses in stool specimens. Although the detection rates of HuCV in children varied between studies and countries, the consensus that NV gastroenteritis is a typical childhood illness has been generally accepted. This is supported by the finding that detection rates of NVs in children with diarrhea were significantly higher than in those without diarrhea (269). Noroviruses were also commonly detected in children hospitalized with acute gastroenteritis (4–53%, mean 15%) (270–273),

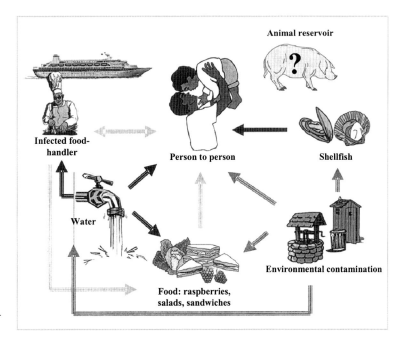

Figure 4.12. Human calicivirus transmission pathways.

observed in emergency rooms (31%) (274), and seen in outpatient clinics (1.3–16%) (275–278), providing an indication that NVs are a cause of severe diarrhea in children. In general, HuCVs have been considered the second most important cause of acute gastroenteritis in children, next only to rotavirus, although the overall clinical symptoms of NV-associated diarrhea are less severe than those of rotavirus.

An important observation made during surveillance of NVs in different countries is that NV-associated acute gastroenteritis is equally important in developing and developed countries. In developing countries such as Argentina, Chile, Mexico, China, Indonesia, and South Africa, the detection rates vary from 3% to 25% and was higher in hospitalized children than outpatients (269,277–283); in developed countries like the United Kingdom, Ireland, France, Spain, Japan, and Australia, the detection rates varied from 4% to 30% and were also highest in hospitalized patients (249,250,270–272,274–276,284–291). This appears to be contradictory to the finding of lower seroprevalence of NVs in children in developed than in developing countries. Possibly, these contradictory results are due to wide variations in detection rates between different countries due to the use of different laboratory methods. For example, in some studies, only small numbers of subjects within highly selected populations were examined; thus the incidence within the entire population could not be determined. The results of these studies are clearly not comparable to others where large numbers of subjects were enrolled. In addition, the major circulating strains in different countries may vary due to the high genetic diversity of HuCVs, and the sensitivity of the commonly used detection methods may also vary. Thus, the key question is: What is the true prevalence of HuCV infection in different countries? High detection rates of NVs were found in some studies but also high rates (up to 50%) of mixed infection with other enteric pathogens were detected, particularly with rotavirus and less frequently with adenovirus or astrovirus, and even mixed infections of NVs and SVs were observed (271,284,287). Therefore, in many of these subjects the cause of disease was not determinable.

In comparison with NVs, the study of SVs has been less advanced. This is probably because SVs mainly infect young children, the illness is milder, and fewer laboratories study SVs. In addition, the methods for diagnosis of SVs remain limited. Thus, our current understanding of SV epidemiology remains preliminary. In contrast to the results on detection of SVs in stool specimens, antibody prevalence studies show that virtually all children are infected with SVs by 5 years of ages, indicating SV infection is widespread. The SV illnesses are sporadic, and most infections appear to be asymptomatic. However, using a highly conserved primer set that detects both NVs and SVs, it was shown that 40% of all HuCV-associated diarrhea in a child cohort study in Mexico was associated with SVs (269). With improved methods and more broadly reactive primers, new unique strains of SVs are continuing to be isolated.

Both NVs and SVs are genetically diverse, and multiple strains with distinct genetic identities usually cocirculate in the same community. In the NV genus, genogroup II (GII) strains have been found to be more prevalent than GI strains, and are more evident in sporadic cases than in outbreaks (284,292). One study showed that GI strains were commonly detected among U.S. Navy personnel (293). Each genogroup also has predominant clusters. The GII/4 cluster (Bristol/Lordsdale, includes Grimsby) is predominant for GII strains, with less frequent detection of GII/1 (Hawaii), GII/3 (Toronto/Mexico), and GII/2 (Snow Mountain/Melksman) strains. The more frequent GI strains reported belong to clusters GI/2 (Southampton), GI/4 (Chiba), and GI/3 (Desert Shield virus) (263,294). The most frequently reported strains of SV belong to the GI/1 cluster (Sapporo/82) (285,295).

Recent reports have included HuCVs on the list of pathogens that cause complications in immunocompromised individuals. Noroviruses have been reported to cause diarrhea in stem-cell transplant recipients (296), severe prolonged secretory diarrhea in an intestinal and liver transplant recipient (297), and chronic diarrhea in HIV-infected children (298).

Immunity to Noroviruses

Immunity to NVs has been a controversial topic because of the unique observations made during human volunteer studies and outbreak investigations that cannot be explained by the general principles of infectious diseases; that is, in volunteer challenge studies, individuals with higher titers of antibodies against NVs were found more susceptible to NV challenge than individuals without or with lower titers of antibodies (197,245,246,299). The recent discovery of HBGA recognition by NVs provides an explanation. Because of the variability in HBGA recognition, a wide spectrum of host ranges for NVs is expected. Thus, in the human volunteer challenge studies, only a portion of individuals are susceptible to the challenge virus based on their blood types and the strain of challenge virus used. Furthermore, because NVs contain many genotypes and possibly serotypes, and the enzyme immunoassays used to measure anti-NV antibodies are not specific for neutralizing antibodies, it is clear why some individuals had antibodies but were not protected against the challenge viruses. Cross-reactive antigenic epitopes have been observed among different NVs, but these epitopes may not induce cross-reactive neutralizing antibodies. Individuals with high levels of antibodies against NVs but who are susceptible to infection are likely to have blood types that match the challenge virus. However, these individuals were also likely to have been previously infected with strains that were antigenically related but serologically distinct from the challenge strain. Thus, the higher level of antibody serves as a marker of susceptibility and past exposure to certain strains but does not necessarily represent protective immunity.

It has been repeatedly demonstrated that individuals were protected against NVs if rechallenged with the same virus within a short time (6 to 14 weeks). However, they were not protected from a rechallenge after a long period following the first challenge. Also, if rechallenged with a NV representing a different genetic type from that used in the first challenge, they were not protected. Thus, NVs may only induce short-term immunity against homologous strains.

The protection against homologous strains also has been observed in outbreaks of NV-associated acute gastroenteritis. In a study of immune responses during an outbreak of acute gastroenteritis on a battleship, serum samples collected from patients as well as asymptomatic crew members were tested for antibody levels against a panel of recombinant capsid antigens representing different genetic clusters of NVs, including the strain isolated during the outbreak (300). The results showed that individuals have variable levels of antibodies against different strains and only individuals with titers >1 : 3200

to the circulating strain prior to exposure were protected. Antibodies against closely related strains within the same genogroup (GI) but in different genetic clusters did not provide protection. This study indicates that acquired immunity against NVs is highly specific.

Vaccine Development for Noroviruses

The need for a vaccine against HuCVs has not been as clearly identified as it has for rotavirus. However, HuCVs are the single most important pathogen causing both epidemic and endemic nonbacterial acute gastroenteritis in all ages in both developed and developing countries, which seems to be sufficient justification for the development of a broadly protective vaccine against these viruses for the general population, including children. Such a vaccine would be particularly useful for the high-risk populations, such as the elderly, travelers to endemic areas, food handlers, crew members of cruise ships, and military personnel. A good vaccine also would be important against a bioterrorism attack.

The research to develop a vaccine for HuCVs has been continuous since the molecular cloning of NVs. One type of vaccine developed was made from recombinant HuCV capsids. Selection of this approach was based on the following: (1) the high yield and ease of production of the recombinant capsid antigens in baculovirus cultures; (2) the baculovirus-expressed recombinant capsid antigens form VLPs that are morphologically and antigenically similar to authentic viral particles and are easily purified by conventional biochemical methods, and (3) the NV VLPs are stable after freezing and lyophilization and over a wide range of pHs (217,301).

Most studies on recombinant NV vaccines have been performed on the prototype Norwalk virus VLPs. Similar approaches also have been developed using other expression systems, such as transgenic plant vectors (302) and Venezuelan equine encephalitis (VEE) virus replicons (303). The main advantage of the transgenic plant vectors such as the potato, tomato, or banana, is that they may provide an "edible" vaccine that allows repetitive vaccination at low cost. The first Norwalk virus recombinant capsid expressed in such vectors was in potatoes. Feeding of the transgenic potato expressing Norwalk viral capsid successfully induced immune responses in mice and human volunteers (205,304). The NV capsid proteins produced in the VEE vector also formed VLPs, but the safety of such a vaccine is of concern because the vector originated from a pathogenic virus.

The study of Norwalk virus recombinant capsid antigens as a candidate vaccine has been performed in preclinical trails in mice and in phase I studies of human volunteers. The results showed that Norwalk VLPs expressed in baculovirus are safe and immunogenic in mice by different routes of immunization (219,305). It also has been shown that the vaccine given orally was safe and immunogenic in humans (217–219,306). It stimulated both IgA and IgG responses as indicated by increases in antibody-secreting cells following immunization. It also induced a modest cellular immune response. This vaccine also was found effective following administration without adjuvant (218). A recent study showed that increased immunity was observed by addition of cholera toxin CT-E29H as an adjuvant (307), which is believed to target the intestinal Peyer's patches.

A major challenge in the development of an NV vaccine is the lack of an animal model for efficacy studies. This also prevents studies on the pathogenesis and mechanism of immunity following a natural infection. Another challenge is the high diversity of genetic and antigenic types within this virus family. *Norovirus* can cause protective immunity following natural infection, but such protection is likely to be homotypic (300). Therefore, a multivalent vaccine that represents the major neutralization epitopes of HuCVs is likely to be necessary if the vaccine is to be widely used and effective.

Control and Prevention of Calicivirus Infections

Because NVs are mainly transmitted by person-to-person and by contaminated environmental surfaces, good personal hygiene is the most important weapon against NV infection, e.g., washing of hands before meals and after each toilet use. In family settings, disinfection of contaminated areas with regular household bleach or other commercial disinfectant solution is valuable. For outbreak situations in the community, finding and removing the sources of the infection and possible mode of transmission are also important. Infected food handlers are the most common sources of food-borne outbreaks. It has been found that NVs commonly cause sub-

clinical infections, and the duration of virus shedding in stools of patients is longer than originally determined, thus increasing the difficulty in controlling food-borne outbreaks caused by infected food handlers.

Astroviruses

Astroviruses are an important cause of pediatric gastroenteritis. They are members of the Astroviridae family and were named for their distinct morphology when examined via negative staining by EM. As encapsidated, single-stranded, positive-sense RNA viruses, they resemble the families Picornaviridae and Caliciviridae but also manifest distinct differences that justify the establishment of their own viral family (309).

History of Astroviruses

Astroviruses were first observed in 1975 by Madeley and Cosgrove (310) in stools of infants with vomiting and diarrhea. Their morphology was distinct from that of other small round viruses such as noroviruses that had been previously identified as agents of gastroenteritis. Over the ensuing years similar viruses were observed in the feces of a variety of species. In the early 1980s, cell culture propagation of astroviruses was achieved via the addition of trypsin to the culture media (311). This discovery facilitated the development of serotyping reagents, monoclonal antibodies suitable for the development of EIAs, and ultimately the cloning and sequencing of astrovirus.

Properties of the Astrovirus Particle

The astrovirus virion is a small 27- to 34-nm-diameter particle when measured in negatively stained EM preparations (Fig. 4.13). The star-like morphology for which the virus is named can be seen in approximately 10% to 15% of virions in such preparations and may represent partially degraded particles (312). Purified astrovirus has a buoyant density of 1.35 to 1.38 in cesium chloride gradients. Computer-enhanced reconstruction of cryo-EM images of human astrovirus particles reveals a rippled solid capsid with 30 dimeric spike-like projections (313).

The capsid consists of two to three proteins of approximately 24- to 36-kd molecular mass

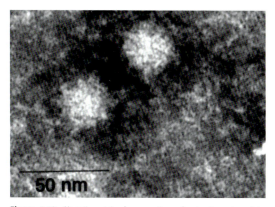

Figure 4.13. Negative stained preparation of astrovirus viewed by electron microscopy. Bar = 50 nm.

determined by sodium dodecyl sulfate (SDS) gel electrophoresis. The capsule proteins are derived from a precursor polypeptide of approximately 97 kd. The genome consists of positive-sense, single-stranded RNA with three ORFs (Fig. 4.14). Open reading frame 1a includes a protease motif, transmembrane helices, and ribosomal frame-shifting and nuclear localization signals. Open reading frame 1b includes an RNA polymerase motif, whereas ORF 2 encodes the capsid precursor polyprotein.

Classification of Astroviruses

The family Astroviridae includes human and animal/avian astroviruses. Mammalian astroviruses are generally associated with localized enteric disease, whereas avian species may be

associated with immunodeficiency (314), interstitial nephritis (315), or hepatitis (316).

Astroviruses isolated from different species are serologically unrelated to each other, although some homology exists at the level of nucleotide sequence. At least eight serotypes of human astrovirus have been described that can be differentiated by a neutralization assay, special EIA, or RT-PCR analysis.

Replication of Astroviruses

Although many details of astrovirus replication are not known, a great deal has been learned in recent years. The astrovirus genome consists of approximately 6800 nucleotides of positive-sense, single-stranded RNA. As already noted, there are three ORFs: ORF 1a, ORF 1b, and ORF 2. Open reading frame 1a encodes a viral protease, which includes transmembrane helices, a functional nuclear localization signal, and autocatalytic sites (317). The in vitro translated product of ORF 1a has serine protease activity, which is abrogated by mutation to critical residues (318). The first AUG codon for ORF 1b falls within the 3′ terminal region of ORF 1a. Initiation occurs by a ribosomal frame-shift mechanism (311,319). Open reading frame 1b encodes the viral RNA polymerase. Open reading frame 2 encodes the structural precursor protein of approximately 87 kd. Open reading frame 2 is also found within infected cells as subgenomic RNA, which facilitates the production of large amounts of capsid protein to package progeny virions (309).

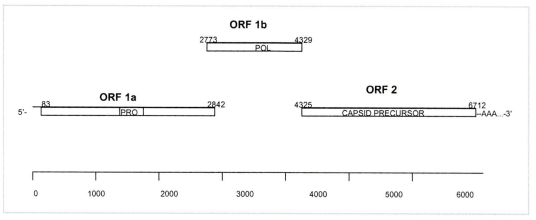

Figure 4.14. Genome organization of human astrovirus serotype 1. Nucleotide numbers vary slightly among strains of astrovirus. Open reading frame (ORF) 1b is initiated via a ribosomal frame shift mechanism located in the 3′ region of ORF 1a. The protease (PRO), polymerase (POLY), and capsid precursor regions are depicted.

In cell culture, astroviruses propagate only in the presence of trypsin (311). Lysosomotrophic agents such ammonium chloride and dansylca-daverine are potent inhibitors of astrovirus replication, suggesting that the endocytic pathway is important in viral entry (320). Electron microscopy has shown astroviruses being taken up by cells in coated pits and transported to large smooth vesicles consistent with endosomes (320). Immunofluorescent studies of astrovirus infected cells show antigen production in the cytoplasm by 7 hours after infection (321). These studies have reported small intranuclear foci of fluorescence during viral infection. Electron micrographs of infected cells at later time points show accumulations of progeny virions in crystalline arrays.

Pulse-chase experiments show that after initial translation as a 87-kd protein, the ORF 2 product is converted intracellularly to a 70-kd protein that forms the viral capsid (322). The processing consists of proteolytic cleavage of the carboxy terminus of the capsid precursor (323). This initial processing appears to facilitate both capsid formation and release of virus from cells (324). It is blocked by caspase inhibitors. The virions encapsidated with the 70-kd protein are minimally infectious, although they bind to cells (322). Subsequent trypsin treatment cleaves the capsid protein to smaller peptides of approximately 34, 27, and 25 kd (322,323). This cleavage is associated with greatly enhanced infectivity of the virions (322). The dependence of the virus on extracellular trypsin likely plays a role in astrovirus tropism for the intestinal epithelium. In cell culture, human astroviruses induce apoptosis (324).

Clinical Illness Induced by Astroviruses

In mammalian species, astrovirus is a cause of gastroenteritis. Human strains of astrovirus are primarily associated with pediatric disease manifested by vomiting, diarrhea, fever, and malaise. The illness is typically milder than rotavirus disease. In a cohort of Mexican children, astrovirus diarrhea was associated with 4.3 stools/day compared with 7.1 stools/day with rotavirus infection (325). Diarrhea is watery, without blood or purulence. Astrovirus gastroenteritis has an incubation period of 1 to 3 days and lasts 4 to 5 days. Viral shedding as detected by RT-PCR may continue for several weeks (326).

Immunocompromised individuals may have prolonged symptoms. Many infections, particularly in older children or adults, are asymptomatic.

Astrovirus infection was originally diagnosed by EM of negatively stained stool samples (310). In recent years, EIAs have been developed that are sensitive and specific. One such assay is commercially available. The RT-PCR assays have been shown to offer even greater sensitivity and may also be designed to determine serotype (327).

Astrovirus Pathogenesis

Knowledge of astrovirus pathogenesis is limited by the lack of a conventional small mammalian model. There are some limited observational studies in humans and a few large animal studies. In an early case report (328), astrovirus particles were observed in small intestinal biopsies of some children with chronic diarrhea. Viral particles were observed in villus epithelial cells of the infected patients.

Human adult volunteer studies have shown that although most subjects underwent seroconversion after oral inoculation, very few developed a gastroenteritis-like picture (329,330). No intestinal histology was obtained during these infections. Histopathology from a 4-year-old bone marrow transplant recipient with prolonged astrovirus diarrhea showed villus blunting, cuboidal villus epithelium, and increased lamina propria inflammation (331). Immunofluorescence showed astrovirus antigen in villus-tip enterocytes, and EM revealed virions within these cells in paracrystalline arrays.

In gnotobiotic lambs, astrovirus infection is associated with mild villus atrophy and crypt hyperplasia with viral replication in villus epithelial cells and scattered subepithelial macrophages (332–334). In bovine infections, a predilection for the dome epithelium over Peyer's patches has been noted (335). Bovine astroviruses appear to be relatively nonpathogenic in calves but do seem to exacerbate rotavirus infection. One-day-old chickens and turkeys develop gastroenteritis during astrovirus infection (336), which is associated with decreased intestinal disaccharidase activity. Ducks infected with their distinct astrovirus manifest hepatitis (316). Some avian viruses seem to be associated with thymic atrophy, enteropathy, and suppressed immune responses (337), whereas others have been associated with interstitial nephritis and growth retardation (315).

In summary, astroviruses are known to infect villus epithelia, resulting in cell death with modest inflammatory changes. Further study is warranted to determine how these changes induce diarrhea and vomiting.

Astrovirus Epidemiology

Astrovirus gastroenteritis occurs in young children throughout the world. Transmission is by the fecal—oral route with evidence for food- and waterborne outbreaks (338). Serotype 1 is the most common serotype isolated in most surveys. Estimates of the incidence of astrovirus disease have increased considerably in recent years in association with improved methods of diagnosis such as EIA and RT-PCR (339). Depending on the setting, astrovirus may account for 3% to 10% of pediatric gastroenteritis (275,340–348). In temperate climates, astrovirus infection rates are increased in winter months in a similar fashion to those of rotavirus and caliciviruses.

Astroviruses have been shown in several studies to be more likely to infect and cause disease in younger infants (<6 months of age) than rotavirus (325,343,344,347,349). This suggests that maternally derived immunity may be less effective against astrovirus than rotavirus. Antibodies to astrovirus are acquired by 70% to 90% of children by school age (350–352). Serotype 1 antibodies are most common in these studies. The incidence of astrovirus infection in a birth cohort of young children was similar to that of rotavirus, that is, approximately 0.2 cases/person-year (325).

Outbreaks among children in day-care centers (326), schools (338), and hospitals (331) are well described. Reverse transcriptase PCR studies in such settings have shown that viral shedding may precede and follow the symptomatic portion of the infection.

Outbreaks of adult astrovirus gastroenteritis may occur in nursing homes (330,353,354) or among military recruits (355). These outbreaks are considerably less common than those attributed to caliciviruses and are usually associated with less common serotypes such as serotypes 3 and 5. An outbreak in Japan appeared to be associated with contaminated food (338), and another was associated with a wading pool (356).

Astroviruses have also been reported to be important pathogens for compromised hosts. Some studies of patients with immunodeficiency (357,358), HIV (359), and transplants (344,360,361) have suggested significant numbers of symptomatic infections occur in these populations. Such infections may be significantly prolonged. Other surveys of HIV-infected patients with diarrhea have not found significant numbers of astrovirus infections (362).

Immunity to Astrovirus

The observation that most symptomatic astrovirus infections occur in young infants suggests that acquired immunity is protective. A longitudinal study of Egyptian infants showed evidence of homotypic protection from infection (325).

Volunteer studies in adults have also suggested that the presence of serum antibody against astrovirus predicts protection from disease (329,330). In one case, administration of immune globulin was associated with clearance of chronic, symptomatic astrovirus infection in an adult with immunodeficiency (363) and a bone marrow transplant recipient (364), but similar therapy has failed in others (360).

Neutralizing epitopes have been identified and roughly mapped using escape mutants generated after neutralization with murine monoclonal antibodies (365). At least one such monoclonal antibody has broad cross-reactivity among different human serotypes in neutralization assays, suggesting a possibility of some degree of heterotypic immunity.

Cellular immunity is also involved in protection from astrovirus disease. Human leukocyte antigen (HLA)-restricted CD4 T cells specific for astrovirus have been isolated from intestinal mucosa of adults (366). T-cell immunodeficiency (358) and chemotherapy, which disproportionally reduces CD4 T-cell counts, have been associated with prolonged astrovirus shedding (363). CD8 and CD16 T-cell populations have also been correlated with reductions in astrovirus shedding (360).

Control and Prevention of Astroviruses

As in all enterically transmitted infections, hygienic standards can play an important role in the institutional control of astrovirus infection. In hospitals, patients with diarrhea should be isolated, and meticulous precautions observed to avoid nosocomial spread. Astroviruses are rendered noninfective by chlorination (327) and by treatment of contaminated surfaces with methanol. Asymptomatic shedding of astrovirus complicates the control of astrovirus infections/outbreaks.

Ultimately an astrovirus vaccine would offer the best hope for control of astrovirus infections. Because many of the more severe infections occur in very young infants, such a vaccine would need to induce protective immunity at an early age. Perhaps further knowledge about this common infection will eventually lead to the development of such a vaccine.

References

1. Gordon I, Ingraham HS, Korns RF. Transmission of epidemic gastroenteritis to human volunteers by oral administration of fecal filtrates. J Exp Med 1947;86: 409–422.
2. Gordon I, Ingraham HS, Korns RF, Trussell RE. Gastroenteritis in man due to a filtrable agent. NY State J Med 1949;49:1918–1920.
3. Kojima S, Fukumi H, Kusama H, et al. Studies on the causative agent of the infectious diarrhea: records of the experiments on human volunteers. Jpn Med J 1948;1:467–476.
4. Reimann HA, Price AH, Hodges JH. The causes of epidemic diarrhea, nausea and vomiting (viral dysentery?) Proc Soc Exp Biol Med 1945;59:8–9.
5. Yamamoto A, Zennyogi H, Yanagita K, Kato S. Research into the causative agent of epidemic gastroenteritis which prevailed in Japan in 1948. Jpn Med J 1948;1:379–384.
6. Adams WR, Kraft LM. Epizootic diarrhea of infant mice: identification of the etiologic agent. Science 1963;141:359–360.
7. Mebus C, Underdahl N, Rhodes M, Twiehaus M. Calf diarrhea (scours): reproduced with a virus from a field outbreak. Res Bull 1969;233:1–16.
8. Malherbe H. Harwin R. The cytopathic effects of vervet monkey viruses. S Afr Med J 1963;37:407 411.
9. Kapikian AZ, Wyatt RG, Dolin R, et al. Visualization by immune electron microscopy of a 27–nm particle associated with acute infectious nonbacterial gastroenteritis. J Virol 1972;10:1075–1081.
10. Bishop RF, Davidson GP, Holmes IH, Ruck BJ. Virus particles in epithelial cells of duodenal mucosa from children with acute non-bacterial gastroenteritis. Lancet 1973;2:1281–1283.
11. Ward RL, Clark HF, Offit PA, Glass GI. Live vaccine strategies to prevent rotavirus disease. In: Levine MM, Kaper JB, Rappuoli R, Liu MA, Good MF, eds. New Generation Vaccines, 3rd ed. New York: Marcel Dekker, 2004:607–620.
12. Tucker AW, Haddix AC, Bresee JS, et al. Cost-effectiveness analysis of a rotavirus immunization program for the United States. JAMA 1998;279:1371–1376.
13. Kapikian AZ, Hoshino Y, Chanock RM. Rotaviruses. In: Knipe DM, Howley PM, Griffin DE, et al., eds. Fields Virology, 4th ed. Philadelphia: Lippincott Williams & Watkins, 2001:1787–1833.
14. Murphy AM, Albrey MB, Hay PJ. Rotavirus infections in neonates. Lancet 1975;2:452–453.
15. Bishop R, Barnes G, Cipriani E, et al. Clinical immunity after neonatal rotavirus infection: a prospective longitudinal study in young children. N Engl J Med 1983;309:72–76.
16. Bhan MK, Lew JF, Sazawal S, et al. Protection conferred by neonatal rotavirus infection against subsequent rotavirus diarrhea. J Infect Dis 1993;168:282–287.
17. Haffejee IE. Neonatal rotavirus infections. Rev Infect Dis 1991;13:957–962.
18. Nakajima H, Nakagomi O, Kamisawa T, et al. Winter seasonality and rotavirus diarrhoea in adults. Lancet 2001;357:1950.
19. Mikami T, Nakagomi T, Tsutsui R, et al. An outbreak of gastroenteritis during school trip caused by serotype G2 group A rotavirus. J Med Virol 2004;73:460–464.
20. Research priorities for diarrhoeal diseases vaccines: memorandum from WHO meeting. Bull WHO 1991; 69:667–676.
21. Shaw AL, Rothnagel R, Chen D, et al. Three-dimensional visualization of the rotavirus hemagglutinin structure. Cell 1993;74:693–701.
22. Prasad BVV, Chiu W. Structure of rotavirus. Curr Top Microbiol Immunol 1994;185:9–29.
23. Prasad BVV, Burns JW, Marietta E, et al. Localization of VP4 neutralization sites in rotavirus by three-dimensional cryo-electron microscopy. Nature 1990; 343:476–479.
24. Yeager M, Berriman JA, Baker TS, et al. Three-dimensional structure of the rotavirus haemagglutinin VP4 by cryo-electron microscopy and difference map analysis. EMBO J 1994;13:1011–1018.
25. Mattion NM, Mitchell DB, Both GW, Estes MK. Expression of rotavirus proteins encoded by alternative open reading frames of genome segment 11. Virology 1991; 181:295–304.
26. Bridger JC. Detection by electron microscopy of caliciviruses, astroviruses and rotavirus-like particles in the faeces of piglets with diarrhoea. Vet Rec 1980;107:532.
27. Saif LJ, Bohl EH, Theil KW, et al. Rotavirus-like, calicivirus-like, and 23–nm virus-like particles associated with diarrhea in young pigs. J Clin Microbiol 1980;12: 105–111.
28. Hung T, Wang C, Fang Z. et al. Waterborne outbreak of rotavirus diarrhea in adult in China caused by a novel rotavirus. Lancet 1984;26:1139–1142.
29. Wang S, Cai S, Chen J, Li R, Jiang R. Etiologic studies of the 1983 and 1984 outbreaks of epidemic diarrhea in Guangxi. Intervirology 1985;24:140–146.
30. Phan TG, Nishimura S, Okame M, et al. Virus diversity and an outbreak of group C rotavirus among infants and children with diarrhea in Maizuru City, Japan during 2002–2003. J Med Virol 2004;74:173–179.
31. Ramig RF, Ward RL. Genomic segment reassortment in rotaviruses and other reoviridae. Adv Virus Res 1991;39:163–207.

32. Yolken R, Arango-Jaramillo S, Eiden J, et al. Lack of genomic reassortant following infection of infant rats with group A and group B rotavirus. J Infect Dis 1988;158:1120–1123.

33. Ward RL, Knowlton DR. Genotypic selection following coinfection of cultured cells with subgroup 1 and subgroup 2 human rotaviruses. J Gen Virol 1989;70: 1691–1699.

34. Nakagomi O, Nakagomi T. Interspecies transmission of rotaviruses studied from the perspective of genogroup. Microbiol Immunol 1993;37:337–348.

35. Nakagomi O, Nakagomi T. Genetic diversity and similarity among mammalian rotaviruses in relation to interspecies transmission of rotavirus. Arch Virol 1991;120:43–55.

36. Nakagomi O, Nakagomi T. Molecular evidence for naturally occurring single VP7 gene substitution reassortant between human rotaviruses belonging to two different genogroups. Arch Virol 1991;119:67–81.

37. Ward RL, Nakagomi O, Knowlton DR, et al. Evidence for natural reassortants of human rotaviruses belonging to different genogroups. J Virol 1990;64:3219–3225.

38. Hoshino Y, Wyatt RG, Greenberg HB, et al. Serotypic similarity and diversity of rotaviruses of mammalian and avian origin as studied by plaque-reduction neutralization. J Infect Dis 1984;149:694–702.

39. Wyatt RG, Greenberg HB, James WD, et al. Definition of human rotavirus serotypes by plaque reduction assay. Infect Immun 1982;37:110–115.

40. Ward R, Knowlton D, Schiff G, et al. Relative concentrations of serum neutralizing antibody to VP3 and VP7 proteins in adults infected with a human rotavirus. J Virol 1988;62:1543–1549.

41. Ward RL, McNeal MM, Sander DS, et al. Immunodominance of the VP4 neutralization protein of rotavirus in protective natural infections of young children. J Virol 1993;67:464–468.

42. Perez-Schael I, Blanco M, Vilar M, et al. Clinical studies of a quadrivalent rotavirus vaccine in Venezuelan infants. J Clin Microbiol 1990;28:553–558.

43. Clark HF, Borian FE, Modesto K, et al. Serotype 1 reassortant of bovine rotavirus WC3 strain, strain W179-9, induces a polytypic antibody response in infants. Vaccine 1990;8:327–332.

44. Gorziglia M, Larralde G, Kapikian AZ, et al. Antigenic relationships among human rotaviruses as determined by outer capsid protein VP4. Proc Natl Acad Sci USA 1990;87:7155–7159.

45. Snodgrass DR, Hoshino Y, Fitzgerald TA, et al. Identification of four VP4 serological types (P serotypes) of bovine rotavirus using viral reassortants. J Gen Virol 1992;73:2319–2325.

46. Estes MK, Cohen J. Rotavirus gene structure and function. Microbiol Rev 1989;53:410–449.

47. Gentsch JR, Woods PA, Ramachandran M, et al. Review of G and P typing results from a global collection of rotavirus strains: implications for vaccine development. J Infect Dis 1996;174:S30–S36.

48. Rao CD, Gowda K, Reddy BSY. Sequence analysis of VP4 and VP7 genes of nontypeable strains identifies a new pair of outer capsid proteins representing novel P and G genotypes in bovine rotaviruses. Virology 2000;276:104–113.

49. Liprandi F, Gerder M, Bastidas Z, et al. A novel type of VP4 carried by a porcine rotavirus strain. Virology 2003;314:373–380.

50. McNeal MM, Sestak K, Choi AH-C, et al. Development of a rotavirus shedding model in rhesus macaques using a homologous wild type rotavirus of a new P genotype. J Virol 2005;79:944–954.

51. Barnes GL, Unicomb L, Bishop RF. Severity of rotavirus infection in relation to serotype, monotype and electropherotype. J Paediatr Child Health 1992;28: 54–57.

52. Bern C, Unicomb L, Gentsch JR, et al. Rotavirus diarrhea in Bangladeshi children: correlation of disease severity with serotypes. J Clin Microbiol 1992;30: 3234–3238.

53. Raul-Velazquez F, Calva JJ, Lourdes-Guerrero M, et al. Cohort study of rotavirus serotype patterns in symptomatic and asymptomatic infections in Mexican children. Pediatr Infect Dis J 1993;12:54–61.

54. Greenberg HB, Valdesuso J, Van Wyke K, et al. Production and preliminary characterization of monoclonal antibodies directed at two surface proteins of rhesus rotavirus. J Virol 1983;47:267–275.

55. Taniguchi K, Urasawa T, Urasawa S, et al. Production of subgroup-specific monoclonal antibodies to an enzyme-linked immunosorbent assay for subgroup determination. J Med Virol 1984;14:115–125.

56. Kalica AR, Greenberg HB, Espejo RT, et al. Distinctive ribonucleic acid patterns of human rotavirus subgroups 1 and 2. Infect Immunol 1981;33:958–961.

57. Ludert JE, Mason BB, Angel J, et al. Identification of mutations in the rotavirus protein VP4 that alter sialic-acid-independent infection. J Gen Virol 1998; 79:725–729.

58. Mendez E, Arias CF, López S. Binding to sialic acids is not an essential step for the entry of animal rotaviruses to epithelial cells in culture. J Virol 1993;67:5253–5259.

59. Hewish MJ, Takada Y, Coulson BS. Integrins α2β1 and α4β1 can mediate SA11 rotavirus attachment and entry into cells. J Virol 2000;74:228–236.

60. Graham KL, Halasz P, Tan Y. et al. Integrin-using rotaviruses bind alpha2beta1 integrin alpha2 I domain via VP4 DGE sequence and recognize alphaX-beta2 and alphavbeta3 by using VP7 during cell entry. J Virol 2003;77:9969–9978.

61. Rolsma MD, Kuhlenschmidt TB, Gelberg HB, Kuhlenschmidt MS. Structure and function of a ganglioside receptor for porcine rotavirus. J Virol 1998;72: 9079–9091.

62. Zárate S, Espinosa R, Romero P, Guerrero CA, Arias CF, López S. Integrin alpha2beta1 mediates the cell attachment of the rotavirus neuraminidase-resistant variant nar3. Virology 2000;278:50–54.

63. Coulson BS, Londrigan SL, Lee DJ. Rotavirus contains integrin ligand sequences and a disintegrin-like domain that are implicated in virus entry into cells. Proc Natl Acad Sci USA 1997;94:5389–5394.

64. Isa P, López S, Segovia L, Arias CF. Functional and structural analysis of the sialic acid-binding domain of rotaviruses. J Virol 1997;71:6749–6756.

65. Zárate S, Cuadras MA, Espinosa R, et al. The interaction of rotaviruses with hsc70 during cell entry is mediated by VP5. J Virol 2003;77:7254–7260.

66. Guerrero CA, Méndez E, Zárate S, Isa P, López S, Arias CF. Integrin alpha(v)beta(3) mediates rotavirus cell entry. Proc Natl Acad Sci USA 2000;97:14644–14649.

67. Chen DY, Estes MK, Ramig RF. Specific interactions between rotavirus outer capsid proteins VP4 and VP7

determine expression of a cross-reactive, neutralizing VP4 specific epitope. J Virol 1992;66:432–439.

68. Méndez E, Arias CF, López S. Interactions between the two surface proteins of rotavirus may alter the receptor-binding specificity of the virus. J Virol 1996;70: 1218–1222.

69. Pralle A, Keller P, Florin E-L, Simons K, Horber JKH. Sphinolipid-cholesterol rafts diffuse as small entities in the plasma membrane of mammalian cells. J Cell Biol 2000;148:997–1007.

70. Brown DA. Seeing is believing: visualization of rafts in model membranes. Proc Natl Acad Sci USA 2001;98:10517–10518.

71. Isa P, Realpe M, Romero P, López S, Arias CF. Rotavirus RRV associates with lipid membrane microdomains during cell entry. Virology 2004;322:370–381.

72. Quan CM, Doane FW. Ultrastructural evidence for the cellular uptake of rotavirus by endocytosis. Intervirology 1983;20:223–231.

73. Ludert JE, Michelangeli F, Gil F, et al. Penetration and uncoating of rotaviruses in cultured cells. Intervirology 1987;27:95–101.

74. Suzuki H, Kitaoka S, Konno T, et al. Two modes of human rotavirus entry into MA104 cells. Arch Virol 1985;85:25–34.

75. Kaljot KT, Shaw RD, Rubin DH, Greenberg HB. Infectious rotavirus enters cells by direct cell membrane penetration, not by endocytosis. J Virol 1988;62: 1136–1144.

76. Fukuhara N, Yoshie O, Kitaoka S, et al. Evidence for endocytosis-independent infection of human rotavirus. Arch Virol 1987;97:93–99.

77. Keljo DJ, Kuhn M, Smith A. Acidification of endosomes in not important for the entry of rotavirus into the cell. J. Pediatr Gastroenterol Nutr 1988;7:257–263.

78. Martin S, Lorrot M, El Azher MA, Vasseur M. Ionic strength- and temperature-induced K_{Ca} shifts in the uncoating reaction of rotavirus strains RF and SA11: correlation with membrane permeabilization. J Virol 2002;76:552–559.

79. Estes MK. Rotaviruses and their replication. In: Knipe DM, Howley PM, Griffin DE, et al., eds. Fields Virology, 4th ed. Philadelphia: Lippincott Williams & Wilkins, 2001:1747–1785.

80. Denisova E, Dowling W, LaMonica R, et al. Rotavirus Capsid Protein VP5* permeabilizes membranes. J Virol 1999;73:3147–3153.

81. Cuadras MA, Arias CF, Lopez S. Rotaviruses induce an early membrane permeabilization of MA104 cells and do not require a low intracellular Ca^{2+} concentration to initiate their replication cycle. J Virol 1997;71: 9065–9074.

82. Liprandi F, Moros Z, Gerder M, et al. Productive penetration of rotavirus in cultured cells induces coentry of the translation inhibitor alpha-sarcin. Virology 1997;237:430–438.

83. Lawton JA, Estes MK, Prasad BV. Three-dimensional visualization of mRNA release from actively transcribing rotavirus particles. Nat Struct Biol 1997;4: 118–121.

84. Silvestri LS, Taraporewala ZF, Patton JT. Rotavirus replication: plus-sense templates for double-stranded RNA synthesis are made in viroplasms. J Virol 2004; 78:7763–7774.

85. Vende P, Piron M, Castagne N, Poncet D. Efficient Translation of rotavirus mRNA requires simultaneous interaction of NSP3 with the eukaryotic translation initiation factor eIF4G and the mRNA 3' end. J Virol 2000;74:7064–7071.

86. Piron M, Vende P, Cohen, Poncet D. Rotavirus RNA-binding protein NSP3 interacts with eIF4GI and evicts the poly (A) binding protein from eIF4F. EMBO J 1998;17:5811–5821.

87. Padilla-Noriega L. Paniagua O, Guzman-Leon S. Rotavirus protein NSP3 shuts off host protein synthesis. Virology 2002;298:1–7.

88. Fabbretti E, Afrikanova I, Vascotto F, Burrone O. Two non-structural rotavirus proteins, NSP2 and NSP5, form viroplasm-like structures in vivo. J Gen Virol 1999;80:333–339.

89. Berois M, Sapin C, Erk I, Poncet D, Cohen J. Rotavirus nonstructural protein NSP5 interacts with major core protein VP2. J Virol 2003;77:1757–1763.

90. Mohan KVK, Muller J, Atreya CD. The N- and C-terminal regions of rotavirus NSP5 are the critical determinants for the formation of viroplasm-like structures independent of NSP2. J Virol 2003;77: 12184–12192.

91. Taraporewala ZF, Patton JT. Nonstructural proteins involved in genome packaging and replication of rotaviruses and other members of the reoviridae. Virus Res 2004;101:57–66.

92. Patton JT, Kearney K, Taraporewala Z. Rotavirus genome replication: role of the RNA—binding protein. In: Desselberger U, Gray J, eds. Viral Gastroenteritis, vol. 9. The Netherlands: Elsevier Science, 2003:165–183.

93. Taniguchi K, Kojima K, Urasawa S. Nondefective rotavirus mutants with an NSP1 gene which has a deletion of 500 nucleotides, including a cysteine-rich zinc finger motif-encoding region (nucleotides 156 to 248) or which has a nonsense codon at nucleotides 153 to 155. J Virol 1996;70:4125–4130.

94. Graff JW, Mitzel DN, Weisend CM, Flenniken ML, Hardy ME. Interferon regulatory factor 3 is a cellular partner of rotavirus NSP1. J Virol 2002;76:9545–9550.

95. Gonzalez RA, Espinosa R, Romero P, López S, Arias CF. Relative localization of viroplasmic and endoplasmic reticulum-resident rotavirus proteins in infected cells. Arch Virol 2000;145:1963–1973.

96. Petrie BL, Greenberg HB, Graham DY, Estes MK. Ultrastructural localization of rotavirus antigens using colloidal gold. Virus Res 1984;1:133–152.

97. Delmas O, Durand-Schneider AM, Cohen J, Colard O, Trugnan G. Spike protein VP4 assembly with maturing rotavirus requires a postendoplasmic reticulum event in polarized Caco-2 cells. J Virol 2004;78: 10987 10994

98. Perez JF, Chemello ME, Liprandi F, Ruiz MC, Michelangeli F. Oncosis in MA104 cells induced by rotavirus infection through an increase in intracellular Ca^{2+} concentration. Virology 1998;252:17–27.

99. Superti F. Ammendolia MG, Tinari A, et al. Induction of apoptosis in HT-29 cells infected with SA-11 rotavirus. J Med Virol 1996;50:325–334.

100. Jourdan N, Maurice M, Delautier D, Quero AM, Servin AL, Trugnan G. Rotavirus is released from the apical surface of cultured human intestinal cells through nonconventional vesicular transport that bypasses the golgi apparatus. J Virol 1997;71:8268–8278.

101. Cuadras MA. Greenberg HB. Rotavirus infectious particles use lipid rafts during replication for transport to

the cell surface in vitro and in vivo. Virology 2003; 313:308–321.

102. Musalem C, Espejo RT. Release of progeny virus from cells infected with simian rotavirus SA11. J Gen Virol 1985;66:2715–2724.

103. Rodriguez WJ, Kim HW, Arrobio JO, et al. Clinical features of acute gastroenteritis associated with human reovirus-like agent in infants and young children. J Pediatr 1977;91:188–193.

104. Bernstein DI, Ward RL. Rotaviruses. In: Feigin RD, Cherry JD, Demmler GJ, Kaplan SL, eds. Textbook of Pediatric Infectious Diseases, 5th ed. Philadelphia: Saunders, 2004:2110–2133.

105. Honeyman MC, Coulson BS, Stone NL, et al. Association between rotavirus infection and pancreatic islet autoimmunity in children at risk of developing type 1 diabetes. Diabetes 2000;49:1319–1324.

106. Blomqvist M, Juhela S, Erkkila S, et al. Rotavirus infections and development of diabetes-associated autoantibodies during the first 2 years of life. Clin Exp Immunol 2002;128:511–515.

107. Murphy TV, Gargiullo PM, Massoudi MS, et al. Intussusception among infants given an oral rotavirus vaccine. N Eng J Med 2001;344:564–572.

108. Parashar UD, Holman RC, Cummings KC, et al. Trends in intussusception-associated hospitalizations and deaths among U.S. infants. Pediatrics 2000;106: 1413–1421.

109. Rennels MB, Parashar UD, Holman RC, et al. Lack of an apparent association between intussusception and wild or vaccine rotavirus infection. Pediatr Infect Dis J 1998;17:924–925.

110. Mebus CA, Stair EL, Underdahl NR, et al. Pathology of neonatal calf diarrhea induced by a reo-like virus. Vet Pathol 1974;8:490–505.

111. Pearson GR, McNulty MS. Ultrastructural changes in small intestinal epithelium of neonatal pigs infected with pig rotavirus. Arch Virol 1979;59:127–136.

112. Torres-Medina A. Effect of combined rotavirus and Escherichia coli in neonatal gnotobiotic calves. Am J Vet Res 1984;45:643–651.

113. Boshuizen JA, Reimerink HJ, Korteland-van Male AM, et al. Changes in small intestinal homeostasis morphology and gene expression during rotavirus infection of infant mice. J Virol 2003;77:13005–13016.

114. Suzuki H, Konno T. Reovirus-like particles in jejunal mucosa of a Japanese infant with acute infectious nonbacterial gastroenteritis. Tohoku J Exp Med 1975;115: 199–221.

115. Holmes IH, Ruck BJ, Bishop RF, et al. Infantile enteritis viruses: morphogenesis and morphology. J Virol 1975;16:937–943.

116. Davidson GP, Gall DG, Petric M, et al. Human rotavirus enteritis induced in conventional piglets: intestinal structure and transport. J Clin Invest 1977;60:1402–1409.

117. Graham DY, Sackman JW, Estes MK. Pathogenesis of rotavirus-induced diarrhea: preliminary studies in miniature swine piglet. Dig Dis Sci 1984;29:1028–1035.

118. Collins J, Starkey WG, Wallis TS, et al. Intestinal enzyme profiles in normal and rotavirus-infected mice. J Pediatr Gastroenterol Nutr 1988;7:264–272.

119. Ball JM, Peng T, Zeng CQY, Morris AP, Estes MK. Age-dependent diarrhea induced by a rotaviral nonstructural glycoprotein. Science 1996;272:101–104.

120. Morris AP, Scott JK, Ball JM, et al. NSP4 elicits age-dependent diarrhea and Ca^{2+}-mediated I^- influx into intestinal crypts of CF mice. Am J Physiol 1999;277: G431–G444.

121. Tian P, Estes MK, Hu Y, et al. The rotavirus nonstructural glycoprotein NSP4 mobilizes Ca^{2+} from the endoplasmic reticulum. J Virol 1995;69:5763–5772.

122. Tian P, Hu Y, Schilling WP, et al. The nonstructural glycoprotein of rotavirus affects intracellular calcium levels. J Virol 1994;68:251–257.

123. Dong Y, Zeng CQY, Ball JM, Estes MK, Morris AP. The rotavirus enterotoxin NSP4 mobilizes intracellular calcium in human intestinal cells by stimulating phospholipase C-mediated inositol 1,4,5-triphosphate production. Proc Natl Acad Sci USA 1997;94:3960–3965.

124. Browne EP, Bellamy AR, Taylor JA. Membrane-destabilizing activity of rotavirus NSP4 is mediated by a membrane-proximal amphipathic domain. J Gen Virol 2000;81:1955–1959.

125. Tian P, Ball JM, Zeng CQY, Estes MK. The rotavirus nonstructural glycoprotein NSP4 possesses membrane destabilization activity. J Virol 1996;70:6973–6981.

126. Brunet J-P, Cotte-Lafitte J, Linxe C, et al. Rotavirus infection induces an increase in intracellular calcium concentration in human intestinal epithelial cells: role in microvillar actin alteration. J Virol 2000;74: 2323–2332.

127. Brunet J-P, Jourdan N, Cotte-Lafitte J, et al. Rotavirus infection induces cytoskeleton disorganization in human intestinal epithelial cells: implication of an increase in intracellular calcium concentration. J Virol 2000;74:10801–10806.

128. Perez JF, Ruiz M-C, Chemello ME, Michelangeli F. Characterization of a membrane calcium pathway induced by rotavirus infection in cultured cells. J Virol 1999;73:2481–2490.

129. Lundgren O, Peregrin AT, Persson K, et al. Role of the enteric nervous system in the fluid and electrolyte secretion of rotavirus diarrhea. Science 2000;287:491–495.

130. Horie Y, Nakagomi O, Koshimura Y, et al. Diarrhea induction by rotavirus NSP4 in the homologous mouse model system. Virology 1999;262:398–407.

131. Zhang M, Zeng CQY, Dong Y, et al. Mutations in rotavirus nonstructural glycoprotein NSP4 are associated with altered virus virulence. J Virol 1998;72:3666–3672.

132. Angel J, Tang B, Feng N, Greenberg HB, Bass D. Studies of the role for NSP4 in the pathogenesis of homologous murine rotavirus diarrhea. J Infect Dis 1998;177: 455–458.

133. Lee C-N, Wang Y-L, Kao C-L, et al. NSP4 gene analysis of rotaviruses recovered from infected children with and without diarrhea. J Clin Microbiol 2000;38:4471–4477.

134. Ward RL, Mason BB, Bernstein DI, et al. Attenuation of a human rotavirus vaccine candidate did not correlate with mutations in the NSP4 protein gene. J Virol 1997; 71:6267–6270.

135. Offit PA, Blavat G, Greenberg HB, et al. Molecular basis of rotavirus virulence role of gene segment 4. J Virol 1986;57:46–49.

136. Broome RL, Vo PT, Ward RL, et al. Murine rotavirus genes encoding outer capsid proteins VP4 and VP7 are

not major determinants of host range restriction and virulence. J Virol 1993;67:2448–2455.

137. Bridger JC, Tauscher GI, and Desselberger U. Viral determinants of rotavirus pathogenicity in pigs: evidence that the fourth gene of a porcine rotavirus confers diarrhea in the homologous host. J Virol 1998;72:6929–6931.

138. Hoshino Y, Saif LJ, Kang S-Y, et al. Identification of group A rotavirus genes associated with virulence of a porcine rotavirus and host range restriction of a human rotavirus in the gnotobiotic piglet model. Virology 1995;209:274–280.

139. McNeal MM, Ward RL. Long-term production of rotavirus antibody and protection against reinfection following a single infection of neonatal mice with murine rotavirus. Virology 1995;211:474–480.

140. Bridger JC. A definition of bovine rotavirus virulence. J Gen Virol 1994;75:2807–2812.

141. Kirstein CG, Clare DA, Lecce JG. Development of resistance of enterocytes to rotavirus in neonatal agammaglobulinemic piglets. J Virol 1985;55:567–573.

142. Riepenhoff-Talty M, Lee PC, Carmody PJ, et al. Age-dependent rotavirus-enterocyte interactions. Proc Soc Exp Biol Med 1982;170:146–154.

143. Varshney KC, Bridger JC, Parson KR, et al. The lesions of rotavirus infection in 1– and 10–day-old gnotobiotic calves. Vet Pathol 1995;32:619–627.

144. Lebenthal E, Lee PC. Development of functional response in human exocrine pancreas. Pediatrics 1980;66:556–560.

145. Zheng BJ, Lo SKF, Tam JSL, et al. Prospective study of community-acquired rotavirus infection. J Clin Microbiol 1989;27:2083–2090.

146. Ward RL, Bernstein DI, Shukla R, et al. Effects of antibody to rotavirus on protection of adults challenged with a human rotavirus. J Infect Dis 1989;159:79–88.

147. Chiba S, Yokoyama T, Nakata S, et al. Protective effect of naturally acquired homotypic and heterotypic rotavirus antibodies. Lancet 1986;2:417–421.

148. Hjelt K, Graubelle PC, Paerregaard A, et al. Protective effect of pre-existing rotavirus-specific immunoglobulin A against naturally acquired rotavirus infection in children. J Med Virol 1987;21:39–47.

149. Bernstein DI, Smith VE, Sander DS, Pax KA, Schiff GM, Ward RL. Evaluation of WC3 rotavirus vaccine and correlates of protection in healthy infants. J Infect Dis 1990;162:1055–1062.

150. Valazques FR, Matson DO, Guerrero ML, et al. Serum antibody as a marker of protection against natural rotavirus infection and disease. J Infect Dis 2000;182:1602–1609.

151. Matson DO, O'Ryan ML, Herrera I, Pickering LK, Estes MK. Fecal antibody responses to symptomatic and asymptomatic rotavirus. J Infect Dis 1993;167:577–583.

152. Coulson BS, Grimwood K, Hudson IL, Barnes GL, Bishop RF. Role of coproantibody in clinical protection of children during reinfection with rotavirus. J Clin Microbiol 1992;30:1678–1684.

153. Graham A, Kudesia G, Allen AM, et al. Reassortment of human rotavirus possessing genome rearrangements with bovine rotavirus: evidence of host cell selection. J Gen Virol 1987;68:115–122.

154. Gombold JL, Ramig RF. Analysis of reassortment of genome segments in mice mixedly infected with rotaviruses SA11 and RRV. J Virol 1986;57:110–116.

155. Bridger JC, Dhaliwal W, Adamson MJV, Howard CR. Determinants of rotavirus host range restriction—a heterologous bovine NSP1 gene does not affect replication kinetics in the pig. Virology 1998;245:47–52.

156. Haffejee IE. The epidemiology of rotavirus infections: a global perspective. J Pediatr Gastroent Nutr 1995;20:275–286.

157. Ward RL, Knowlton DR, Pierce MJ. Efficiency of human rotavirus propagation in cell culture. J Clin Microbiol 1984;19:748–753.

158. Keswick BH, Pickering LK, Dupont HL, et al. Survival and detection of rotaviruses on environmental surfaces in day care centers. Appl Environ Microbiol 1983;46:813–816.

159. Estes MK, Graham DY, Smith EM, et al. Rotavirus stability and inactivation. J Gen Virol 1979;43:403–409.

160. Butz AM, Fosarelli P, Dick J, et al. Prevalence of rotavirus on high-risk fomites in daycare facilities. Pediatrics 1993;92:202–205.

161. Shaw R, Merchant A, Groene W, Cheng EH. Persistence of intestinal antibody response to heterologous rotavirus infection in a murine model beyond 1 year. J Clin Microbiol 1993;31:188–191.

162. Feng N, Burns JW, Bracy L, Greenberg HB. Comparison of mucosal and systemic humoral immune responses and subsequent protection in mice orally inoculated with a homologous or a heterologous rotavirus. J Virol 1994;68:7766–7773.

163. McNeal MM, Broome RL, Ward RL. Active immunity against rotavirus infection in mice is correlated with viral replication and titers of serum rotavirus IgA following vaccination. Virology 1994;204:642–650.

164. Chiba S, Yokoyama T, Nakata S, et al. Protective effect of naturally acquired homotypic and heterotypic rotavirus antibodies. Lancet 1986;2:417–421.

165. Hoshino Y, Saif LJ, Sereno MM, Chanock RM, Kapikian AZ. Infection immunity of piglets to either VP3 or VP7 outer capsid protein confers resistance to challenge with a virulent rotavirus bearing the corresponding antigen. J Virol 1988;62:744–748.

166. Ward RL, McNeal MM, Sheridan JF. Evidence that active protection following oral immunization of mice with live rotavirus is not dependent on neutralizing antibody. Virology 1992;188:57–66.

167. Flores J, Perez-Schael I, Gonzales M, et al. Protection against severe rotavirus diarrhoea by rhesus rotavirus vaccine in Venezuelan infants. Lancet 1987;1:882–884.

168. Santosham M, Letson GW, Wolff M, et al. A field study of the safety and efficacy of two candidate rotavirus vaccines in a Native American population. J Infect Dis 1991;163:483–487.

169. Ward RL, Knowlton DR, Zito ET, Davidson BL, Rappaport R, Mack ME. Serological correlates of immunity in a tetravalent reassortant rotavirus vaccine trial. J Infect Dis 1997;176:570–577.

170. Burns JW, Siadat-Pajouh M, Krishnaney AA, Greenberg HB. Protective effect of rotavirus VP6–specific IgA monoclonal antibodies that lack neutralizing activity. Science 1996;272:104–107.

171. Feng N, Lawton JA, Gilbert J, et al. Inhibition of rotavirus biogenesis by a non-neutralizing, rotavirus VP6–specific IgA mAb. J Clin Invest 2002;109:1203–1213.

172. O'Neal CM, Crawford SE, Estes ME, Conner ME. Rotavirus VLPs administered mucosally induce protective immunity. J Virol 1997;71:8707–8717.

173. Choi AHC, Basu M, McNeal MM, Clements JD, Ward RL. Antibody-independent protection against rotavirus infection of mice stimulated by intranasal immunization with chimeric VP4 or VP6 protein. J Virol 1999;73:7574–7581.

174. Offit P, Dudzik K. Rotavirus-specific cytotoxic T lymphocytes passively protect against gastroenteritis in suckling mice. J Virol 1990;64:6325–6328.

175. Dharakul T, Rott L, Greenberg H. Recovery from chronic rotavirus infection in mice with severe combined immunodeficiency: virus clearance mediated by adoptive transfer of immune CD8+ T lymphocytes. J Virol 1990;64:4375–4382.

176. Ward RL, McNeal MM, Sheridan JF. Development of an adult mouse model for studies on protection against rotavirus. J Virol 1990;64:5070–5075.

177. Feng N, Vo PT, Chung D, Hoshino Y, Greenberg HB. Heterotypic protection following oral immunization with live heterologous rotaviruses in the mouse model. J Infect Dis 1997;175:330–341.

178. Moser CA, Cookinham S, Coffin SE, Clark HF, Offit PA. Relative importance of rotavirus-specific effector and memory B cells in protection against challenge. J Virol 1998;72:1108–1114.

179. Franco MA, Greenberg HB. Role of B cells and cytotoxic T lymphocytes in clearance of and immunity to rotavirus infection in mice. J Virol 1995;69:7800–7806.

180. McNeal MM, Barone KS, Rae MN, Ward RL. Effector functions of antibody and CD8+ cells in resolution of rotavirus infection and protection against reinfection in mice. Virology 1995;214:387–397.

181. Coffin SE, Clark SL, Bos NA, Brubaker JO, Offit PA. Migration of antigen-presenting B cells from peripheral to mucosal lymphoid tissues may induce intestinal antigen-specific IgA following parental immunization. J Immunol 1999;163:3064–3070.

182. O'Neal CM, Harriman GR, Conner ME. Protection of the villus epithelial cells in the small intestine infection does not require immunoglobulin A. J Virol 2000;74:4102–4109.

183. Kuklin NA, Rott L, Feng N, et al. Protective intestinal anti-rotavirus B cell immunity is dependent on α4β7 integrin expression but does not require IgA antibody production. J Immunol 2001;166:1894–1902.

184. Choi AHC, Basu M, McNeal MM, et al. Functional mapping of protective domains and epitopes in the rotavirus VP6 protein. J Virol 2000;74:11574–11580.

185. McNeal, MM, VanCott JL, Choi AHC, et al. CD4 T cells are the only lymphocytes needed to protect mice against rotavirus shedding after intranasal immunization with a chimeric VP6 protein and the adjuvant LT(R192G). J Virol 2002;76:560–568.

186. Bernstein DI, Sack DA, Rothstein E, et al. Efficacy of live, attenuated, human rotavirus vaccine 89–12 in infants: a randomized placebo-controlled trial. Lancet 1999;354:287–290.

187. Ciarlet M, Crawford SE, Barone C, Bertolotti-Ciarlet A, Estes MK, Conner ME. Subunit rotavirus vaccine administered parenterally to rabbits induces active protective immunity. J Virol 1998;72:9233–9246.

188. Coste A, Sirard JC, Johansen K, Cohen J, Kraehenbuhl JP. Nasal immunization of mice with virus-like particles protects offspring against rotavirus diarrhea. J Virol 2000;74:8966–8971.

189. Yuan L, Iosef C, Azevedo MSP, et al. Protective immunity and antibody-secreting cell responses elicited by combined oral attenuated Wa human rotavirus and intranasal Wa 2/6–VLPs with mutant Escherichia coli heat-labile toxin in gnotobiotic pigs. J Virol 2001;75:9229–9238.

190. Ward RL, Clemens JD, Sack DA, et al. Culture-adaptation of group A rotaviruses causing diarrheal illnesses in Bangladesh during 1985–1986. J Clin Microbiol 1991;29:1915–1923.

191. Green KY, Ando T, Balayan MS, et al. Taxonomy of the caliciviruses. J Infect Dis 2000;181:S322–330.

192. Liu BL, Clarke IN, Caul EO, Lambden PR. Human enteric caliciviruses have a unique genome structure and are distinct from the Norwalk-like viruses. Arch Virol 1995;140:1345–1356.

193. Adler J, Zickl R. Winter vomiting disease. J Infect Dis 1969;119:668–673.

194. Dolin R. Norwalk agent-like particles associated with gastroenteritis in human beings. J Am Vet Med Assoc 1978;173:615–619.

195. Dolin R, Reichman RC, Roessner KD, et al. Detection by immune electron microscopy of the Snow Mountain agent of acute viral gastroenteritis. J Infect Dis 1982;146:184–189.

196. Thornhill TS, Wyatt RG, Kalica AR, Dolin R, Chanock RM, Kapikian AZ. Detection by immune electron microscopy of 26- to 27–nm viruslike particles associated with two family outbreaks of gastroenteritis. J Infect Dis 1977;135:20–27.

197. Wyatt RG, Dolin R, Blacklow NR, et al. Comparison of three agents of acute infectious nonbacterial gastroenteritis by cross-challenge in volunteers. J Infect Dis 1974;129:709–714.

198. Chiba S, Sakuma Y, Kogasaka R, et al. An outbreak of gastroenteritis associated with calicivirus in an infant home. J Med Virol 1979;4:249–254.

199. Jiang X, Graham DY, Wang KN, Estes MK. Norwalk virus genome cloning and characterization. Science 1990;250:1580–1583.

200. Jiang X, Wang M, Wang K, Estes MK. Sequence and genomic organization of Norwalk virus. Virology 1993;195:51–61.

201. De Leon R, Matsui SM, Baric RS, et al. Detection of Norwalk virus in stool specimens by reverse transcriptase-polymerase chain reaction and nonradioactive oligoprobes. J Clin Microbiol 1992;30:3151–3157.

202. Jiang X, Wang J, Graham DY, Estes MK. Detection of Norwalk virus in stool by polymerase chain reaction. J Clin Microbiol 1992;30:2529–2534.

203. Jiang X, Wang M, Graham DY, Estes MK. Expression, self-assembly, and antigenicity of the Norwalk virus capsid protein. J Virol 1992;66:6527–6532.

204. Baric RS, Yount B, Lindesmith L, et al. Expression and self-assembly of Norwalk virus capsid protein from Venezuelan equine encephalitis virus replicons. J Virol 2002;76:3023–3030.

205. Mason HS, Ball JM, Shi JJ, Jiang X, Estes MK, Arntzen CJ. Expression of Norwalk virus capsid protein in transgenic tobacco and potato and its oral immunogenicity in mice. Proc Natl Acad Sci USA 1996;93:5335–5340.

206. Graham DY, Jiang X, Tanaka T, Opekun AR, Madore HP, Estes MK. Norwalk virus infection of volunteers:

new insights based on improved assays. J Infect Dis 1994;170:34–43.

207. Monroe SS, Stine SE, Jiang X, Estes MK, Glass RI. Detection of antibody to recombinant Norwalk virus antigen in specimens from outbreaks of gastroenteritis. J Clin Microbiol 1993;31:2866–2872.

208. Huang P, Farkas T, Marionneau S, et al. Noroviruses bind to human ABO, Lewis, and secretor histo-blood group antigens: identification of 4 distinct strain-specific patterns. J Infect Dis 2003;188:19–31.

209. Marionneau S, Ruvoen N, Le Moullac-Vaidye B, et al. Norwalk virus binds to histo-blood group antigens present on gastroduodenal epithelial cells of secretor individuals. Gastroenterology 2002;122:1967–1977.

210. Gray JJ, Cunliffe C, Ball J, Graham DY, Desselberger U, Estes MK. Detection of immunoglobulin M (IgM), IgA, and IgG Norwalk virus-specific antibodies by indirect enzyme-linked immunosorbent assay with baculovirus-expressed Norwalk virus capsid antigen in adult volunteers challenged with Norwalk virus. J Clin Microbiol 1994;32:3059–3063.

211. Green KY, Lew JF, Jiang X, Kapikian AZ, Estes MK. Comparison of the reactivities of baculovirus-expressed recombinant Norwalk virus capsid antigen with those of the native Norwalk virus antigen in serologic assays and some epidemiologic observations. J Clin Microbiol 1993;31:2185–2191.

212. Jiang X, Cubitt D, Hu J, et al. Development of an ELISA to detect MX virus, a human calicivirus in the Snow Mountain agent genogroup. J Gen Virol 1995;76:2739–2747.

213. Jiang X, Wang J, Estes MK. Characterization of SRSVs using RT-PCR and a new antigen ELISA. Arch Virol 1995;140:363–374.

214. Parker S, Cubitt D, Jiang JX, Estes M. Efficacy of a recombinant Norwalk virus protein enzyme immunoassay for the diagnosis of infections with Norwalk virus and other human "candidate" caliciviruses. J Med Virol 1993;41:179–184.

215. Parker SP, Cubitt WD. Measurement of IgA responses following Norwalk virus infection and other human caliciviruses using a recombinant Norwalk virus protein EIA. Epidemiol Infect 1994;113:143–151.

216. Parker SP, Cubitt WD, Jiang X. Enzyme immunoassay using baculovirus-expressed human calicivirus (Mexico) for the measurement of IgG responses and determining its seroprevalence in London, UK. J Med Virol 1995;46:194–200.

217. Ball JM, Estes MK, Hardy ME, et al. Recombinant Norwalk virus-like particles as an oral vaccine. Arch Virol 1996;12:243–249.

218. Ball JM, Graham AR, Opekun MA, et al. Recombinant Norwalk virus-like particles given orally to volunteers: phase I study. Gastroenterology 1999;117:40–48.

219. Ball JM, Hardy ME, Atmar RL, Conner ME, Estes MK. Oral immunization with recombinant Norwalk virus-like particles induces a systemic and mucosal immune response in mice. J Virol 1998;72:1345–1353.

220. Prasad BV, Hardy ME, Dokland T, Bella J, Rossmann MG, Estes MK. X-ray crystallographic structure of the Norwalk virus capsid. Science 1999;286:287–290.

221. Prasad BV, Rothnagel R, Jiang X, Estes MK. Three-dimensional structure of baculovirus-expressed Norwalk virus capsids. J Virol 1994;68:5117–5125.

222. Prasad BV, Hardy ME, Estes MK. Structural studies of recombinant Norwalk capsids. J Infect Dis 2000;181:S317–321.

223. Tan M, Hegde RS, Jiang X. The P domain of norovirus capsid protein forms dimer and binds to histo-blood group antigen receptors. J Virol 2004;78:6233–6242.

224. Tan M, Huang P, Meller J, Zhong W, Farkas T, Jiang X. Mutations within the P2 domain of norovirus capsid affect binding to human histo-blood group antigens: evidence for a binding pocket. J Virol 2003;77:12562–12571.

225. Greenberg HB, Valdesuso JR, Kalica AR, et al. Proteins of Norwalk virus. J Virol 1981;37:994–999.

226. Matson DO, Zhong WM, Nakata S, et al. Molecular characterization of a human calicivirus with sequence relationships closer to animal caliciviruses than other known human caliciviruses. J Med Virol 1995;45:215–222.

227. Dastjerdi AM, Green J, Gallimore CI, Brown DW, Bridger JC. The bovine Newbury agent-2 is genetically more closely related to human SRSVs than to animal caliciviruses. Virology 1999;254:1–5.

228. Guo M, Chang KO, Hardy ME, Zhang Q, Parwani AV, Saif LJ. Molecular characterization of a porcine enteric calicivirus genetically related to Sapporo-like human caliciviruses. J Virol 1999;73:9625–9631.

229. Sugieda M, Nagaoka H, Kakishima Y, Ohshita T, Nakamura S, Nakajima S. Detection of Norwalk-like virus genes in the caecum contents of pigs. Arch Virol 1998;143:1215–1221.

230. van der Poel WH, Vinje J, van der Heide R, Herrera MI, Vivo A, Koopmans MP. Norwalk-like calicivirus genes in farm animals. Emerg Infect Dis 2000;6:36–41.

231. Katayama K, Shirato-Horikoshi H, Kojima S, et al. Phylogenetic analysis of the complete genome of 18 Norwalk-like viruses. Virology 2002;299:225–239.

232. Ando T, Noel JS, Fankhauser RL. Genetic classification of "Norwalk-like viruses." J Infect Dis 2000;181:S336–348.

233. Schuffenecker I, Ando T, Thouvenot D, Lina B, Aymard M. Genetic classification of "Sapporo-like viruses." Arch Virol 2001;146:2115–2132.

234. Karst SM, Wobus CE, Lay M, Davidson J, Virgin HWT. STAT1–dependent innate immunity to a Norwalk-like virus. Science 2003;299:1575–1578.

235. Jiang X, Matson DO, Cubitt WD, Estes MK. Genetic and antigenic diversity of human caliciviruses (HuCVs) using RT-PCR and new EIAs. Arch Virol 1996;12:251–262.

236. Jiang X. Development of serological and molecular tests for the diagnosis of calicivirus infections. In: Desselberger U, Gray J, eds. Viral Gastroenteritis, 1st ed. Amsterdam: Elsevier Science BV, 2003:505–522.

237. Jiang X, Wilton N, Zhong WM, et al. Diagnosis of human caliciviruses by use of enzyme immunoassays. J Infect Dis 2000;181:S349–S359.

238. White LJ, Ball JM, Hardy ME, Tanaka TN, Kitamoto N, Estes MK. Attachment and entry of recombinant Norwalk virus capsids to cultured human and animal cell lines. J Virol 1996;70:6589–6597.

239. Duizer E, Schwab KJ, Neill FH, Atmar RL, Koopmans MP, Estes MK. Laboratory efforts to cultivate noroviruses. J Gen Virol 2004;85:79–87.

240. Flynn WT, Saif LJ. Serial propagation of porcine enteric calicivirus-like virus in primary porcine kidney cell cultures. J Clin Microbiol 1988;26:206–212.

241. Guo M, Hayes J, Cho KO, Parwani AV, Lucas LM, Saif L. Comparative pathogenesis of tissue culture-adapted and wild-type Cowden porcine enteric calicivirus (PEC) in gnotobiotic pigs and induction of diarrhea by intravenous inoculation of wild-type PEC. J Virol 2001;75:9239–9251.

242. Guo M, Saif LJ. Pathogenesis of enteric calicivirus infections. In: Desselberger U, Gray J, eds. Viral Gastroenteritis, 1st ed. Amsterdam: Elsevier Science BV, 2003:489–503.

243. Chang KO, Sosnovtsev SV, Belliot G, et al. Bile acids are essential for porcine enteric calicivirus replication in association with down-regulation of signal transducer and activator of transcription 1. Proc Natl Acad Sci USA 2004;101:8733–8738.

244. Wobus CE, Karst SM, Thackray LB, et al. Replication of norovirus in cell culture reveals a tropism for dendritic cells and macrophages. PLoS Biol 2004;2:e432.

245. Dolin R, Levy AG, Wyatt RG, Thornhill TS, Gardner JD. Viral gastroenteritis induced by the Hawaii agent. Jejunal histopathology and serologic response. Am J Med 1975;59:761–768.

246. Parrino TA, Schreiber DS, Trier JS, Kapikian AZ, Blacklow NR. Clinical immunity in acute gastroenteritis caused by Norwalk agent. N Engl J Med 1977; 297:86–89.

247. Thornhill TS, Kalica AR, Wyatt RG, Kapikian AZ, Chanock RM. Pattern of shedding of the Norwalk particle in stools during experimentally induced gastroenteritis in volunteers as determined by immune electron microscopy. J Infect Dis 1975;132: 28–34.

248. Rockx B, De Wit M, Vennema H, et al. Natural history of human calicivirus infection: a prospective cohort study. Clin Infect Dis 2002;35:246–253.

249. Pang XL, Joensuu, J, Vesikari T. Human calicivirus-associated sporadic gastroenteritis in Finnish children less than two years of age followed prospectively during a rotavirus vaccine trial. Pediatr Infect Dis J 1999;18:420–426.

250. Pang XL, Zeng SQ, Honma S, Nakata S, Vesikari T. Effect of rotavirus vaccine on Sapporo virus gastroenteritis in Finnish infants. Pediatr Infect Dis J 2001;20:295–300.

251. Flynn WT, Saif LJ, Moorhead PD. Pathogenesis of porcine enteric calicivirus-like virus in four-day-old gnotobiotic pigs. Am J Vet Res 1988;49:819–825.

252. Hardy ME, White LJ, Ball JM, Estes MK. Specific proteolytic cleavage of recombinant Norwalk virus capsid protein. J Virol 1995;69:1693–1698.

253. Ruvoen-Clouet N, Ganiere JP, Andre-Fontaine G, Blanchard D, Le Pendu J. Binding of rabbit hemorrhagic disease virus to antigens of the ABH histo-blood group family. J Virol 2000;74:11950–11954.

254. Hutson AM, Atmar RL, Marcus DM, Estes MK. Norwalk virus-like particle hemagglutination by binding to histo-blood group antigens. J Virol 2003;77: 405–415.

255. Hutson AM, Atmar RL, Graham DY, Estes MK. Norwalk virus infection and disease is associated with ABO histo-blood group type. J Infect Dis 2002;185: 1335–1337.

256. Hennessy E, Green AD, Connor MP, Darby R, MacDonald P. Norwalk virus infection and disease is associated with ABO histo-blood group type. J Infect Dis 2003;188:176–177.

257. Lindesmith L, Moe C, Marionneau S, et al. Human susceptibility and resistance to Norwalk virus infection. Nature Med 2003;9:548–553.

258. Blacklow NR, Cukor G, Bedigian MK, et al. Immune response and prevalence of antibody to Norwalk enteritis virus as determined by radioimmunoassay. J Clin Microbiol 1979;10:903–909.

259. Nakata S, Chiba S, Terashima H, Yokoyama T, Nakao T. Humoral immunity in infants with gastroenteritis caused by human calicivirus. J Infect Dis 1985;152: 274–279.

260. Evans MR, Meldrum R, Lane W, et al. An outbreak of viral gastroenteritis following environmental contamination at a concert hall. Epidemiol Infect 2002;129: 355–360.

261. Marks PJ, Vipond IB, Carlisle D, Deakin D, Fey RE, Caul EO. Evidence for airborne transmission of Norwalk-like virus (NLV) in a hotel restaurant. Epidemiol Infect 2000;124:481–487.

262. Marks PJ, Vipond IB, Regan FM, Wedgewood K, Fey RE, Caul EO. A school outbreak of Norwalk-like virus: evidence for airborne transmission. Epidemiol Infect 2003;131:727–736.

263. Cubitt WD, Green KY, Payment P. Prevalence of antibodies to the Hawaii strain of human calicivirus as measured by a recombinant protein based immunoassay. J Med Virol 1998;54:135–139.

264. Gray JJ, Jiang X, Morgan-Capner P, Desselberger U, Estes MK. Prevalence of antibodies to Norwalk virus in England: detection by enzyme-linked immunosorbent assay using baculovirus-expressed Norwalk virus capsid antigen. J Clin Microbiol 1993;31:1022–1025.

265. Jiang X, Matson DO, Velazquez FR, et al. Study of Norwalk-related viruses in Mexican children. J Med Virol 1995;47:309–316.

266. Jing Y, Qian Y, Huo Y, Wang LP, Jiang X. Seroprevalence against Norwalk-like human caliciviruses in Beijing, China. J Med Virol 2000;60:97–101.

267. Numata K, Nakata S, Jiang X, Estes MK, Chiba S. Epidemiological study of Norwalk virus infections in Japan and Southeast Asia by enzyme-linked immunosorbent assays with Norwalk virus capsid protein produced by the baculovirus expression system. J Clin Microbiol 1994;32:121–126.

268. Parker SP, Cubitt WD, Jiang XJ, Estes MK. Seroprevalence studies using a recombinant Norwalk virus protein enzyme immunoassay. J Med Virol 1994; 42:146–150.

269. Farkas T, Jiang X, Guerrero ML, et al. Prevalence and genetic diversity of human caliciviruses (HuCVs) in Mexican children. J Med Virol 2000;62:217–223.

270. Foley B, O'Mahony J, Morgan SM, Hill C, Morgan JG. Detection of sporadic cases of Norwalk-like virus (NLV) and astrovirus infection in a single Irish hospital from 1996 to 1998. J Clin Virol 2000;17:109–117.

271. Kirkwood CD, Bishop RF. Molecular detection of human calicivirus in young children hospitalized with acute gastroenteritis in Melbourne, Australia, during 1999. J Clin Microbiol 2001;39:2722–2424.

272. Marie-Cardine A, Gourlain K, Mouterde O, et al. Epidemiology of acute viral gastroenteritis in children hospitalized in Rouen, France. Clin Infect Dis 2002; 34:1170–1178.

273. Subekti DS, Tjaniadi P, Lesmana M, et al. Characterization of Norwalk-like virus associated with gastroenteritis in Indonesia. J Med Virol 2002;67:253–258.

274. Roman E, Negredo A, Dalton RM, Wilhelmi I, Sanchez-Fauquier A. Molecular detection of human calicivirus among Spanish children with acute gastroenteritis. J Clin Microbiol 2002;40:3857–3859.

275. Bon F, Fascia P, Dauvergne M, et al. Prevalence of group A rotavirus, human calicivirus, astrovirus, and adenovirus type 40 and 41 infections among children with acute gastroenteritis in Dijon, France. J Clin Microbiol 1999;37:3055–3058.

276. Iritani N, Seto Y, Kubo H, et al. Prevalence of Norwalk-like virus infections in cases of viral gastroenteritis among children in Osaka City, Japan. J Clin Microbiol 2003;41:1756–1759.

277. Martinez N, Espul C, Cuello H, et al. Sequence diversity of human caliciviruses recovered from children with diarrhea in Mendoza, Argentina, 1995–1998. J Med Virol 2002;67:289–298.

278. O'Ryan ML, Mamani N, Gaggero A, et al. Human caliciviruses are a significant pathogen of acute sporadic diarrhea in children of Santiago, Chile. J Infect Dis 2000;182:1519–1522.

279. Bereciartu A, Bok K, Gómez J. Identification of viral agents causing gastroenteritis among children in Buenos Aires, Argentina. J Clin Virol 2002;25:197–203.

280. Bonrud P, Volmer A, Dosch T, et al. Leads from the MMWR. Viral gastroenteritis—South Dakota and New Mexico. JAMA 1988;259:1459–1460.

281. Qiao H, Nilsson M, Abreu ER, et al. Viral diarrhea in children in Beijing, China. J Med Virol 1999;57:390–396.

282. Subekti D, Lesmana M, Tjaniadi P, et al. Incidence of Norwalk-like viruses, rotavirus and adenovirus infection in patients with acute gastroenteritis in Jakarta, Indonesia. FEMS Immunol Med Microbiol 2002;33:27–33.

283. Wolfaardt M, Taylor MB, Booysen HF, Engelbrecht L, Grabow WO, Jiang X. Incidence of human calicivirus and rotavirus infection in patients with gastroenteritis in South Africa. J Med Virol 1997;51:290–296.

284. Buesa J, Collado B, Lopez-Andujar P, et al. Molecular epidemiology of caliciviruses causing outbreaks and sporadic cases of acute gastroenteritis in Spain. J Clin Microbiol 2002;40:2854–2859.

285. de Wit MA, Koopmans MP, Kortbeek LM, et al. Sensor, a population-based cohort study on gastroenteritis in the Netherlands: incidence and etiology. Am J Epidemiol 2001;154:666–674.

286. McIver CJ, Hansman G, White P, Doultree JC, Catton M, Rawlinson WD. Diagnosis of enteric pathogens in children with gastroenteritis. Pathology 2001;33:353–358.

287. Oh DY, Gaedicke G, Schreier E. Viral agents of acute gastroenteritis in German children: prevalence and molecular diversity. J Med Virol 2003;71:82–93.

288. Sakai Y, Nakata S, Honma S, Tatsumi M, Numata-Kinoshita K, Chiba S. Clinical severity of Norwalk virus and Sapporo virus gastroenteritis in children in Hokkaido, Japan. Pediatr Infect Dis J 2001;20:849–853.

289. Schnagl RD, Barton N, Patrikis M, Tizzard J, Erlich J, Morey F. Prevalence and genomic variation of Norwalk-like viruses in Central Australia in 1995–1997. Acta Virol 2000;44:265–271.

290. Simpson R, Aliyu S, Iturriza-Gomara M, Desselberger U, Gray J. Infantile viral gastroenteritis: On the way to closing the diagnostic gap. J Med Virol 2003;70:258–262.

291. Traore O, Belliot G, Mollat C, et al. RT-PCR identification and typing of astroviruses and Norwalk-like viruses in hospitalized patients with gastroenteritis: evidence of nosocomial infections. J Clin Virol 2000;17:151–158.

292. Fankhauser RL, Monroe SS, Noel JS, et al. Epidemiologic and molecular trends of "Norwalk-like viruses" associated with outbreaks of gastroenteritis in the United States. J Infect Dis 2002;186:1–7.

293. Thornton SV, Davies DV, Chapman F, et al. Detection of Norwalk-like virus infection aboard two U.S. Navy ships. Mil Med 2002;167:826–830.

294. Vinje J, Vennema L, Maunula L, et al. International collaborative study to compare reverse transcriptase PCR assays for detection and genotyping of noroviruses. J Clin Microbiol 2003;41:1423–1433.

295. Okada M, Shinozake K, Ogawa T, Kaiho I. Molecular epidemiology and phylogenetic analysis of Sapporo-like viruses. Arch Virol 2002;147:1445–1551.

296. Chakrabarti S, Collingham KE, Stevens RH, et al. Isolation of viruses from stools in stem cell transplant recipients: a prospective surveillance study. Bone Marrow Transplant 2000;25:277–282.

297. Kaufman SS, Chatterjee NK, Fuschino ME, et al. Calicivirus enteritis in an intestinal transplant recipient. Am J Transplant 2003;3:764–768.

298. Cegielski JP, Msengi AE, Miller SE. Enteric viruses associated with HIV infection in Tanzanian children with chronic diarrhea. Pediatr AIDS HIV Infect 1994;5:296–299.

299. Wardley RC, Povey RC. The clinical disease and patterns of excretion associated with three different strains of feline caliciviruses. Res Vet Sci 1977;23:7–14.

300. Farkas T, Thornton SA, Wilton N, Zhong W, Altaye M, Jiang X. Homologous versus heterologous immune responses to Norwalk-Like viruses among crew members after acute gastroenteritis outbreaks on 2 US Navy vessels. J Infect Dis 2003;187:187–193.

301. Estes MK, Ball JM, Guerrero RA, et al. Norwalk virus vaccines: challenges and progress. J Infect Dis 2000;181:S367–373.

302. Richter L, Mason HS, Arntzen CJ. Transgenic plants created for oral immunization against diarrheal diseases. J Travel Med 1996;3:52–56.

303. Harrington PR, Yount B, Johnston RE, Davis N, Moe C, Baric RS. Systemic, mucosal, and heterotypic immune induction in mice inoculated with Venezuelan equine encephalitis replicons expressing Norwalk virus-like particles. J Virol 2002;76:730–742.

304. Tacket CO, Mason HS, Losonsky G, Estes MK, Levine MM, Arntzen CJ. Human immune responses to a novel Norwalk virus vaccine delivered in transgenic potatoes. J Infect Dis 2000;182:302–305.

305. Guerrero R, Ball J, Estes M. Immunogenicity in mice of recombinant Norwalk virus-like particles administered by mucosal routes. J Pediatr Gastroenterol Nutr 1998;26:547.

306. Estes MK, Ball JM, Crawford SE, et al. Virus-like particle vaccines for mucosal immunization. Adv Exp Med Biol 1997;412:387–395.

307. Periwal SB, Kourie KR, Ramachandaran N, et al. A modified cholera holotoxin CT-E29H enhances systemic and mucosal immune responses to recombinant Norwalk virus-virus like particle vaccine. Vaccine 2003;21:376–385.

308. Prasad BV, Hardy ME, Jiang X, Estes MK. Structure of Norwalk virus. Arch Virol 1996;12:237–242.

309. Monroe SS, Jiang B, Stine SE, et al. Subgenomic RNA sequence of human astrovirus supports classification of Astroviridae as a new family of RNA viruses. J Virol 1993;67:3611–3614.

310. Madeley CR, Cosgrove BP. Viruses in infantile gastroenteritis. Lancet 1975;2:124.

311. Lee TW, Kurtz JB. Serial propagation of astrovirus in tissue culture with the aid of trypsin. J Gen Virol 1981; 57:421–424.

312. Risco C, Carrascosa JL, Pedregosa AM, et al. Ultrastructure of human astrovirus serotype 2. J Gen Virol 1995;76:2075–2080.

313. Matsui M, Greenberg HB. Astroviruses. In: Knipe DM, Howley PM, eds. Fields Virology. Philadelphia: Lippincott Williams & Wilkins, 2001:875–916.

314. Qureshi MA, Saif YM, Heggen-Peay CL, et al. Induction of functional defects in macrophages by a poultry enteritis and mortality syndrome-associated turkey astrovirus. Avian Dis 2001;45:853–861.

315. Imada T, Yamaguchi S, Mase M, et al. Avian nephritis virus (ANV) as a new member of the family Astroviridae and construction of infectious ANV cDNA. J Virol 2000;74:8487–8493.

316. Gough RE, Collins MS, Borland E, et al. Astrovirus-like particles associated with hepatitis in ducklings. Vet Rec 1984;114:279.

317. Jiang B, Monroe SS, Koonin EV, et al. RNA sequence of astrovirus: distinctive genomic organization and a putative retrovirus-like ribosomal frameshifting signal that directs the viral replicase synthesis. Proc Nat Acad Sci USA 1993;90:10539–10543.

318. Kiang D, Matsui SM. Proteolytic processing of a human astrovirus nonstructural protein. J Gen Virol 2002;83:25–34.

319. Lewis TL, Matsui SM. An astrovirus frameshift signal induces ribosomal frameshifting in vitro. Arch Virol 1995;140:1127–1135.

320. Donelli G, Superti F, Tinari A, et al. Mechanism of astrovirus entry into Graham 293 cells. J Med Virol 1992;38:271–277.

321. Aroonprasert D, Fagerland JA, Kelso NE, et al. Cultivation and partial characterization of bovine astrovirus. Vet Microbiol 1989;19:113–125.

322. Bass DM, Qiu S. Proteolytic processing of the astrovirus capsid. J Virol 2000;74:1810–1814.

323. Mendez E, Fernandez-Luna T, Lopez S, et al. Proteolytic processing of a serotype 8 human astrovirus ORF2 polyprotein. J Virol 2002;76:7996–8002.

324. Mendez E, Salas-Ocampo E, Arias CF. Caspases mediate processing of the capsid precursor and cell release of human astroviruses. J Virol 2004;78:8601–8608.

325. Naficy AB, Rao MR, Holmes JL, et al. Astrovirus diarrhea in Egyptian children. J Infect Dis 2000;182:685–690.

326. Mitchell DK, Van R, Morrow AL, et al. Outbreaks of astrovirus gastroenteritis in day care centers. J Pedriatr 1993;123:725–732.

327. Traore O, Belliot G, Mollat C, et al. RT-PCR identification and typing of astroviruses and Norwalk-like viruses in hospitalized patients with gastroenteritis: evidence of nosocomial infections. J Clin Virol 2000;17:151–158.

328. Phillips AD, Rice S, Walker-Smith JA. Astrovirus within the human small intestinal mucosa. Gut 1982;23: A923–A924.

329. Kurtz JB, Lee TW, Craig JW, et al. Astrovirus infection in volunteers. J Med Virol 1979;3:221–230.

330. Midthun K, Greenberg HB, Kurtz JB, et al. Characterization and seroepidemiology of a type 5 astrovirus associated with an outbreak of gastroenteritis in Marin County, California. J Clin Microbiol 1993;31: 955–962.

331. Sebire, NJ, Malone M, Shah N, et al. Pathology of astrovirus associated diarrhoea in a paediatric bone marrow transplant recipient. J Clin Pathol 2004;57: 1001–1003.

332. Gray EW, Angus KW, Snodgrass DR. Ultrastructure of the small intestine in astrovirus-infected lambs. J Gen Virol 1980;49:71–82.

333. Hall GA. Comparative pathology of infection by novel diarrhoea viruses. Ciba Fdn Sym 1987;128:192– 217.

334. Snodgrass DR, Angus KW, Gray EW, et al. Pathogenesis of diarrhoea caused by astrovirus infections in lambs. Arch Virol 1979;60:217–226.

335. Woode GN, Pohlenz JF, Gourley NE, et al. Astrovirus and Breda virus infections of dome cell epithelium of bovine ileum. J Clin Microbiol 1984;19:623–630.

336. Baxendale W, Mebatsion T. The isolation and characterisation of astroviruses from chickens. Avian Path 2004;33:364–370.

337. Koci MD, Moser LA, Kelley LA, et al. Astrovirus induces diarrhea in the absence of inflammation and cell death. J Virol 2003;77:11798–11808.

338. Oishi I, Yamazaki K, Kimoto T, et al. A large outbreak of acute gastroenteritis associated with astrovirus among students and teachers in Osaka, Japan. J Infect Dis 1994;170:439–443.

339. Glass RI, Noel J, Mitchell D, et al. The changing epidemiology of astrovirus-associated gastroenteritis: a review. Arch Virol 1996;12:287–300.

340. Herrmann JE, Taylor DN, Echeverria P, et al. Astroviruses as a cause of gastroenteritis in children. N Engl J Med 1991;324:1757–1760.

341. Cruz JR, Bartlett AV, Herrmann JE, et al. Astrovirus-associated diarrhea among Guatemalan ambulatory rural children. J Clin Microbiol 1992;30:1140– 1144.

342. Unicomb LE, Banu NN, Azim T, et al. Astrovirus infection in association with acute, persistent and nosocomial diarrhea in Bangladesh. Pedriatr Infect Dis J 1998;17:611–614.

343. Shastri S, Doane AM, Gonzales J, et al. Prevalence of astroviruses in a children's hospital. J Clin Microbiol 1998;36:2571–2574.

344. Rodriguez-Baez N, O'Brien R, Qiu SQ, et al. Astrovirus, adenovirus, and rotavirus in hospitalized children: prevalence and association with gastroenteritis. J Pediatr Gastroenterol Nutr 2002;35:64–68.

345. Schnagl RD, Belfrage K, Farrington R, et al. Incidence of human astrovirus in central Australia (1995 to 1998) and comparison of deduced serotypes detected from 1981 to 1998. J Clin Microbiol 2002;40:4114– 4120.

346. Dalton RM, Roman ER, Negredo AA, et al. Astrovirus acute gastroenteritis among children in Madrid, Spain. Pedriatr Infect Dis J 2002;21:1038–1041.

347. Liu CY, Shen KL, Wang SX, et al. Astrovirus infection in young children with diarrhea hospitalized at Beijing Children's Hospital. Chin Med J 2004;117:353– 356.

348. Phan TG, Okame M, Nguyen TA, et al. Human astrovirus, norovirus (GI, GII), and sapovirus infections in Pakistani children with diarrhea. J Med Virol 2004;73:256–261.

349. Espul C, Martinez N, Noel JS, et al. Prevalence and characterization of astroviruses in Argentinean children with acute gastroenteritis. J Med Virol 2004;72: 75–82.

350. Mitchell DK, Matson DO, Cubitt WD, et al. Prevalence of antibodies to astrovirus types 1 and 3 in children and adolescents in Norfolk, Virginia. Pediatr Infect Dis J 1999;18:249–254.

351. Koopmans, MP, Bijen MH, Monroe SS, et al. Age-stratified seroprevalence of neutralizing antibodies to astrovirus types 1 to 7 in humans in the Netherlands. Clin Diag Lab Immunol 1998;5:33–37.

352. Kriston S, Willcocks MM, Carter MJ, et al. Seroprevalence of astrovirus types 1 and 6 in London, determined using recombinant virus antigen. Epidemiol Infect 1996;117:159–164.

353. Lewis DC, Lightfoot NF, Cubitt WD, et al. Outbreaks of astrovirus type 1 and rotavirus gastroenteritis in a geriatric in-patient population. J Hosp Infect 1989; 14:9–14.

354. Gray JJ, Wreghitt TG, Cubitt WD, et al. An outbreak of gastroenteritis in a home for the elderly associated with astrovirus type 1 and human calicivius. J Med Virol 1987;23:377–381.

355. Belliot G, Laveran H, Monroe SS. Outbreak of gastroenteritis in military recruits associated with serotype 3 astrovirus infection. J Med Virol 1997;51:101–106.

356. Maunula L, Kalso S, Von Bonsdorff CH, et al. Wading pool water contaminated with both noroviruses and astroviruses as the source of a gastroenteritis outbreak. Epidemiol Infect 2004;132:737–743.

357. Noel J and Cubitt D. Identification of astrovirus serotypes from children treated at the Hospitals for Sick Children, London 1981–93. Epidemiol Infect 1994;113:153–159.

358. Wood DJ, David TJ, Chrystie IL, et al. Chronic enteric virus infection in two T-cell immunodeficient children. J Med Virol 1988;24:435–444.

359. Grohmann GS, Glass RI, Pereira HG, et al. Enteric viruses and diarrhea in HIV-infected patients. Enteric opportunistic infections working group. N Engl J Med 1993;329:14–20.

360. Cubitt WD, Mitchell DK, Carter MJ, et al. Application of electron microscopy, enzyme immunoassay, and RT-PCR to monitor an outbreak of astrovirus type 1 in a paediatric bone marrow transplant unit. J Med Virol 1999;57:313–321.

361. Cox GJ, Matsui SM, Lo RS, et al. Etiology and outcome of diarrhea after marrow transplantation: a prospective study. Gastroenterology 1994;107:1398–1407.

362. Liste MB, Natera I, Suarez JA, et al. Enteric virus infections and diarrhea in healthy and human immunodeficiency virus-infected children. J Clin Microbiol 2000;38:2873–2877.

363. Coppo P, Scieux C, Ferchal F, et al. Astrovirus enteritis in a chronic lymphocytic leukemia patient treated with fludarabine monophosphate. Annals Hematol 2000;79:43–45.

364. Yuen KY, Woo PC, Liang RH, et al. Clinical significance of alimentary tract microbes in bone marrow transplant recipients. Diag Microbiol Infect Dis 1998;30: 75–81.

365. Bass DM, Upadhyayula U. Characterization of human serotype 1 astrovirus-neutralizing epitopes. J Virol 1997;71:8666–8671.

366. Molberg O, Nilsen EM, Sollid LM, et al. CD4+ T cells with specific reactivity against astrovirus isolated from normal human small intestine (see comment). Gastroenterology 1998;114:115–122.

5

Oral Manifestations of Viral Diseases

Denis P. Lynch

The oral cavity is a unique biologic ecosystem involving both hard and soft tissues. The soft tissues of the mouth, oral pharynx, and adjacent salivary glands are particularly susceptible to various viral infections, some of which are site specific. Oral mucosa varies from simple stratified squamous epithelium, for example, labial mucosa, to highly specialized tissue, for example, dorsal tongue. Glandular tissue may be serous, mucous, or mixed. Furthermore, the oral cavity manifests many secondary bacterial, fungal, and viral infections, as well as neoplasms, as a result of immunosuppression due to infection with human immunodeficiency virus (HIV). As such, the mouth acts as a biologic barometer relative to the progression of such virally mediated immunosuppression.

The infectivity of oral viral disease varies dramatically from organism to organism. The sequelae of oral transmission also vary from the inconsequential to the potentially fatal. This is of concern not only to dental health care workers, but to any health care worker whose responsibilities involve oral examination or the handling of oral tissues. Dermatologists, in particular, are more likely than most physicians to become involved in an extended examination of the oral cavity and adjacent tissues, the diagnosis of oral soft tissue lesions, and their subsequent treatment.

The purpose of this chapter is fourfold. First, an overview of significant oral diseases and lesions with a viral etiology is presented, with an emphasis on their epidemiology. The pathophysiology of such infections is presented in other chapters dealing with specific viruses. Second, the clinical oral manifestations of such infections are described, with an emphasis on the differential diagnosis of specific oral viral lesions. Third, the methods used in the diagnosis of oral viral lesions are presented. Fourth, a summary of current therapeutic management strategies is presented, along with their relationship to long-term prognosis.

The chapter concludes with an overview of the oral manifestations of HIV infection. This overview includes significant nonviral infections and neoplasms that are seen in the oral cavities of individuals who are immunosuppressed as a result of their HIV infection.

Herpesvirus

Herpesviruses comprise the largest family of viruses with oral manifestations (Table 5.1) (1). Eight types of herpesvirus are known to be pathogenic in humans, with varying significance relative to oral disease. Herpes simplex virus type 1 (HSV-1) is the most common oral and perioral viral infection. Herpes simplex virus type 2 (HSV-2) is normally a genital infection; however, oral lesions have been reported.

The oral manifestations of primary varicella-zoster virus (VZV) infection (or herpes 3) or chickenpox are overshadowed by its cutaneous manifestations. Recurrent VZV infection, that is, herpes zoster or shingles, can present with both facial and intraoral clinical signs and symptoms. Epstein-Barr virus (EBV or herpes 4) has been

Table 5.1. Viral diseases with oral manifestations

Viral family	Virus	Disease
Herpesvirus	Herpes simplex 1 (herpes 1)	Primary herpetic gingivostomatitis
		Recurrent intraoral herpes simplex
		Herpes labialis (fever blister or cold sore)
	Herpes simplex 2 (herpes 2)	Indistinguishable from herpes simplex 1
	Varicella-zoster (herpes 3)	Varicella (chicken pox)
		Herpes zoster (shingles)
	Epstein-Barr (herpes 4)	Mononucleosis
		Burkitt's lymphoma
		Nasopharyngeal carcinoma
		Oral hairy leukoplakia
	Cytomegalovirus (herpes 5)	Sialadenopathy
		Oral aphthae
	Kaposi's sarcoma herpes virus (herpes 8)	Kaposi's sarcoma
Papillomavirus	Papilloma virus	Oral warts
		Condyloma acuminatum
		Focal epithelial hyperplasia
		(Heck's disease)
Paramyxovirus	Measles	Measles
Mumps	Mumps	Parotitis
Picornavirus	Coxsackie virus	Hand-foot-and-mouth disease
		Herpangina

associated with infectious mononucleosis, Burkitt's lymphoma, nasopharyngeal carcinoma, and hairy leukoplakia. Cytomegalovirus (CMV) infection (or herpes 5) is known to result in sialadenopathy; however, CMV-associated ulcers have recently been reported in HIV-infected, immunosuppressed individuals. Human herpesvirus type 6 (herpes 6) is the causative agent of exanthem subitum (roseola infantum); however, its relationship to specific oral lesions remains unclear. Human herpesvirus type 7 (herpes 7) has no known oral mucosal manifestations. There is persuasive evidence that human herpesvirus type 8 (herpes 8) is the etiologic agent for Kaposi's sarcoma in patients with acquired immune deficiency syndrome (AIDS).

Reactivation of the virus may occur in conjunction with various contributing factors, the most frequently cited relative to herpes labialis being exposure to ultraviolet light (Table 5.2) (6). All of these factors place some type of stress on the infected patient; however, the individual response to such stresses is quite variable.

When reactivated, HSV-1 travels distally down the nerve axon to the epithelium, where it replicates, resulting in acantholysis and a typical cluster of small vesicles. For reasons that are not entirely clear, secondary infections are limited in scope, with rare systemic manifestations. Interestingly, recurrent lesions frequently recur in the same anatomic location. Reactivation factors are listed in (Table 5.2).

Herpes Simplex Virus Type 1

The most common type of herpesvirus infection in humans is due to HSV-1 (2–5). Its oral lesions are most prominent on the gingiva during the initial infection, thus the common term *primary herpetic gingivostomatitis*; however, the vermilion borders of the lips and any intraoral mucosal site may be involved. Although HSV-1 infections result in acantholysis with vesicle formation, intact intraoral vesicles are rare, due to the friction associated with speaking, eating, and swallowing.

Table 5.2. Frequently reported factors resulting in reactivation of latent herpes simplex virus, type 1

Ultraviolet light (sunlight)
Emotional stress
Endocrine fluctuations (menstrual and pregnancy)
Fever
Physical trauma
Immunosuppression
Upper respiratory infection
Allergy
Gastrointestinal disturbances

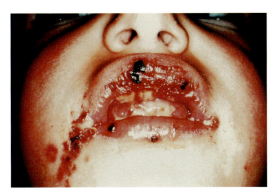

Figure 5.1. Primary herpes simplex virus type 1 (HSV-1). The bloody, crusted, ruptured vesicles on the vermilion border of the lips are reminiscent of erythema multiforme in an adult.

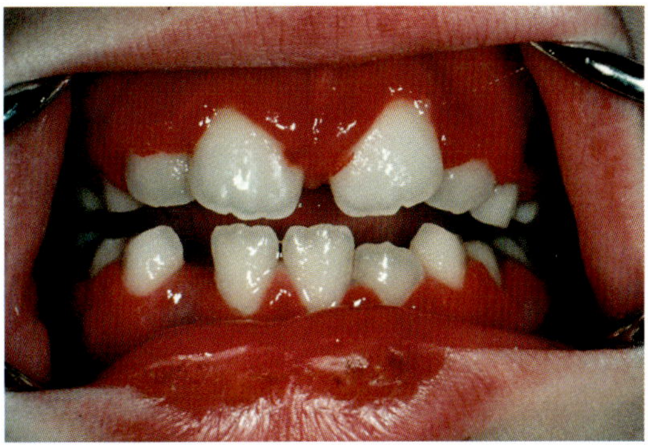

Figure 5.3. Primary HSV-1. The gingival vesicles and subsequent ulcers are the clinical hallmarks of primary herpetic gingivostomatitis. (*Source*: Courtesy of J. Robert Newland, D.D.S., M.S., University of Texas, Houston Health Science Center, Dental Branch.)

Oral Manifestations

In otherwise healthy children, primary herpetic gingivostomatitis presents a fairly typical clinical picture (7). Normally, infants and children develop a moderate fever with accompanying headache, malaise, dysphagia, occasional arthralgia, and cervical lymphadenopathy (8). Vesicles may be apparent on the vermilion borders of the lips and perioral skin (Fig. 5.1); however, intraoral lesions, especially involving the gingiva, are often more prominent (Figs. 5.2 to 5.4) (9). Unlike recurrent intraoral herpetic lesions, primary infections affect both keratinized and nonkeratinized oral mucosa; however, the pharynx is normally spared. The lesions resolve uneventfully within 10 to 14 days, at which time the virus has migrated to regional nerve ganglia and become dormant (10).

Recurrent lesions, especially those on the vermilion borders of the lips, are associated with a distinct, individualized prodrome, most frequently described as a sensation of "tingling," "tightness", "burning," or "itching." Within 24 hours of the prodrome, multiple vesicles appear, rapidly coalesce, and rupture to form a typical "fever blister" or "cold sore" (Fig. 5.5). The frequency of reactivation is increased in immunosuppressed individuals.

Recurrent intraoral lesions are limited to keratinized mucosa, that is, the attached gingiva (Figs. 5.6 and 5.7) and hard palate (Figs. 5.8 and 5.9), which distinguishes them from recurrent aphthous ulcers, which occur only on nonkeratinized mucosa (11, 12). Prodromal symptoms are less frequently reported with recurrent intraoral lesions, and the detection of intact vesicles is distinctly uncommon. The clinical course of intraoral lesions parallels that of vermilion

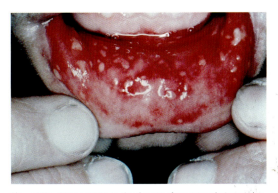

Figure 5.2. Primary HSV-1. In the absence of concurrent lesions on keratinized gingiva, e.g., hard palate and gingiva, ruptured vesicles on the lower lip mucosa are clinically indistinguishable from herpetiform aphthous ulcers, which do not have a viral etiology. (*Source*: Courtesy of J. Robert Newland, D.D.S., M.S., University of Texas, Houston Health Science Center, Dental Branch.)

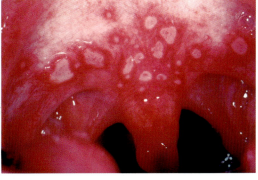

Figure 5.4. Primary HSV-1. Primary herpetic gingivostomatitis may also occur in adults and involve the soft palate, resulting in dysphagia.

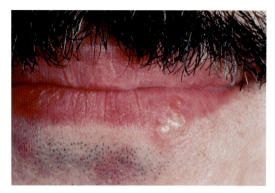

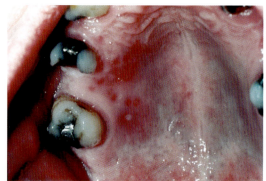

Figure 5.5. Recurrent HSV-1. Typically, recurrent herpetic lesions of the vermilion border of the lower lip and adjacent skin are preceded a pro-drome of itching, burning, stinging, or tightness, followed by an herpeti-form eruption of small vesicles, which subsequently coalesce and rupture.

Figure 5.8. Recurrent HSV-1. Recurrent intraoral HSV-1 infections may affect the palate and occur without a history of herpes labialis.

border and perioral lesions, with complete reso-lution of the lesions within 14 days.

Diagnosis

The diagnosis of primary herpetic gingivostom-atitis is frequently made on the basis of clinical signs and symptoms, especially when there is compelling evidence to suggest exposure to the virus, for example, a parent or sibling with a recent history of a recurrent lesion. A fourfold rise in antibody titer to HSV-1 over a 2-week period is also considered confirmatory; however, by the time the diagnosis has been confirmed, the episode has resolved. Due to the necessity of instituting treatment in a timely manner, this procedure is not normally performed for an uncomplicated case of presumptive primary her-petic gingivostomatitis in an otherwise healthy individual.

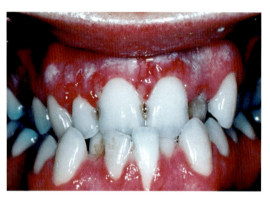

Figure 5.6. Recurrent HSV-1. Sporadically recurrent episodes of vesicles and ulcers of the gingiva characterize recurrent intraoral HSV-1 infection. Such lesions may resemble the gingival lesions of pemphigus vulgaris. (*Source*: Courtesy of J. Robert Newland, D.D.S., M.S., University of Texas, Houston Health Science Center, Dental Branch.)

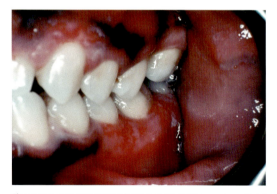

Figure 5.7. Recurrent intraoral HSV-1. Incidental trauma from dental procedures may precipitate recurrent intraoral HSV-1 infections. Additional recurrences may be prevented by the administration of systemic acyclovir, prior to and following such procedures.

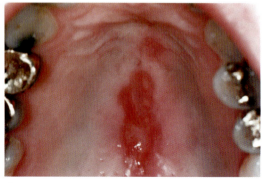

Figure 5.9. Recurrent HSV-1. Ruptured vesicles of recurrent intraoral HSV-1 lesions rapidly coalesce to form irregular, shallow erosions covered by pseudomembrane.

The cytopathologic diagnosis of herpetic lesions (Tzanck test) has not found widespread use in dental offices, although it is a rapid method of detecting cytopathic change in infected epithelial cells (13). Although not diagnostic, the presence of such changes is highly suggestive of a virally induced epithelial acantholysis.

Viral culture is the "gold standard" for the diagnosis of oral infections due to HSV-1; however, it is not widely used because of associated costs and delay in diagnosis. Specific situations, however, for example, a persistent vesiculo-ulcerative lesion in an immunosuppressed patient, indicate the use of this diagnostic procedure (14). Polymerase chain reaction techniques can also be used to amplify viral copy numbers in specimens containing low levels of HSV-1 (15).

Differential Diagnosis

Primary herpetic gingivostomatitis may occasionally be confused clinically with impetigo, especially when the lesions are primarily perioral, rather than intraoral. Adult patients with primary herpetic gingivostomatitis are sometimes misdiagnosed as having erythema multiforme (16). This is of particular concern, as such patients are routinely treated with systemic corticosteroids, which may exacerbate the underlying primary viral infection.

Herpes labialis does not present a significant clinical diagnostic challenge, as the patient's history frequently confirms the clinical suspicion (17). Intraoral lesions can be distinguished from recurrent aphthous ulcers by their anatomic location, that is, recurrent intraoral HSV-1 lesions occur only on keratinized mucosa (attached gingiva and hard palate), whereas recurrent aphthous ulcers occur only on nonkeratinized mucosa (labial and buccal mucosa, alveolar gingiva, tongue, floor of mouth, soft palate, and oral pharynx) (18, 19). Table 5.3 provides more information on differential diagnosis.

Treatment and Prognosis

A number of therapeutic agents for primary and, especially, recurrent HSV-1 infections have been proposed (20–28) (Table 5.4). Although numerous nonprescription preparations are available, their actual efficacy has been most often derived from anecdotal reports. Their primary value, for the most part, appears to be one of symptomatic relief and placebo effect rather than virucidal action. An over-the-counter preparation, Abreva™ (10% docosanol), has a reported therapeutic index equivalent to that of prescription creams and ointments used to treat herpes labialis (Table 5.5) (29).

Systemic antiviral therapy, for example, acyclovir, has not been approved by the U.S. Food and Drug Administration (FDA) for otherwise uncomplicated primary or recurrent oral HSV-1 infections; however, anecdotal evidence and preliminary reports indicate that early intervention with systemic acyclovir greatly attenuates the clinical signs and symptoms associated with primary herpetic gingivostomatitis (30).

Systemic acyclovir has also been used in the treatment of recurrent herpes labialis; however,

Table 5.3. Differential diagnoses for recurrent HSV-1 and -2 lesions

Aphthous ulcers: only occur on nonkeratinized mucosa (lips, cheeks, tongue, alveolar mucosa, soft palate and oral pharynx); never preceded by vesicles

Impetigo: often prominently affects perioral and adjacent facial skin

Erythema multiforme: prominent, bloody vesicles and bullae involving the vermilion borders; pathognomonic bull's eye, target and iris lesions often present on visible skin

Herpes labialis: prodrome of tingling, burning or throbbing; coalescing cluster of small, fluid-filled vesicles

Varicella zoster (VZV): intraoral VZV lesions are relatively uncommon in primary infections (chickenpox); recurrent lesions (shingles) present as painful, unilateral vesicles that follow a dermatomal pattern

Table 5.4. Therapeutic agents used in the treatment of oral herpes simplex virus type 1 infections

Agent	Form	Use
Acyclovir (Zovirax™)	Capsules	Systemic
Acyclovir (Zovirax™)	Ointment	Topical
Acyclovir (Zovirax™)	Cream	Topical
Penciclovir (Denavir™)	Cream	Topical
Ganciclovir (Cytovene™)	Capsule	Systemic
	Powder	Intravenous
Valacyclovir (Valtrex™)	Caplet	Systemic
Famciclovir (Famvir™)	Tablet	Systemic
Foscarnet (Foscavir™)	Solution	Intravenous
Idoxuridine (Stoxil™)	Ophthalmic ointment	Topical
Vidarabine (Vira-A™)	Ophthalmic ointment	Topical
Trifluridine (Viroptic™)	Ophthalmic solution	Topical

Table 5.5. Nonprescription remedies used in the treatment of recurrent oral herpes simplex virus type 1 infections

Antiviral agent
Abreva™ (10% docosanol)

Occlusive and anesthetic agents
Orabase™
Orabase™ with benzocaine
Anbesol™ gel
Anbesol™ liquid
Zilactin™ and Zilactin-B™ gel
Zilactin-L™ liquid
Campho-Phenique™ gel

Drying agents
Alcohol
Ether
Chloroform

Ultraviolet light-blocking agents
Chap-Stick™ Sunblock 15 balm
Herpecin-L™ balm
Pre-Sun-15™ lotion
Pre-Sun-15™ lip gel

such treatment is ineffective if instituted at any time other than the first onset of prodromal symptoms (31). Prophylactic use of systemic acyclovir has been reported to be effective in preventing recurrent HSV-1 infections, which precipitate other conditions, for example, erythema multiforme (32). Protocols for managing acyclovir-resistant herpes simplex virus infections have also been developed (33). Symptomatic treatment of HSV-1 is shown in Table 5.6.

Anecdotal reports of the use of topical acyclovir ointment are prevalent, with the greatest success reported when the medication is applied during the prodromal stage (34–38). Acyclovir cream is FDA-approved to treat herpes labialis and is much more effective than the ointment form due to its ability to penetrate the skin and vermilion border. Denavir™ cream (1% penciclovir) was the first FDA-approved topical medication for use in treating recurrent herpes labialis in immunocompetent individuals (39, 40). The potential of drug resistance has been proposed as a reason for limiting the use of antiviral medications to severe cases of both primary and recurrent HSV-1 infections (41–43).

Other therapeutic agents that have been proposed include cyclooxygenase inhibitors (44), chlorhexidine (45), idoxuridine (46), vidarabine (47) and helicase primase inhibitors (48). For many infected individuals, herpes labialis is effectively prevented by the use of lip balms and

other preparations that contain para-amino benzoic acid and other agents that block ultraviolet radiation to the skin (49).

Supportive care and palliative treatment is essential in the treatment of both primary and recurrent HSV-1 (50). Table 5.7 lists several palliative mouth rinses containing coating and anesthetic agents that may be used to relieve the discomfort associated with recurrent intraoral herpetic lesions.

The prognosis of primary HSV-1 is excellent in an otherwise healthy patient, although immunosuppressed patients are prone to more severe disease (51, 52). In healthy patients, the lesions resolve in 10 to 14 days and never recur in such a protean fashion. Unfortunately, recur-

Table 5.6. Treatment of common symptoms of herpes simplex virus-1 (HSV-1)

Primary herpetic gingivostomatitis

General oral discomfort	Topical anesthetics, acetaminophen
	Coating agents (Kaopectate™, Milk of Magnesia™)
Dehydration	Water
	Popsicles and ice chips are soothing
	Gatorade™, Powerade™
Herpes labialis	Preventable with the use of lip balms with para-amino benzoic acid and other UV-blockers
Prodromal symptoms	Abreva™ or Denavir™ applied every 2 hours
	Systemic agents are not usually given to healthy patients
	Cytovene™, Valtrex™, Famvir™, Zovirax™, Foscavir™ are effective in immunosuppressed patients
	Fever blister/cold sore
Fever blister/cold sore	Drying agents (alcohol, ether, chloroform)
	Topical anesthetics, acetaminophen
	Occlusive dressing (Zilactin-L™)
	Avoid direct application with fingertip
	Systemic agents are minimally effective in healthy patients
	Cytovene™, Valtrex™, Famvir™, Zovirax™, Foscavir™ are effective in immunosuppressed patients
Recurrent intraoral herpes 1	Topical anesthetics, acetaminophen
	Coating agents (Kaopectate™, Milk of Magnesia™)
	Occlusive dressings (Orabase™, Zilactin-B™)
Herpetic whitlow	Systemic Zovirax™ or analogous antiviral medication
	Avoided through use of barrier precaution (gloves)
Herpetic conjunctivitis	See Chapter 6 for additional information

Table 5.7. Palliative mouth rinses used in the treatment of oral ulcerations

Topical anesthetic	Coating agent
Benzocaine (Cetacaine™) liquid	Sucralfate (Carafate™) suspension*
Lidocaine (Xylocaine™) viscous	Phillips' Milk of Magnesia™ liquid
Dyclonine (Dyclone™) liquid	Maalox™ suspension
Diphenhydramine (Benadryl™) elixir**	Kaopectate™ liquid
Diphenhydramine (Benalyn™) syrup**	Amphogel™ suspension
Promethazine (Phenergan™) syrup**	Gaviscon™ liquid

A specific anesthetic agent and a specific coating agent are normally combined in a 1:1 (vol:vol) solution for rinsing, followed by expectoration.
* 1.0-g tablet dissolved in 5.0 mL H_2O.
** Antihistamine with topical anesthetic properties.

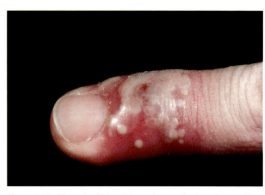

Figure 5.10. Herpetic whitlow. Prior to the use of universal precautions, herpetic whitlow was a significant occupational hazard in dentistry. Fortunately, it has been virtually eliminated through the use of latex gloves for all dental procedures.

rent HSV-1 lesions may arise as often as every few weeks, depending on the constitution of the infected patient and the individual contributing factors. Although the lesions have a certain amount of associated morbidity, complications are rare.

Primary and recurrent HSV-1 infections can present significant complications in individuals who are immunosuppressed secondary to malignancy (53), cancer chemotherapy (54–56), radiation therapy (57), or in preparation for receipt of an organ or tissue transplant (58–62). Oral HSV-1 lesions tend to be more severe and prolonged in such patients. Recurrent episodes, like those in immunosuppressed HIV-positive individuals, are often widespread and mimic primary infections with HSV-1. Acyclovir and related systemic antiviral agents are used prophylactically, as well as therapeutically, in these individuals (63).

Herpetic Whitlow

Several unique manifestations of herpetic lesions affect specific populations of individuals. A well-known occupational hazard of dental health care workers is herpetic whitlow (64–67). This condition results from a herpetic infection of the finger pad or nail bed that recurs periodically, analogous to recurrent HSV-1 lesions of the vermilion border of the lips and perioral skin. A painful, herpetiform cluster of vesicles appears on the skin of the finger or cuticle, which subsequently ruptures, crusts over, and heals (Fig. 5.10). Herpetic whitlow is responsive to acy-

clovir (68). Fortunately, this condition has virtually disappeared in dental health care workers with the advent of widespread use of barrier protection in clinical dentistry, but still occurs as a result of HSV-2 infection of the finger following digital I genital contact.

Herpetic conjunctivitis is another occupational hazard of dentistry, the incidence of which has decreased dramatically, in this case due to the use of protective eye wear (Fig. 5.11). This condition, although less frequent in occurrence than herpetic whitlow, has significant manifestations in that it can lead to blindness, in spite of aggressive therapy (69, 70). Because of the necessity of binocular vision for depth perception, an essential requirement for most dental procedures, complications of herpetic conjunctivitis can have a profound effect on the ability of an affected dentist to practice.

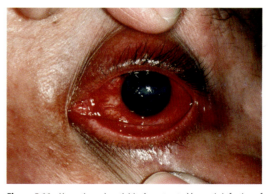

Figure 5.11. Herpetic conjunctivitis. An untreated herpetic infection of the eye can lead to blindness. Like herpetic whitlow, herpetic conjunctivitis in dental health care workers has virtually been eliminated through the use of protective eyewear.

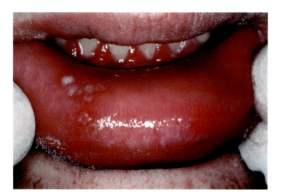

Figure 5.12. Primary HSV-2 lesions, in this case acquired through oral–genital contact, are identical to those seen in primary HSV-1 infections.

Herpes Simplex Virus Type 2

Herpes simplex virus type 2 (HSV-2) has occasionally been reported in the oral cavity as the causative agent of both primary and recurrent disease. The prevalence of HSV-2 in oral lesions is unknown, as the diagnosis is normally made clinically, without the benefit of viral culture or typing. It is presumed that an increase in oral HSV-2 lesions is due to an increase in oral–genital contact; however, this explanation is speculative.

The clinical manifestations of oral HSV-2 infection are identical to those found in oral HSV-1 infections (71) (Fig. 5.12). Oral HSV-2 lesions respond to the same therapeutic modalities as HSV-1 lesions.

Varicella-Zoster Virus [Human Herpesvirus 3 (HHV-3)]

Oral lesions of VZV share many similarities with those of HSV-1 (72). Both viruses result in a primary mucocutaneous eruption, with intraoral lesions presenting as fragile vesicles, which rupture to form shallow ulcers (73).

Following primary infection, VZV, like HSV-1, remains latent in nerve ganglia and can be reactivated. In such cases, the virus travels from the nerve cell body down the axon, causing a characteristic vesicular eruption that follows the distribution of the infected nerve. These lesions, commonly referred to as shingles, may be found in a wide range of individuals from children (74) to the elderly (75).

Although the vast majority of adults have been exposed to VZV in childhood and develop chickenpox, a much smaller number develop recurrent disease. Certain conditions that result in immunosuppression predispose individuals to developing shingles. At highest risk are individuals with hematopoietic or lymphoid malignancies, HIV-positive patients, chemotherapy patients, and transplant patients. Physical trauma to the nerve root has also been documented as an inciting factor. A small percentage of affected patients have an idiopathic episode of recurrent VZV without any apparent precipitating event (76).

Oral Manifestations

Although the oral lesions of primary VZV infection have been described, they are of minor significance and are overshadowed by the presence of cutaneous disease. Intact intraoral vesicles are rare and the subsequent shallow ulcers, while uncomfortable, are not particularly dramatic in terms of number or symptomatology.

Oral lesions of recurrent VZV are pathognomonic (77). Normally there is an antecedent prodrome of pain or paresthesia, followed by a vesicular eruption that extends to, but does not cross, the midline (Fig. 5.13). The vesicles rapidly rupture to form shallow ulcers, which heal within 2 weeks. Frequently there is a postherpetic neuralgia, which is often refractory to nonnarcotic analgesics (78–80). Oral complications

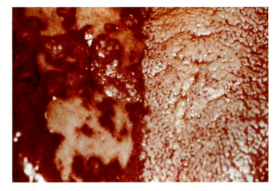

Figure 5.13. Recurrent varicella-zoster virus. An exclusively unilateral distribution of the lesion, with an abrupt cessation at the midline, is characteristic of a recurrent intraoral Varicella zoster infection.

are rare, but tooth exfoliation and mandibular necrosis have been reported (81,82).

Diagnosis

The diagnosis of primary VZV infection is rarely made on the basis of oral lesions for several reasons. First, the intraoral lesions are not prominent. Second, the initial symptoms of a primary infection are nonspecific. Third, the cutaneous manifestations of infection rapidly follow the onset of oral lesions.

Differential Diagnosis

Few lesions can be confused with recurrent intraoral VZV, due to its distinct clinical presentation; however, other unilateral herpetiform vesicular lesions, for example, recurrent intraoral HSV-1 on the hard palate, might be considered in the differential diagnosis. Other differential diagnoses are shown in Table 5.8.

Treatment and Prognosis

For most individuals, supportive therapy is sufficient to treat both primary and recurrent intraoral VZV infections; however, immunosuppressed individuals may require more aggressive therapy. Acyclovir administered either orally or intravenously has been reported to be effective and is the drug of choice in immunosuppressed patients (83–90). It must be used, however, as early as possible during the recurrent episode to be beneficial. Other specific antiviral agents, for example valacyclovir, famciclovir, vidarabine (91) and idoxuridine (92), have also been reported to be successful.

Table 5.8. Differential diagnoses of varicella zoster virus (VZV)

HSV-1: Primary lesions affect the perioral skin, vermilion borders of the lips and all oral mucosal surfaces. Recurrent extraoral lesions are vesicular and preceded by a prodrome of tingling, burning or throbbing. They tend to occur in the same anatomic location. Recurrent intraoral lesions are limited to keratinized mucosa, i.e., attached gingival and hard palate.

Primary varicella (chickenpox): Oral lesions are uncommon and of minimal clinical significance.

Recurrent varicella zoster (shingles): Both extraoral and intraoral lesion manifest as painful vesicular eruptions that occur in a unilateral dermatomal distribution.

Recurrent intraoral VZV lesions are relatively painful, as compared to other recurrent intraoral viral lesions, for example, recurrent HSV-1. Oral VZV lesions have an excellent prognosis; however, in individuals who are predisposed to multiple recurrences there can be a significant morbidity associated with the lesions. Individuals who are immunosuppressed are at much greater risk for extended recurrences that may be unresponsive to conventional therapy.

Although most dental health care workers have had a primary varicella infection in childhood, appropriate infection control measures will prevent the unlikely transmission of this virus in the dental setting (93). Future generations will benefit from vaccination against this infection, essentially eliminating both transmission in the dental setting and recurrent episodes (94).

Epstein-Barr Virus

Epstein-Barr virus (EBV or HHV-4) is a herpesvirus that has been shown to have relationships with multiple infectious and neoplastic processes (95–102). It has been recognized for many years as the causative agent of infectious mononucleosis, usually transmitted by infected saliva. Individuals infected prior to adolescence rarely develop the classic symptoms of infectious mononucleosis (103). Of greatest notoriety over the past decade has been the association of EBV with oral hairy leukoplakia, a unique lesion of oral mucosa associated with HIV infection. Epstein-Barr virus has also been linked with Burkitt's lymphoma in African children and nasopharyngeal carcinoma in Asian populations. Symptoms characteristic of these diseases are shown in Table 5.9.

Oral Manifestations

Oral lesions of infectious mononucleosis characteristically consist of pharyngitis and petechial hemorrhages of the soft palate and oral pharynx, usually in young adults with concomitant fever and cervical lymphadenopathy (104) (Fig. 5.14). Constitutional symptoms associated with the systemic manifestations of infection are helpful in distinguishing the oral lesions of infectious mononucleosis from clinically similar lesions, for example, oral and pharyngeal

Table 5.9. Diseases associated with Epstein-Barr virus (EBV) (in addition to infectious mononucleosis)

Disease	Manifestation	Treatment
Burkitt's lymphoma (non-Hodgkin's)	Common in Ugandan children Jaw enlargement Loose teeth Pain and paresthesia	Chemotherapy Guarded prognosis
Nasopharyngeal carcinoma	Racial predilection (Chinese) Often asymptomatic Granular, velvety, erythematous Anaplastic Cervical metastasis common	Radiation therapy Poor prognosis
Oral hairy leukoplakia	Primarily on lateral tongue Vertically corrugated to "hairy"	Usually not treated Responds to podophyllin resin Responds to antiviral drugs Recurs when therapy is stopped

petechiae secondary to fellatio and violent coughing or sneezing (Fig. 5.15). Oral complications are rare, but subsequent cranial nerve deficit following an acute episode of infectious mononucleosis has been reported (105), as well as cervical abscess (106), parotid mass with facial nerve palsy (107), and lingual tonsillitis (108).

Burkitt's lymphoma is a high-grade, non-Hodgkin's lymphoma, first described in 1958 as a jaw sarcoma endemic in Ugandan children (109). African Burkitt's lymphoma occurs in a younger age group, predominantly in males, with a predilection for jaw involvement (110–112). Most significantly, although over 90% of endemic African Burkitt's lymphoma cases have detectable EBV (113, 114), less than 10% of nonendemic Burkitt's lymphoma cases are similarly infected (115, 116). Epstein-Barr virus–positive Burkitt's lymphoma is characterized by widespread jaw involvement, with jaw expansion, loosening of the teeth, and pain paresthesia (117–119) (Figs. 5.16 and 17). Significant differences exist between the African and American forms (120, 121).

Nasopharyngeal carcinoma is a poorly differentiated, anaplastic, nonkeratinizing carcinoma that arises in the fossa of Rosenmüller (122). It is found most commonly in Chinese, Eskimos, Southeast Asian natives, and Arabs from North Africa and Kuwait (123). Epstein-Barr virus is frequently found in association with this malignancy and appears to be involved in its pathogenesis (124, 125). Nasopharyngeal carcinomas account for nearly one in five malignancies in mainland Chinese and is the third leading cancer

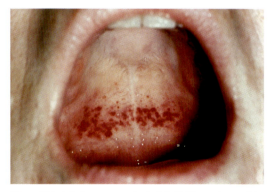

Figure 5.14. Infectious mononucleosis. Petechial hemorrhages of the soft palate and oral pharynx, in conjunction with constitutional symptoms of low-grade fever, malaise, and cervical lymphadenopathy are characteristic of infectious mononucleosis. (*Source*: Courtesy of J. Robert Newland, D.D.S., M.S., University of Texas, Houston Health Science Center, Dental Branch.)

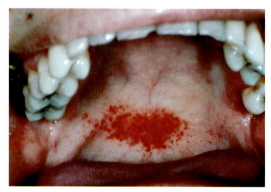

Figure 5.15. Traumatic hemorrhages. Asymptomatic petechial hemorrhages of the soft palate are suggestive of traumatic injury, in this case fellatio. (*Source*: Courtesy of J. Robert Newland, D.D.S., M.S., University of Texas, Houston Health Science Center, Dental Branch.)

Figure 5.16. Burkitt's lymphoma. Endemic Burkitt's lymphoma is characterized by asymptomatic, unilateral mid-face expansion and radiographically evident bone destruction. Immunohistochemistry confirms a B-cell lineage and evidence of Epstein-Barr virus.

in southeast Asia (126). The lesions are remarkably asymptomatic, with cervical metastasis being the primary presenting complaint, followed by nasal and aural symptoms (127). Cranial nerve involvement has also been reported (128). Clinically, the lesions are erythematous with a granular or velvety appearance.

Oral hairy leukoplakia (OHL) is a unique HIV-related lesion, characterized by an EBV-induced epithelial hyperplasia, primarily on the lateral tongue. This lesion is discussed in greater detail in the section on oral manifestations of HIV infection.

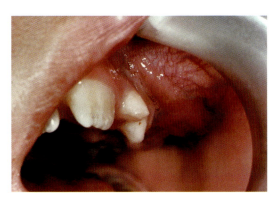

Figure 5.17. Burkitt's lymphoma. Intraorally, Burkitt's lymphoma results in obliteration of the maxillary vestibule with loosening of the teeth.

Diagnosis

The diagnosis of infectious mononucleosis is rarely pursued exclusively on the basis of oral signs and symptoms. Lymphocytosis and the presence of atypical lymphocytes are highly suggestive of infectious mononucleosis in the presence of other associated clinical signs and symptoms. Rapid serologic tests, for example, heterophile antibody, are available that can provide the diagnosis expediently; however, EBV-specific antibody is confirmatory (129). Burkitt's lymphoma requires an incisional biopsy, as does nasopharyngeal carcinoma, for diagnosis (130–136).

Treatment and Prognosis

The routine oral–pharyngeal lesions of infectious mononucleosis require no treatment and resolve spontaneously with the regression of the infection (137, 138). Symptoms may be more severe in older adults (139). Infectious mononucleosis is, in most cases, self-limiting; however, the incidence and prevalence of chronic EBV infection continues to be a source of great debate (140). Acyclovir has been proposed for use in the early stages of the disease.

Cytomegalovirus

Traditionally, cytomegalovirus has been associated with salivary gland disease in immunocompetent hosts (141,142) and birth defects in infected fetuses (143–145). Cytomegalovirus has been implicated in a number of destructive periodontal conditions (146). In recent years, however, it has assumed a much more prominent role in individuals who are immunosuppressed secondary to HIV infection. Oral manifestations of cytomegalovirus infection in HIV-infected individuals are discussed in the section on oral manifestations of HIV infection.

Human Herpesvirus-6

Human herpesvirus-6 (HHV-6), first discovered in 1986, was found in 1988 to be the causative agent of roseola infantum or exanthem subitum, although more recently it is has been tentatively associated with other diseases, for example, multiple sclerosis (147–157). No specific antiviral therapy is currently available for HHV-6 infection (158).

Although it appears to be closely related to cytomegalovirus (159), there are no specific oral manifestations related to HHV-6 infection. Most adults have been exposed to HHV-6 in infancy and shed virus in their saliva (160–162).

Human Herpesvirus-7

There are no reported oral manifestations associated with human herpesvirus-7 (HHV-7) infection, although the virus is shed in saliva (163,164).

Human Herpesvirus-8

The most recently discovered human herpesvirus (HHV-8) has also been called Kaposi's sarcoma–related herpes virus (KSHV) due to its relationship with Kaposi's sarcoma in AIDS patients. This lesion is discussed in greater detail in the section on oral manifestations of HIV infections. Human herpesvirus 8 has also been shown to be associated with primary effusion lymphoma and multicentric Castleman's disease, both of which are more commonly found in HIV-infected individuals (165), as well as multiple myeloma (166) and various lymphoproliferative disorders (167–169).

Human Papilloma Virus

Papillary and verrucal proliferations of oral epithelium are generically referred to as oral warts. Squamous papillomas comprise the largest group of papillary lesions of the oral cavity, accounting for approximately 2.5% of all oral lesions (170). Various subtypes of human papilloma virus (HPV) have been associated with both oral squamous papillomas and oral verruca vulgaris (HPV-2, -6, -11, and -57). Human papilloma virus 2 is associated with cutaneous verruca vulgaris; HPV-6 and -11 have been found in condyloma acuminatum, and HPV-11 has also been found in association with laryngeal and conjunctival papillomas. Focal epithelial hyperplasia, or Heck's disease, has been found to harbor HPV-13 and -32. Of greatest interest is the detection of HPV-16 and -18 in association with dysplastic and neoplastic conditions of squamous epithelium. Of the over 100 subtypes of HPV, at least 13 have been associated with lesions of oral squamous epithelium (Table 5.10).

Table 5.10. Differential diagnoses of human papilloma virus (HPV)

Squamous papilloma: most common oral HPV-related lesion; usually solitary and pedunculated; common on labial and palatal mucosa, as well as the uvula

Verruca vulgaris: usually solitary and sessile; skin lesions may lead to autoinoculation of labial gingival and anterior tongue

Condyloma acuminatum: usually due to oral–genital contact; social history is helpful in determining risk behaviors; more easily transmitted than other oral HPV-related lesions; common on lingual frenum, anterior tongue, and soft palate as multiple, soft, sessile, pink, fleshy masses

Focal epithelial hyperplasia (Heck's disease): multiple dome-shaped or papular, pink soft tissue masses, usually involving the labial, lingual and buccal mucosa; vertical transmission may mimic clinically similar genodermatoses

Oral Warts

Human papilloma virus (HPV) is a member of the papillomavirus family (171,172). All of the over various subtypes are DNA viruses. They do not have an envelope but share many common antigenic determinants (173–175).

The route of infection of HPV in oral mucosa is thought to be direct contact, although, in most cases, it is unusual to determine a specific route of infection. Maternal transmission has been suggested as a likely route in children (176). Viral replication occurs in the infected epithelial cell nucleus, but detection of HPV in basal keratinocytes is difficult due to low copy numbers of the virus (177). Human papilloma virus has also been detected in normal oral squamous epithelial cells (178–180).

Oral Manifestations

Oral warts can be found on the vermilion border of the lips as well as any intraoral mucosal site (181) (Fig. 5.18). There is a predilection for the hard and soft palates, as well as the uvula, with over one third of all intraoral lesions occurring on these sites (182). The lesions are generally

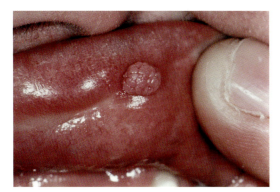

Figure 5.18. Verruca vulgaris. Patients with oral mucosal verruca vulgaris frequently have digital lesions, which they bite. Autoinoculation results in labial mucosal lesions.

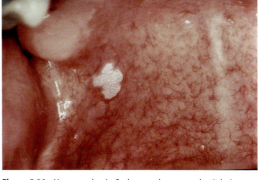

Figure 5.20. Verruca vulgaris. Oral mucosal verruca vulgaris lesions are frequently asymptomatic; however, patients may be aware of their presence, especially if the affected tissue is mobile.

small, with a maximum diameter of less than 0.5 cm. Intraoral squamous papillomas are normally pedunculated (Fig. 5.19), while verruca vulgaris lesions tend to be sessile (Fig. 5.20). Both lesions are normally solitary and asymptomatic, unless secondarily traumatized.

Diagnosis

With rare exception, oral squamous papillomas and verruca vulgaris lesions have a fairly characteristic clinical presentation (183). Both of these lesions require an excisional biopsy for diagnosis. Routine hematoxylin and eosin (H&E)-stained sections are sufficient for diagnosis (184). Human papilloma virus can also be demonstrated with transmission electron microscopy (185). Specific subtyping for

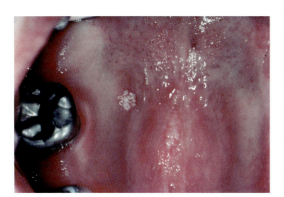

Figure 5.19. Squamous papilloma. While some squamous papillomas of the oral cavity are negative for the presence of human papilloma virus (HPV), many reveal the presence of HPV-2, -6, -11, or -57 by in situ DNA hybridization.

definitive viral identification requires immunodiagnostic techniques, for example, in situ DNA hybridization (186–193). Interestingly, not all squamous papillomas stain positively for the presence of HPV (194). Differential diagnoses are shown in Table 5.10.

Treatment and Prognosis

Oral warts are most commonly treated with conservative surgical excision. Electrodesiccation and laser ablation have also been used successfully. Cryosurgery or thermal ablation is not recommended for lesions directly over bone or adjacent to teeth. Chemotherapeutic and immunotherapeutic treatment, for example podophyllin resin and interferon, have not been extensively evaluated for oral warts; however, the relative ease and low morbidity associated with surgical excision of oral lesions makes that method preferable.

Oral warts have an excellent prognosis, as HPV has a relatively low infectivity. Although the exact placement of excisional surgical margins is necessarily imprecise, significant recurrences have not been reported following conservative excision. Obviously, if there are cutaneous lesions noted as the presumptive origin of infection, for example, digital warts, which are bitten or chewed, they must also be treated.

Condyloma Acuminatum

Condyloma acuminatum, or venereal warts, are moderately infectious viral lesions associated primarily with HPV-6 and -11 (195). The lesions occur almost exclusively on moist squamous mucosa. Over the past decade there has been a marked increase in the incidence of oral lesions in both the heterosexual and homosexual populations (196–198). The presence of oral lesions in children is highly suggestive of sexual abuse (199). Although oral–genital contact is the primary mechanism of oral inoculation, the possibility of self-inoculation also exists.

Oral Manifestations

Unlike oral squamous papillomas and verruca vulgaris, oral condyloma acuminatum characteristically appears as multiple, small, soft, sessile masses that appear several months following infection. The lesions are pink in color and ultimately coalesce to form exophytic papillary growths with varying amounts of keratinization (Figs. 5.21 and 5.22). The lesions may be quite widespread in the mouth; however, they are normally self-limiting (200–202).

Diagnosis

An excisional biopsy is required for the definitive diagnosis of oral condyloma acuminatum. In situ DNA hybridization can be used to confirm the presence of specific HPV, that is, HPV-6 or HPV-11. Like other oral warts, viral detection is uncommon in basal cells, with viral replication paralleling keratinocyte maturation.

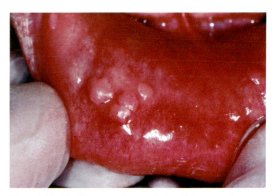

Figure 5.22. Condyloma acuminatum. Oral condyloma acuminata may be multiple, as well as solitary. (*Source*: Courtesy of J. Robert Newland, D.D.S., M.S., University of Texas, Houston Health Science Center, Dental Branch.)

Treatment and Prognosis

As with other oral warts, surgical excision is the treatment of choice; however, recurrences are reportedly more common with oral condyloma acuminatum. Presumably, this is due to more extensive viral infection of cells in the normal-appearing, perilesional tissue. For this reason, surgical excision of intraoral condyloma acuminatum requires a wider surgical margin to decrease the risk of recurrence, usually several millimeters beyond the base of the lesion.

Cryosurgery, laser ablation, and electrodesiccation have also been used successfully in the treatment of condyloma acuminatum; however, caution is recommended for the use of these procedures on lesions that are overlying bone or adjacent to teeth (203). Interferon and imiquinod appear to be promising areas of immunotherapy for condyloma acuminatum; however, their use in the treatment of intraoral lesions, to date, has been minimal (204,205).

Focal Epithelial Hyperplasia (Heck's Disease)

Focal epithelial hyperplasia (FEH), or Heck's disease, was first described in 1965 in Eskimos and American Indians (206) and later in South Africans, Mexicans, and Central Americans (207). Originally the disorder was thought to have a variety of etiologies, the most common one being that it represented a genodermatosis, as it was described in successive generations of affected individuals. It is now generally accepted

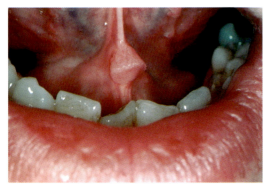

Figure 5.21. Condyloma acuminatum. Intraoral condyloma acuminatum is currently more prevalent due to an increase in oral–genital contact in both heterosexual and homosexual populations.

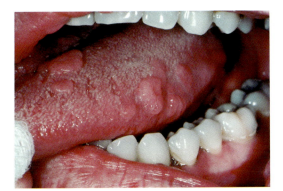

Figure 5.23. Focal epithelial hyperplasia (Heck's disease). Intraoral lesions of focal epithelial hyperplasia may resemble other mucosal HPV infections, including condyloma acuminatum.

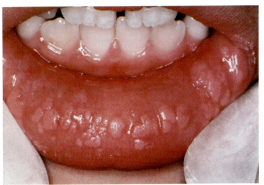

Figure 5.25. Focal epithelial hyperplasia (Heck's disease). Focal epithelial hyperplasia may present with a vertical transmission of the virus. This patient is the daughter of the woman shown in Figures 5.23 and 5.24. Two other siblings, as well as the grandmother were similarly affected.

that the disorder represents an infection with HPV-13 and, perhaps, -32, which can be transmitted vertically, thus mimicking a heritable defect (208,209). Although the disease historically has been described in children, both genders of adults are equally affected (210). Although the disease is more prevalent in some ethic groups, it has been described in other groups.

Oral Manifestations

Focal epithelial hyperplasia is characterized clinically by the presence of multiple, pink, soft tissue masses, which may be either dome-shaped or papular. They are most commonly found on the buccal mucosa, tongue, and labial mucosa, although any oral mucosa may be affected (211–213) (Figs. 5.23 to 5.25).

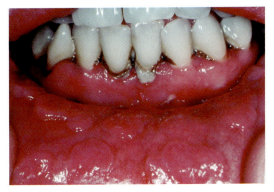

Figure 5.24. Focal epithelial hyperplasia (Heck's disease). Widespread lesions of focal epithelial hyperplasia are difficult to treat definitively, due to the amount of mucosal involvement.

Diagnosis

The clinical appearance of FEH can range from subtle to dramatic. Less striking cases can be confused clinically with multiple immature fibrous hyperplasias or oral condyloma acuminatum. The routine histopathologic findings are suggestive of a virally induced epithelial hyperplasia and immunodiagnostic techniques, for example, in situ DNA hybridization, are necessary to confirm the presence of associated HPV, for example, HPV-13 and -32 (214–217).

Treatment and Prognosis

Although the lesions of FEH are virally induced, their infectivity is relatively low. While solitary lesions can be removed for functional or aesthetic reasons (218), the normally widespread nature of the disease precludes surgical excision of all lesions. Of particular interest is the apparent spontaneous regression of lesions in some individuals in the absence of any therapeutic intervention (219). It has been speculated that such regressions represent an expression of delayed viral recognition, with subsequent activation of cell-mediated immunity.

Relationship of Human Papilloma Virus to Oral Carcinoma

As with EBV, a relationship between HPV and mucosal carcinogenesis has been suggested (220–236). Although HPV has been found in association with oral malignancies (237–241) and putative premalignant conditions (242), the precise role, if any, that it plays in the genesis of oral carcinoma remains unclear (243–247). A recent meta-analysis of 94 articles indicated that HPV is detected with increasing frequency in oral malignancies and premalignancies, compared to normal oral mucosa (248).

Paramyxovirus

The paramyxoviruses are a heterogeneous family of RNA viruses that include measles virus, mumps virus, parainfluenza virus, and respiratory syncytial virus of humans, as well as several other strains that are pathogenic only in animals. The only two viruses in this family that have significant oral manifestations are measles and mumps virus.

Measles

Measles, or rubeola, is a highly contagious viral infection caused by a paramyxovirus known simply as measles virus. It is related to orthomyxoviruses, which cause mumps and influenza. Measles virus is spread through contaminated airborne droplets through the respiratory tract. It is not related to the togavirus, which causes German measles or rubella (249), although it shares similar, though more prolonged, clinical symptomatology.

Oral Manifestations

Measles has a characteristic clinical course, characterized by a 7- to 10-day incubation period followed by a typical prodrome of fever, malaise, conjunctivitis, photophobia, and cough (250). Within 48 hours small erythematous macules with white necrotic centers appear on the buccal mucosa (Koplik's spots) and precede the characteristic maculopapular skin rash by 1 to 2 days (251). Mucosal ulcerations, gingivitis, and pericoronitis have also been reported (252). Interestingly, the infection has a seasonal variation, being most prominent in the winter and spring seasons.

Diagnosis

The diagnosis of measles is rarely, if ever, made on the basis of initial oral findings. The preceding clinical signs and symptoms are suggestive of a primary viral infection; however, they are nonspecific. Anti-measles antibody can be detected in saliva (253). Koplik's spots are indistinguishable from other oral aphthae; however, the index of clinical suspicion should be raised when the clinical picture of measles is present in an individual who has not been immunized.

Treatment and Prognosis

There is no specific treatment for measles beyond supportive care. Koplik's spots have variable symptomatology; however, palliative mouth rinses are effective in controlling any oral discomfort. Complications such as encephalitis, thrombocytopenic purpura, and secondary infections are rare in otherwise healthy individuals.

Mumps

The etiologic agent of this disorder is simply called mumps virus. Mumps is the most common salivary gland disease in humans. Infected patients are most commonly seen in the winter and spring; however, cases of mumps can be found throughout the year. Virus transmission is by direct contact with infected aerosolized salivary droplets.

Although the disease is conceptualized as a parotid gland infection, mumps actually represents a systemic viral infection with involvement of other glandular tissues, as well as hepatic, renal, pancreatic, and nervous system involvement (254).

Oral Manifestations

Infected children will normally manifest a non-specific prodrome of fever, chills, malaise, and headache. The initial presence of preauricular pain makes mumps highly suspect, and the subsequent parotid swelling is essentially confirmatory. Bilateral swelling is found in approximately three fourths of infected individuals. Salivary gland swelling resolves over a period of 7 to 10 days, with a concomitant decrease of symptoms (198–256).

Salivary gland ducts, especially Stensen's duct in the parotid, become compromised due to stromal edema. Any activity that stimulates salivary gland secretion, for example, eating, can result in acute discomfort, which characterizes the infection.

The clinical signs and symptoms of mumps can occasionally be mimicked by other viral infections, for example, Coxsackie A virus, echovirus, and cytomegalovirus. The acute nature of mumps and the typical age of the patient effectively rules out the possibility of a neoplastic etiology for the salivary gland enlargement. Unilateral involvement of the parotid gland in mumps may mimic an occlusive sialadenopathy, especially when edema surrounding Stensen's duct restricts salivary flow. Mumps should be strongly considered in children who have not been vaccinated and present with either a unilateral or bilateral parotid swelling (257).

Treatment and Prognosis

The management of mumps is primarily symptomatic in nature, combined with bed rest and analgesics, as indicated (258). Corticosteroids have been used occasionally in severe cases. Complications of encephalitis, myocarditis, and nephritis have been reported in children. Orchitis, and occasionally oophoritis, are the most serious complications of mumps in adults.

Coxsackie Virus

Coxsackie viruses, discovered in Coxsackie, New York, are members of the picornavirus family. Two conditions with significant oral manifestations that are caused by Coxsackie viruses are hand, foot, and mouth disease and herpangina. Both infections are normally transmitted by infected salivary droplets.

Hand, Foot, and Mouth Disease

Hand, foot, and mouth disease (HFM) is a highly contagious viral infection that is normally transmitted through infected saliva; however, oral–fecal routes of transmission have also been

reported (259–261). The most frequently reported causative agent is Coxsackie virus A-16, although other Coxsackie subtypes have also been reported (262–265). The infection normally occurs as an epidemic primarily affecting children under the age of 5 years (266). The infection is occasionally fatal. Interestingly, fatal cases of HFM are characterized by a lack of oral ulcers (267).

Oral Manifestations

Following a brief incubation period, infected individuals demonstrate a low-grade fever, malaise, lymphadenopathy, and stomatodynia (268). Mucosal lesions initially appear as vesicles that rapidly rupture to form shallow ulcers. The mucosal ulcers are covered by pseudomembrane and are surrounded by an erythematous halo, mimicking recurrent aphthous ulcers (269,270) (Fig. 5.26).

The palate, tongue, and buccal mucosa are favored sites; however, lesions have been reported on all intraoral mucosal sites. Concurrent with, or shortly after the onset of oral lesions, erythematous maculopapular lesions appear on the skin of the hands and feet (Fig. 5.27). The cutaneous lesions eventually vesiculate and rupture with later encrustation of the resulting ulcers (271).

Diagnosis

The diagnosis of HFM is normally made on the basis of clinical signs and symptoms; however,

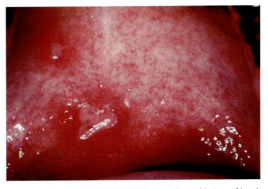

Figure 5.26. Hand, foot, and mouth disease. The oral lesions of hand, foot, and mouth disease resemble aphthous ulcers; however, they are accompanied by a nonspecific maculopapular, vesicular rash on the hands and feet.

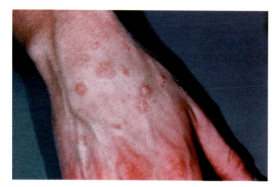

Figure 5.27. Hand, foot, and mouth disease. The cutaneous lesions of hand, foot, and mouth disease are useful in differentiating oral lesions from recurrent aphthous ulcers. (*Source*: Courtesy of J. Robert Newland, D.D.S., M.S., University of Texas, Houston Health Science Center, Dental Branch.)

antibody titers are useful in confirming the clinical suspicion. Virus can also be cultured from intact vesicles. The subsequent appearance of cutaneous lesions in specific anatomic sites is very helpful in narrowing the differential diagnosis (272).

Treatment and Prognosis

The treatment of HFM is normally symptomatic, as the disease is self-limiting. Analgesics are appropriate for the treatment of fever, and palliative mouth rinses significantly reduce the attendant morbidity. Ablation of oral ulcers associated with HFM with low-level laser therapy has also been reported to be effective (273). Little is known about subsequent immunity to reinfection.

Herpangina

Herpangina is an acute viral infection caused by one of several Coxsackie A viruses (-1 to -6, -8, -10, and -22). Like hand, foot, and mouth disease, herpangina infections are transmitted through contaminated droplets of saliva; however, the fecal–oral route of transmission has also been described. Characteristically, the infection occurs endemically in the summer and fall months, and more frequently in children than adults (274–277).

Oral Manifestations

Following exposure, infected individuals have a brief, nonspecific prodrome of fever and malaise, followed by an erythematous pharyngitis and dysphagia. These clinical signs and symptoms are followed by the appearance of a patchy vesicular eruption involving the soft palate, tonsillar pillars, and fauces (Fig. 5.28). Characteristically, the remaining oral mucosa is unaffected. The vesicles rapidly rupture to form shallow ulcers covered by pseudomembrane and surrounded by an erythematous halo, mimicking recurrent aphthous ulcers. The lesions resolve within 1 week without any significant additional symptoms (278).

Diagnosis

The diagnosis of herpangina is normally made clinically, based on the patient's presenting signs and symptoms. Detectable serum antibodies are present and the virus can be cultured from intact vesicles.

The oral lesions of herpangina may be confused with both recurrent aphthous ulcers and primary oral HSV-1 infections. The limitation of lesions primarily to the posterior mouth and pharynx is not characteristic of recurrent aphthous ulcers, nor are the constitutional symptoms that accompany the lesions. Although primary HSV-1 lesions are found in association with fever and malaise, the lack of gingival and labial

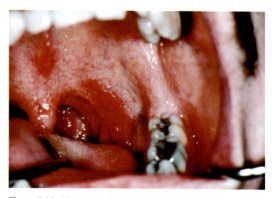

Figure 5.28. Herpangina. Herpangina normally affects the posterior oral cavity and oral pharynx, without significant involvement of the buccal, lingual, or labial mucosa. (*Source*: Courtesy of J. Robert Newland, D.D.S., M.S., University of Texas, Houston Health Science Center, Dental Branch.)

involvement makes such a clinical diagnosis unlikely.

Treatment and Prognosis

Due to the mild nature of the infection and lack of specific antiviral therapy, treatment is supportive and symptomatic. Patients respond well to palliative mouth rinses and gargles. Steroid-containing preparations should be avoided. Presumably, immunity to reinfection exists; however, due to the number of strains that have been reported to result in herpangina, it is possible to become reinfected with a similar, but antigenically distinct, strain of Coxsackie virus.

Human Immunodeficiency Virus

The oral manifestations associated with HIV infection and AIDS have played a prominent role in the HIV/AIDS epidemic since the first reported cases of Kaposi's sarcoma in 1981 (Table 5.11) (279,280). None of the oral lesions associated with HIV infection result directly from HIV infection per se, but rather from subsequent immunosuppression (281). Such immunosuppression not only results in decreased resistance to infection, but also may play a role in decreased viral suppression and enhanced neoplastic transformation.

Because the primary etiologic agents of HIV-associated oral lesions are often not viral, this portion of the chapter deals with bacterial and fungal, as well as viral, diseases of the oral cavity that are seen in individuals who are HIV-positive (282–289). They are grouped by their association with HIV infection, according to the consensus criteria adopted by the European Community (EC) Clearinghouse on Oral Problems Related to HIV Infection, World Health Organization (WHO) Collaborating Center on Oral Manifestations of the Immunodeficiency Virus, and the U.S. Workshop on Oral Manifestations of HIV Infection in September 1992 (290).

A number of excellent reviews on the oral manifestations of HIV infection and AIDS have been published (291–320), as well as large case series on the oral manifestations of HIV infection and AIDS in high risk groups, including homosexual and bisexual males (321,322), injecting drug users (323), and hemophiliacs

Table 5.11. Revised classification of oral lesions associated with HIV infection

Group 1: lesions strongly associated with HIV infection
 Candidiasis
  Erythematous candidiasis
  Pseudomembranous candidiasis
 Viral
  Hairy leukoplakia
 Periodontal disease
  Linear gingival erythema
  Necrotizing (ulcerative) gingivitis
  Necrotizing (ulcerative) periodontitis
 Kaposi's sarcoma
 Non-Hodgkin's lymphoma
Group 2: lesions less commonly associated with HIV infection
 Bacterial infections
  Mycobacterium avium-intercellulare
  Mycobacterium tuberculosis
 Viral infections
  Herpes simplex virus
  Human papillomavirus
  Condyloma acuminatum
  Focal epithelial hyperplasia
  Verruca vulgaris
  Varicella-zoster virus
  Herpes zoster
  Varicella
 Salivary gland disease
  Xerostomia due to decreased salivary flow rate
  Unilateral or bilateral enlargement of major salivary glands
 Others
  Melanotic hyperpigmentation
  Necrotizing (ulcerative) stomatitis
  Thrombocytopenic purpura
  Ulceration not otherwise specified
Group 3: other lesions seen in HIV infection
 Bacterial infections
  Actinomyces israelii
  Escherichia coli
  Klebsiella pneumonia
  Epithelioid (bacillary) angiomatosis
  Cat-scratch disease
 Drug reactions
  Ulcerative
  Erythema multiforme
  Lichenoid
  Toxic epidermolysis
 Fungal infections
  Cryptococcus neoformans
  Geotrichum candidum
  Histoplasma capsulatum
  Mucoraceae (mucormycosis/zygomycosis)
  Aspergillus flavus
 Neurologic disturbances
  Facial palsy
  Trigeminal neuralgia
 Recurrent aphthous stomatitis
 Viral infections
  Cytomegalovirus
  Molluscum contagiosum

(324), as well as in other specific populations, including women (325), children (326–332), Africans (333,334), and other ethnic groups (335–339).

Group 1: Lesions Strongly Associated with HIV Infection

Candidiasis

Oral candidiasis is the most common oral fungal infection in humans (340–344), as well as the most common presenting oral infection in HIV-positive men and women (345–350). The vast majority of cases are caused by *Candida albicans* (351); however, other oral isolates have been described (352–355). Although some cases may be initially asymptomatic, patients eventually complain of a nonspecific stomatopyrosis, often accompanied by stomatodynia and dysphagia (356,357). Oral candidiasis in HIV-infected individuals, especially with a progression to esophageal candidiasis, is a negative prognostic sign (358–360). Fortunately, the incidence of many HIV-associated diseases and conditions, including oral candidiasis, decreases dramatically following the institution of highly active antiretroviral therapy (HAART) (361).

Many HIV-positive patients harbor *C. albicans* and are otherwise asymptomatic (362–364). Various host factors, primarily immunosuppression, disrupt the normal oral ecologic microbial balance and frequently result in candidiasis (365–368).

Oral candidiasis traditionally manifests itself in its acute pseudomembranous form, which is characterized by white, curd-like dislodgeable plaques that, when removed, reveal an erythematous, occasionally hemorrhagic, base (Figs. 5.29 and 5.30). Frequently, however, infected patients present with erythematous candidiasis. This variant is characterized by the identical symptoms as those seen with the acute pseudomembranous type, with the exception of visible superficial fungal colonies or plaques (369) (Figs. 5.31 and 5.32). This diffuse oral erythema, normally with a lack of obvious fungal organisms, occasionally obscures the appropriate clinical impression. Median rhomboid glossitis, a unique form of *C. albicans* infection, can be found without any other associated oral candidal lesions (Fig. 5.33).

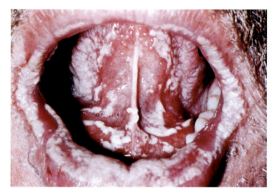

Figure 5.29. Pseudomembranous candidiasis. Severe cases of pseudomembranous candidiasis may be apparent extraorally. Superficial fungal colonies can be dislodged to reveal an eroded, erythematous mucosal base. (*Source*: Courtesy of J. Robert Newland, D.D.S., M.S., University of Texas, Houston Health Science Center, Dental Branch.)

The diagnosis of oral candidiasis is relatively straightforward. Potassium hydroxide preparations are normally sufficient to make the diagnosis; however, other rapid diagnostic methodologies have been applied, for example, latex agglutination utilizing CandidaSure™ (Life Sign First Care, Somerset, NJ) or immunofluorescent staining using calcofluor white (370). If speciation is required, cultures can be taken; however, the relatively slow growth rate of *Candida* sp. may interfere with the institution of appropriate therapy. A simplified, in-office culture system for oral candidiasis is also available (Oricult-N™, Orion Diagnostica, Espoo, Finland) (371).

The ability to treat effectively oral candidiasis in HIV-positive individuals is somewhat proportional to their immune suppression (372), and treatment efficacy varies from patient to patient

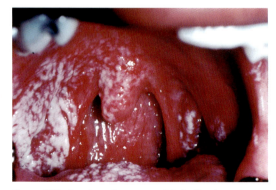

Figure 5.30. Pseudomembranous candidiasis. Pseudomembranous candidiasis of the posterior oral cavity and oral pharynx frequently results in dysphagia.

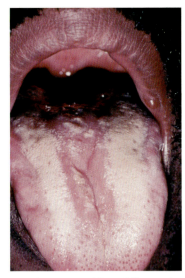

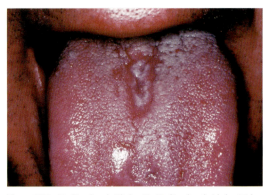

Figure 5.33. Median rhomboid glossitis. Median rhomboid glossitis has a characteristic clinical appearance; however, the diamond-shaped, depapillated area occasionally persists despite elimination of the organism from the mouth.

Figure 5.31. Erythematous candidiasis. Unlike pseudomembranous candidiasis, the erythematous variant does not exhibit any superficial colonies and presents clinically as patchy mucosal erythema. When it occurs on the tongue, there is frequently depapillation of the dorsal surface.

(373,374). Oral troches, for example, clotrimazole or nystatin, and antifungal mouth rinses are often effective in individuals who are relatively immunocompetent (375–378). Systemic antifungals, for example, ketoconazole, fluconazole, and itraconazole, are effective in more severe or refractory cases, or in those cases where patients are not compliant with the treatment regimen required of troches, that is, five times daily (379–385). Various antimicrobial and antifungal

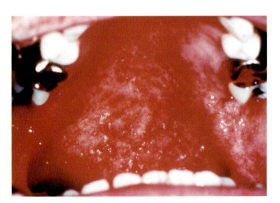

Figure 5.32. Erythematous candidiasis. Erythematous candidiasis of the palate often presents as diffuse erythema with associated stomatopyrosis and dysphagia. Occasionally, discrete, superficial fungal colonies are visible. (*Source*: Courtesy of J. Robert Newland, D.D.S., M.S., University of Texas, Houston Health Science Center, Dental Branch.)

agents have been used in the prophylaxis of oral candidiasis (386–389). Other formulations of antifungal medications, for example, intravenous, oral suspensions, creams, and ointments, are also useful in specific clinical situations, for example, severe infections, pediatric patients, perioral infections, and on the tissue-bearing surfaces of removable dental appliances.

A number of strains of *C. albicans* that are resistant to various azole-based antifungal medications have been reported (390–396). Such resistance may be due to infection with multiple strains (397,398), although there is evidence that *C. albicans* strains are maintained in individual patients (399,400).

The presence of oral candidiasis is a good indicator that immunosuppression exists to some degree in HIV-positive individuals (401,402). In all but the most severely immunosuppressed individuals, the prognosis for oral candidal infections is good (403–405); however, the frequency of recurrence is high, due to the ubiquitous nature of the organism. Unfortunately, in the absence of *Candida* prophylaxis, recurrence is the norm rather than the exception.

Hairy Leukoplakia

The unique lesion of oral hairy leukoplakia (OHL) was first described by Greenspan and coworkers in the early 1980s in a number of homosexual males in San Francisco (406–408). The lesion was subsequently described in other risk groups for HIV infection, for example,

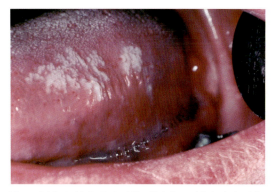

Figure 5.34. Oral hairy leukoplakia. Clinically, oral hairy leukoplakia may mimic leukoplakia secondary to tobacco use, tongue chewing, or other dysfunctional habits. The diagnosis is confirmed by in situ DNA hybridization for Epstein-Barr virus.

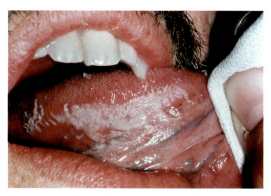

Figure 5.36. Oral hairy leukoplakia. Involvement of the filiform papillae on the dorsal surface of the tongue in oral hairy leukoplakia, when present, imparts a "hairy" appearance, whereas lesions on the ventral tongue lesions are more plaque-like.

hemophiliacs and transfusion recipients (409–412). The initially described cases of OHL occurred on the dorsal tongue and involved the filiform papillae, which gave the lesions a "hairy" appearance. The majority of lesions occur on the lateral surfaces of the tongue and have a corrugated appearance although they can extend distally into the oral pharynx (Figs. 5.34 and 5.35) (413). The OHL lesions on the ventral tongue have a more plaque-type appearance (Fig. 5.36).

Oral hairy leukoplakia is not a specific sign of HIV infection but a manifestation of generalized immunosuppression (414,415). Identical lesions have now been described in HIV-negative transplant patients receiving bone marrow (416–420), kidney (421–423), liver (424), and heart (425), as

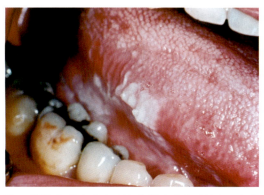

Figure 5.35. Oral hairy leukoplakia. Oral hairy leukoplakia is routinely found on the lateral border of the tongue, where it has a corrugated appearance. (*Source*: Courtesy of J. Robert Newland, D.D.S., M.S., University of Texas, Houston Health Science Center, Dental Branch.)

well as an HIV-negative patient with myelodysplastic syndrome (426). The occurrence of OHL in HIV-negative, immunocompetent individuals is rare (427,428).

Oral hairy leukoplakia is characterized by an epithelial hyperplasia, secondary to EBV replication (429–431). Epstein-Barr virus preferentially infects oral keratinocytes (432) but can be found in the oral tissues of both HIV-positive and HIV-negative individuals (433–435). Human papilloma virus has also been found in some OHL specimens, although its significance is questionable (436). Depending on the clinical situation, the appearance of OHL may be confused with a number of other conditions, especially when HIV seropositivity is not suspected. Frictional hyperkeratoses, tobacco-associated leukoplakias, hyperplastic candidiasis, and plaque-type lichen planus, among others, can all resemble OHL (437–439).

In an individual who is known to be HIV-positive, an incisional, H&E-stained specimen that reveals the koilocytic-like and other characteristic epithelial changes is usually sufficient to render a diagnosis of OHL. Such changes can be seen in other oral mucosal lesions; therefore, in cases where infection by HIV is not suspected or known, confirmation by in situ DNA hybridization is essential (440,441). The ultrastructural features of EBV, such as infected epithelial cells, are also quite characteristic (442,443). Exfoliative cytology, utilizing immunodiagnostic staining, has also been described for use in the diagnosis of OHL (444,445). A noninvasive brush biopsy technique has also been described (446).

Oral hairy leukoplakia does not require any definitive intervention, unless the extent of involvement poses a functional or aesthetic problem (447). The lesions resolve with the administration of oral antiviral agents, for example, acyclovir (448,449), desciclovir (450), dihydroxy-propoxymethyl-guanine (DHPG) (451), zidovudine (452–454), and valacyclovir (455); however, they return when the medication is discontinued. Vitamin A has also been used topically to treat OHL (456). Podophyllin resin has been used successfully (457,458), as has surgery (459). Although the prognosis of OHL is excellent, it is a negative prognostic marker for the overall progression of HIV disease, as it indicates a more depressed immune system (460–462).

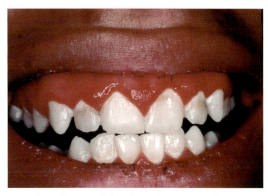

Figure 5.38. Linear gingival erythema. HIV-positive pediatric patients have similar gingival manifestations as those found in adults. (*Source*: Courtesy of Joel H. Berg, D.D.S., M.S., University of Washington, School of Dentistry.)

Periodontal Disease

The gingival and periodontal manifestations of HIV infection continue to be oral hallmarks of HIV infection in both heterosexuals and homosexuals (463–468). Linear gingival erythema, formerly called HIV-associated gingivitis, is characterized by an intensely erythematous marginal gingivitis. It is occasionally accompanied by erythematous plaques of the alveolar gingiva and vestibular mucosa (467) (Figs. 5.37 and 5.38). Unlike marginal gingivitis in HIV-negative individuals, however, the erythema does not resolve following increased oral hygiene measures, for example, tooth brushing and flossing.

A subpopulation of HIV-positive patients with linear gingival erythema will progress to develop necrotizing gingivitis and periodontitis, formerly called HIV-associated gingivitis and periodontitis, respectively (468,469). These individuals exhibit a rapidly progressive destruction of periodontal tissues, including loss of periodontal attachment and resorption of underlying alveolar bone (Fig. 5.39). Occasionally, these destructive processes extend beyond the periodontium and result in a necrotizing stomatitis of periodontal origin (Fig. 5.40). This infection results in massive destruction of the adjacent mucosa and underlying bone and can be life-threatening.

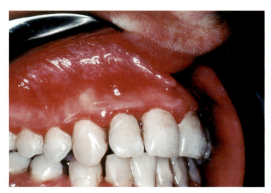

Figure 5.37. Linear gingival erythema. Linear gingival erythema is characterized by a bright red marginal gingivitis, which fails to respond to normal oral hygiene measures. The microbial flora is not dramatically different from that found in healthy individuals.

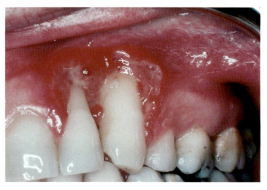

Figure 5.39. Necrotizing periodontitis. Necrotizing periodontitis is characterized by rapid apical migration of the periodontal attachment and resorption of underlying bone.

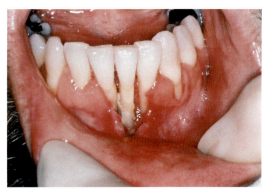

Figure 5.40. Necrotizing stomatitis. Patients with necrotizing stomatitis have a necrotic destruction of the soft tissues, with spontaneous hemorrhage, exposed alveolar bone, sequestration of bone, and acute, severe pain. (*Source*: Courtesy of John M. Wright, D.D.S., M.S., Baylor College of Dentistry, Dallas, Texas.)

In general, there are no significant differences in putative periodontal pathogens between HIV-positive and HIV-negative individuals (470–474), which suggests there is a continuum from declining periodontal health to linear gingival erythema to necrotizing gingivitis and periodontitis (475).

The diagnosis of HIV-related periodontopathies is purely clinical, once HIV-seropositivity has been established. All forms of these diseases are characterized by their refractory response to routine therapeutic measures, for example, increased oral hygiene measures. In addition to local debridement, local and systemic antimicrobial therapy, for example, chlorhexidine and metronidazole, immediate follow-up care, and long-term maintenance are essential to control this condition (476–479).

The prognosis for the HIV-associated periodontopathies is good, in the sense that the progression of the diseases can be arrested, even though previous periodontal destruction cannot be reversed. It is essential, however, that affected patients are scrupulous with their personal oral hygiene and professional dental care.

Kaposi's Sarcoma

Kaposi's sarcoma (KS) was the first AIDS-defining disease described at the outset of the epidemic and continues to be the most prevalent malignancy in AIDS, especially among homosexual men (480,481). It is a malignant neoplasm of, presumptively, small blood vessels (482), although vascular endothelium has been proposed to play a role in the histogenesis of KS (483). The classic lesion of KS was described approximately 100 years ago as an indolent, slow-growing lesion, limited to the skin, commonly found on the lower extremities of men of Mediterranean descent. The KS lesions in HIV-positive individuals progress rapidly, involve internal organs, and account for a significant mortality in affected patients. From the outset of the AIDS epidemic, it had been postulated that KS was a virally induced malignancy, as its spread throughout the United States mimicked that of other infectious epidemics (484,485). In 1994, herpes-like viral sequences were identified in AIDS-associated KS (486). The following year the same DNA sequences were described in KS in patients who were not HIV infected (487). Human herpesvirus 8 has been identified in saliva (488–491), as well as oral KS by several investigators (492) and in oral tissues not affected by KS (493,494). It appears that HHV-8 is necessary, but not solely sufficient, to cause KS (495–498). Human herpesvirus 8 has an estimated global seroprevalence between 10% and 25%; however, it appears to be under immunologic control in otherwise healthy individuals (499,500).

It is of particular importance to dental health care workers that, of all AIDS patients who have visible KS, one third of them will have KS intraorally or on the normally visible skin of the head and neck. Furthermore, KS is the most common oral malignancy in AIDS, although over the past decade its relative incidence in AIDS patients has been declining, due to its sensitivity to immune deficiency and the advent of HAART (501–503). The palate is the most common intraoral site for KS; however, it has been reported on every intraoral mucosal surface, as well as intraosseously (504). Early lesions are macular in appearance, ranging from slightly red to violet (Figs. 5.41 to 5.43), whereas late-stage lesions are more nodular in nature (Figs. 5.44 to 5.48) and may result in underlying bony destruction (505).

An incisional biopsy is essential to confirm the diagnosis of KS, even when the patient's HIV serostatus is confirmed, due to the histopathologic similarities between KS and epithelioid (bacillary) angiomatosis, a bacterial lesion

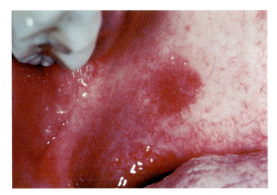

Figure 5.41. Kaposi's sarcoma. Early lesions of Kaposi's sarcoma are asymptomatic, flat, red-to-violet in color, and clinically resemble a mucosal ecchymosis. (*Source:* Courtesy of J. Robert Newland, D.D.S., M.S., University of Texas, Houston Health Science Center, Dental Branch.)

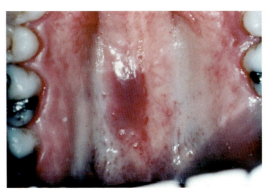

Figure 5.44. Kaposi's sarcoma. More advanced lesions of Kaposi's sarcoma have a nodular appearance and are more likely to be noticed by the patient. (*Source:* Courtesy of J. Robert Newland, D.D.S., M.S., University of Texas, Houston Health Science Center, Dental Branch.)

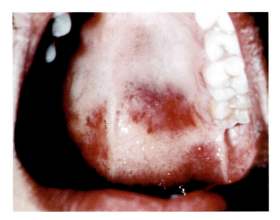

Figure 5.42. Kaposi' sarcoma. The palate is the most common site of intraoral Kaposi's sarcoma.

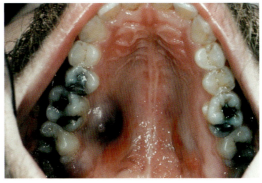

Figure 5.45. Kaposi's sarcoma. Nodular lesions of the palate may enlarge enough to cause functional problems with speech and swallowing.

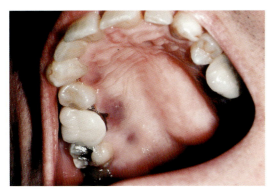

Figure 5.43. Kaposi's sarcoma. Occasionally, palatal Kaposi's sarcoma may resemble trauma secondary to a dental procedure.

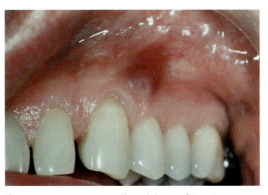

Figure 5.46. Kaposi's sarcoma. The early lesions of Kaposi's sarcoma on the gingiva may be mistaken for gingival cysts. (*Source:* Courtesy of J. Robert Newland, D.D.S., M.S., University of Texas, Houston Health Science Center, Dental Branch.)

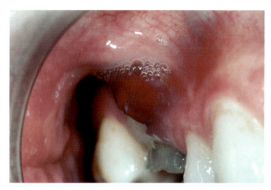

Figure 5.47. Kaposi's sarcoma. The gingiva is the second most common intraoral site for Kaposi's sarcoma. (*Source*: Courtesy of J. Robert Newland, D.D.S., M.S., University of Texas, Houston Health Science Center, Dental Branch.)

Table 5.12. Differential diagnosis of Kaposi's sarcoma
Kaposi's sarcoma: most common on gingival and hard palate, but may involve any oral mucosal site; one-third of patients with intraoral KS will have a cutaneous head and neck lesion, as well; early lesions are more erythematous than violaceous and more macular than nodular; most lesions are asymptomatic unless secondarily traumatized
Bacillary angiomatosis: reddish-brown mucocutaneous papules and nodules; clinically resembles intraoral Kaposi's sarcoma; due to a focal infection by *Bartonella* species; organism demonstrable with special strains; lesion respond to antibiotics
Non-Hodgkin's lymphoma: most common on palate and alveolar ridge; may be limited to soft tissue; jaw involvement common, with loose teeth, pain and paresthesia; variable coloration from pink to purple

that mimics KS in AIDS patients (506,507). Cytomegalovirus has also been noted in oral KS lesions, although its role in the pathogenesis of KS remains unclear (508) (Table 5.12).

Intraoral KS is normally treated, exclusive of other systemic therapy, when the lesions impair function or become an aesthetic problem. Radiation therapy was initially used to decrease tumor volume (509); however, its usefulness is limited because of complications from radiation toxicity and subsequent mucositis. Surgical debulking of large lesions provides temporary relief of functional impairment; however, the lesions frequently reenlarge. Photodynamic

therapy is somewhat successful with few side effects (510). Combined systemic antiviral (zidovudine) and immunotherapy (α-interferon) is effective in some patients (511,512), although anemia and constitutional symptoms preclude its widespread application. Most recently, intralesional vinblastine has been used successfully in inducing tumor sclerosis and palliation (Figs. 5.49 and 5.50) (513,514). Sodium tetradecyl sulfate (Sotradecol™) has also been suggested as a sclerosing agent for localized

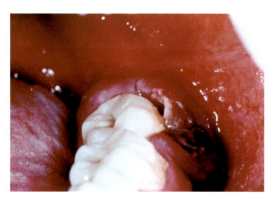

Figure 5.48. Kaposi's sarcoma. Advanced gingival lesions of Kaposi's sarcoma can result in significant periodontal destruction. (*Source*: Courtesy of J. Robert Newland, D.D.S., M.S., University of Texas, Houston Health Science Center, Dental Branch.)

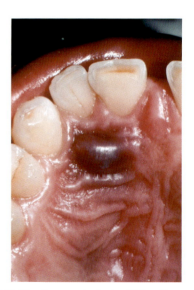

Figure 5.49. Kaposi's sarcoma (pretreatment). Kaposi's sarcoma of the anterior hard palate can become a functional problem, as the lesion may be traumatized by the anterior mandibular teeth. (*Source*: Courtesy of C. Mark Nichols, D.D.S., Bering Dental Clinic, Houston, Texas.)

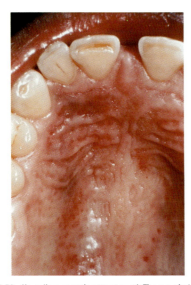

Figure 5.50. Kaposi's sarcoma (posttreatment). The same lesion seen in Figure 5.49 following three intralesional injections of vinblastine over a 6-week period. (*Source*: Courtesy of C. Mark Nichols, D.D.S., Bering Dental Clinic, Houston, Texas.)

intraoral KS lesions (515). Treatment of intraoral KS should be coordinated with the management of other oral manifestations of HIV-related oral disease (516,517).

The morbidity of intraoral KS far exceeds its ultimate mortality; thus, the prognosis is relatively good when compared to opportunistic infections. In fact, overall mortality from AIDS-related KS appears to be decreasing, due, in part, to the increasing use of HAART (518). Patients in whom KS has regressed have been noted to have immunoglobulin G (IgG) and IgA antibodies directed against HHV-8; however, the significance of this finding remains unclear (519). Unfortunately, it is unusual to achieve long-term remissions of intraoral KS lesions, and recurrences as well as new lesions can be expected over time.

Non-Hodgkin's Lymphoma

Non-Hodgkin's lymphomas (NHL) comprise the second most common malignancy in AIDS patients (520–526). The vast majority of affected patients have extranodal involvement, with central nervous system, bone marrow, bowel, and mucocutaneous sites being most common.

Approximately 5% of NHLs in AIDS patients have the presenting lesion in the mouth (527). Oral NHLs are primarily of B-cell lineage, most are high grade, and many are found to contain EBV DNA (528–530).

Oral NHLs have been reported to involve primarily the palate and alveolar ridge, although they can involve other intraoral sites, usually by tumor extension (531–533) (Fig. 5.51). The clinical presentation of the lesions can range from necrotizing (ulcerative) gingivitis to voluminous tumor masses, both solitary and multifocal (534–538). Patients may be initially asymptomatic; however, loose teeth and paresthesia are common when there is bone involvement. Early lesions may be confused with nonneoplastic gingival lesions, while advanced disease shares clinical features with metastatic carcinoma, melanoma, or malignant histiocytosis.

The diagnosis of NHLs requires both routine histopathology to grade the tumor and specific immunohistochemistry procedures to characterize and subclassify the neoplasm, for example, the presence of EBV as well as B and T cell markers.

Acquired immune deficiency syndrome patients with NHL have a rapidly progressive, declining clinical course. The lesions respond well initially to traditional chemotherapy; however, their remissions are not durable. Although radiation therapy has also been used, it is felt to be inadequate in the absence of combined chemotherapy (539).

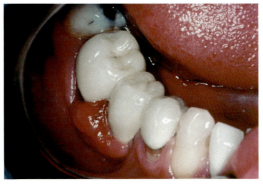

Figure 5.51. Non-Hodgkin's lymphoma. Non-Hodgkin's lymphoma is the second most common intraoral malignancy in AIDS and, in its early stages, may mimic benign reactive lesions of the gingiva, e.g., pyogenic granuloma.

Group 2: Lesions Less Commonly Associated with HIV Infection

Bacterial Infections

Both *Mycobacterium avium-intercellulare* complex (MAC) and *Mycobacterium tuberculosis* (TB) are found with increasing frequency in HIV-positive individuals (540,541), including children (542,543). Approximately 50% of all HIV-positive individuals have at least one episode of mycobacterial infection (544,545).

Mycobacterium avium-intercellulare complex infections are acquired through a pulmonary route of exposure and are felt to be unavoidable environmental exposures in immunosuppressed patients (546–549). The organism rapidly colonizes the pulmonary and gastrointestinal mucosa (550–554) and is characterized by multiple drug resistance (555–563). Manifestations of MAC have been reported to include oral ulceration in HIV-infected individuals and should be considered in the differential diagnosis of such lesions.

Although MAC is more common overall, TB is more frequently reported in certain high-risk groups, for example, Haitians and injecting drug users (564). Its prevalence ranges from less than 3% in U.S.-born AIDS patients to 13% in Haitian-born AIDS patients, with an overall U.S. prevalence of 2.5% (565). Approximately one fourth of all extrapulmonary cases of TB in the United States occur in individuals who are HIV-positive (566,567).

In most cases, TB occurs earlier in the course of HIV infection than MAC (568) and presents a

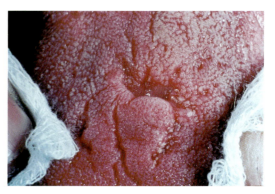

Figure 5.53. Tuberculosis. Tuberculous ulcers of the tongue are indurated, with irregular borders. (*Source*: Courtesy of J. Robert Newland, D.D.S., M.S., University of Texas, Houston Health Science Center, Dental Branch.)

significant risk to dental and other health care workers who do not use barrier protection and appropriate infection control measures (569). Like MAC, TB is normally transmitted through contaminated, aerosolized saliva (570,571). Unfortunately, although TB is both preventable and treatable (572,573), at-risk and infected patients may not practice accepted protocols and are often noncompliant with treatment regimens (574). Of greatest concern in recent years is the emergence of multiple-drug resistant TB, with an increased incidence of associated mortality in both immunosuppressed and immunocompetent individuals (575).

Oral lesions of TB are nonspecific. Clinically, the lesions most commonly appear as indolent ulcers with varying symptomatology (576) (Figs. 5.52 to 5.54). The characteristic granulomatous

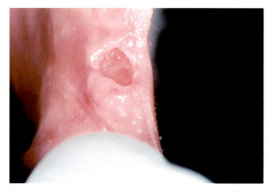

Figure 5.52. Tuberculosis. Tuberculosis may present intraorally as a nonhealing ulcer, resembling squamous cell carcinoma. (*Source*: Courtesy of J. Robert Newland, D.D.S., M.S., University of Texas, Houston Health Science Center, Dental Branch.)

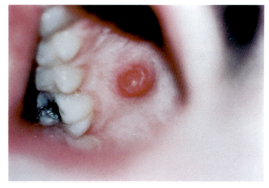

Figure 5.54. Tuberculosis. Tuberculosis may also affect the palate and resemble malignant salivary gland neoplasms. (*Source*: Courtesy of J. Robert Newland, D.D.S., M.S., University of Texas, Houston Health Science Center, Dental Branch.)

response may not be apparent on histopathologic examination and oral cultures for *M. tuberculosis* are unreliable. Strong consideration should be given to the possibility of a mucosal manifestation of TB for immunosuppressed individuals who present with nonresolving oral ulcers.

Melanotic Hyperpigmentation

For reasons that are not well understood, individuals who are HIV-positive occasionally have a diffuse, patchy oral pigmentation that is clinically similar to that seen in patients with endocrine abnormalities or pigmentary disorders (577,578). Brownish-black macules have been reported on the buccal mucosa, lips, gingival, and palate in 6.4% of HIV-positive individuals followed for up to 24 months, as compared to 3.6% in HIV-negative controls (579) (Fig. 5.55). Pigmented skin and nail changes have also been described (580).

Certain therapeutic agents, for example, ketoconazole, clofazimine, and azidothymidine, are known to occasionally result in increased oral pigmentation (581); however, oral mucosal melanocyte stimulation appears to occur in HIV-positive individuals in the absence of such drugs (582). The significance of these findings is debatable (583), in light of the occasional finding of oral pigmentation in HIV-negative individuals. Nevertheless, the presence of oral mucosal pigmentation, in the presence of other suggestive findings, should raise the clinical index of suspicion for HIV infection (584).

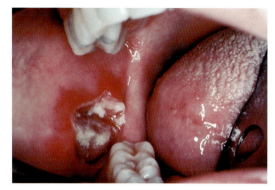

Figure 5.56. Necrotizing stomatitis. Necrotizing stomatitis may be quite destructive, despite the lack of identifiable pathogens.

Necrotizing (Ulcerative) Stomatitis

Necrotizing (ulcerative) stomatitis (NS) is a less common, but more severe extension of necrotizing (ulcerative) periodontitis (585). It is clinically reminiscent of noma or gangrenous stomatitis and differs from necrotizing (ulcerative) periodontitis on the basis of anatomic location (586). It is characterized by extension of the destruction to nonperiodontal soft tissue and bone, with considerable necrosis of the epithelium and underlying connective tissue and sloughing of bone (587,588) (Fig. 5.56). Occasionally, the lesions can become life-threatening. Necrotizing stomatitis is best treated with local debridement, dental prophylaxis, analgesics, and antimicrobial therapy, for example, chlorhexidine rinses and metronidazole (589).

Salivary Gland Disease

Salivary gland disease in HIV-positive and AIDS patients has been reported as both xerostomia and salivary gland enlargement (590–594). These manifestations appear to be more common in children than adults and characteristically resemble the lesions of benign lymphoepithelial disease, for example, Sjögren's syndrome, both clinically and histopathologically (595–604). The majority of affected patients have bilateral enlargement of the parotid glands with decreased salivary flow rates (605–607). HIV can be found in low levels in whole saliva (608–614), and, while salivary protein is decreased, secretory IgA is increased (615). The risk of salivary transmission of HIV is extremely low (616,617).

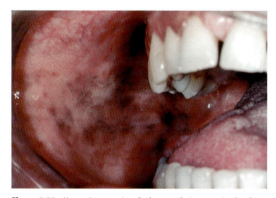

Figure 5.55. Hyperpigmentation. Oral mucosal pigmentation has been described in HIV-positive individuals due to a variety of causes; however, the significance of idiopathic pigmentation is unknown.

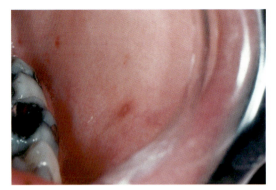

Figure 5.57. Thrombocytopenic purpura. Oral mucosal petechiae may be the first indication of thrombocytopenic purpura.

Recently, the FDA approved a rapid oral fluid-based test for HIV detection, OraQuick™ (OraSure Technologies, Bethlehem, PA) (618,619).

The treatment of salivary gland disease in HIV-infected patients parallels that of patients with Sjögren's syndrome. Scrupulous oral hygiene is necessary to prevent an increased incidence of dental caries (620,621). Sialogogues, for example, chewing gum, are useful to stimulate residual functional salivary gland parenchyma (622). Pilocarpine has been used to achieve the same result with minimal side effects. A newly approved cholinergic agonist, Evoxac™ (cevimeline) (Daiichi Pharmaceutical Corp., Tokyo, Japan), binds to muscarinic receptors and has a similar effect on compromised salivary glands (623,624).

Thrombocytopenic Purpura

Thrombocytopenic purpura (TP), as a result of increased platelet destruction, is an uncommon finding in HIV-positive and AIDS patients (625–627). It is significant because of the likelihood of oral lesions as an initial presenting clinical sign (628,629) (Fig. 5.57). Furthermore, the presence of vascular lesions of the oral mucosa in HIV-positive individuals requires a definitive workup to rule out the possibility of other clinically similar lesions, for example, Kaposi's sarcoma and epithelioid (bacillary) angiomatosis (630,631). Gingival biopsy has also been proposed as a useful diagnostic tool in the diagnosis of TP (632). Treatment consists of prednisone, immunoglobulin therapy, and splenectomy (633).

Nonspecific Ulcerations

Patients who are HIV positive may have a variety of oral ulcers of indeterminate etiology (634). There is no reason to suspect that HIV-positive individuals are any less likely to manifest oral ulcerations of various etiologies than the population at large. Nevertheless, it is critical to rule out ulcers of infectious origin prior to instituting antiinflammatory or exclusively palliative therapy. Both corticosteroids (635) and thalidomide (636) have been used successfully in treating refractory oral aphthae.

Herpes Simplex Virus

Herpes simplex virus (HSV) infections in HIV-positive children and adults present a diagnostic and therapeutic challenge (637–640). They may occur as solitary infections or in concert with other oral microbial infections commonly found in HIV-positive patients (641,642). The vast majority of lesions in HIV-positive individuals are type 1 and recurrent, although type 2 and primary infections have been reported (643). Unfortunately, recurrent HSV-1 lesions in immunosuppressed individuals often present as widespread, severe disease with prolonged involvement of both nonkeratinized and keratinized mucosa, thus mimicking primary HSV-1 in an immunocompetent individual (644,645) (Fig. 5.58). As such, they are a significant source of morbidity in affected patients.

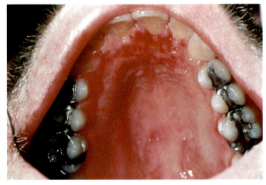

Figure 5.58. Herpes simplex virus type 1. Recurrent intraoral HSV-1 lesions in AIDS patients may resemble primary HSV-1 infections, occurring on keratinized mucosa, i.e., hard palate and attached gingiva, with a protracted clinical course.

The diagnosis of HSV-1 lesions can be made by exfoliative cytology (Tzanck test) (646), direct immunofluorescence (647), immunoperoxidase (648), radioimmunoassay (649), or viral culture (650). Various antiviral agents have been used both prophylactically and therapeutically, including acyclovir, valacyclovir, famcicolvir, ganciclovir, and foscarnet (651–660). Unfortunately, both acyclovir- and foscarnet-resistant HSV-1 infections have been reported in HIV-positive individuals (661–663).

Condyloma Acuminatum

Oral condyloma acuminatum (CA) or venereal warts are commonly found on the tongue, lingual frenum, gingiva, and lips of HIV-positive individuals, presumably due to oral-genital contact (664,665). In situ DNA hybridization is routinely positive for the presence of HPV-6 and-11 (666,667).

The lesions are normally asymptomatic, pink, fleshy, soft tissue masses with a smooth, bosselated surface (Figs. 5.59 and 5.60). Due to their relative ease of virus transmission, the lesions merit removal. This can be accomplished most expeditiously by surgical removal; however, laser surgery and cryosurgery are also effective (668). Podophyllin resin and interferon have been used on nonoral condyloma acuminatum, but their usefulness on oral mucosa remains to be seen.

Focal Epithelial Hyperplasia

Focal epithelial hyperplasia (FEH) has been described in HIV-positive individuals (669). The

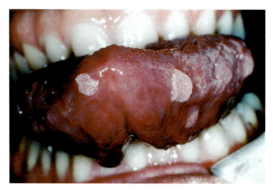

Figure 5.60. Condyloma acuminatum. Multiple intraoral condyloma acuminata may present therapeutic problems due to the amount of mucosal involvement. (*Source*: Courtesy of J. Robert Newland, D.D.S., M.S., University of Texas, Houston Health Science Center, Dental Branch.)

lesions appear as multiple, asymptomatic, flesh colored papules and plaques involving all oral mucosal surfaces (Fig. 5.61). In situ hybridization of the lesions reveals the presence of HPV-13 or -32. No treatment is indicated in the absence of aesthetic considerations or functional impairment.

Oral Warts

Oral verruca vulgaris, squamous papillomas, and warts other than condyloma acuminatum have been described in HIV-positive and AIDS patients (670–672) (Figs. 5.62 and 5.63). Like their counterparts in immunocompetent patients, these lesions are positive for the presence of various HPV subtypes and exhibit koilocytic and other characteristic histopathologic

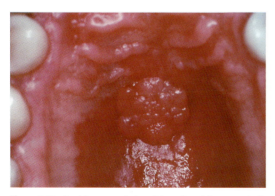

Figure 5.59. Condyloma acuminatum. Intraoral condyloma acuminata are frequently seen in HIV-positive and AIDS patients.

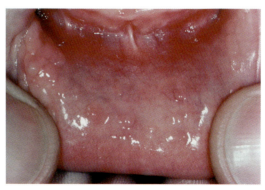

Figure 5.61. Focal epithelial hyperplasia (Heck's disease). Focal epithelial hyperplasia or Heck's disease, caused by HPV-13 and -32, has also been reported in HIV-positive and AIDS patients.

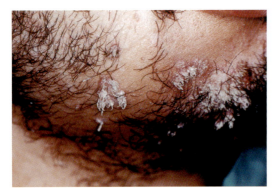

Figure 5.62. Verruca vulgaris. Multiple verruca vulgaris lesions are present on the perioral skin.

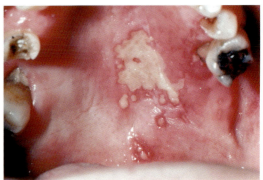

Figure 5.64. Recurrent varicella-zoster virus. Intraoral herpes zoster is characterized by a unilateral distribution, following the distribution of infected nerves.

features of oral warts (673). Occasionally, unusual HPV subtypes are reported, for example, HPV-7, the causative agent of butcher's warts (674).

Unlike condyloma acuminatum, these oral warts appear to be less likely to spread or be transmitted. Solitary or focal clusters of lesions merit removal; however, in the event that multi-focal disease cannot be managed in that manner, the possibility of rapid, widespread extension is minimal. Topical cidofovir, a purine analogue of cytosine, has been reported to be effective in the management of recalcitrant gingival warts in an HIV-positive individual (675).

Varicella-Zoster Virus

Primary varicella-zoster virus (VZV) infection (chickenpox) in immunosuppressed individuals can be fatal (676–678) and recurrent oral VZV

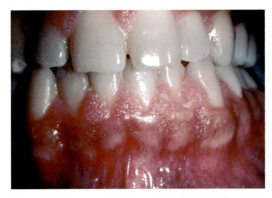

Figure 5.63. Verruca vulgaris. Verruca vulgaris lesions on the gingiva, caused by various HPV subtypes, may cause treatment complications when the lesions are widespread.

(herpes zoster or shingles) is a source of severe morbidity (679–686). The oral manifestations of primary VZV in HIV-positive and AIDS patients is similar to that seen in immunocompetent individuals (687); however, the clinical impression is often obscured by other coexistent infections (688–690). Recurrent oral VZV lesions have a characteristic, unilateral distribution over the area of infected nerve distribution, although multiple nerves may be involved (Fig. 5.64).

Acyclovir, valaryclovir, and famcicolvir are commonly used to treat VZV-infected individuals and has replaced older antiviral agents, for example, vidarabine, as the treatment of choice (691–697). Bromovinyldeoxuridine, foscarnet, ganciclovir, and other acyclovir congeners have also been advocated for use in treating VZV infections (698,699). As with other herpesvirus infections in immunosuppressed individuals, resistance of VZV to acyclovir has been reported (700–703). Various palliative mouth rinses, in combination with analgesics, are also very helpful in the treatment of oral discomfort associated with oral VZV infection.

Group 3: Lesions Seen in HIV Infection

Bacterial Infections

Numerous atypical oral bacterial infections have been described in individuals with HIV infection, including those with *Actinomyces* sp., *Escherichia coli*, and *Klebsiella pneumoniae* (704,705). The clinical manifestations of these infections are not unique, but underscore the necessity of appropriate diagnostic methods, including culture, to ensure the timely institution of appropriate therapy.

Epithelioid (Bacillary) Angiomatosis and Cat Scratch Disease

Epithelioid (bacillary) angiomatosis (EA) is a pseudoneoplastic infectious disease that is characterized by a mucocutaneous eruption of reddish-brown papules and nodules of vascular origin (706–708). The etiologic agent of EA is a gram-negative bacillus of the genus *Bartonella* (formerly *Rochalimaea*), which is similar, if not identical, to the causative organism of cat scratch disease and is demonstrable by use of the Warthin-Starry silver stain (709–718).

The lesions have been reported in HIV-positive and AIDS patients and resemble Kaposi's sarcoma, both clinically and histopathologically (719–725). As an infectious process, EA responds favorably to erythromycin and other antibiotics (726–730).

The accurate diagnosis of this lesion is important for several reasons. First, misdiagnosis of this lesion in an HIV-positive individual may result in an inappropriate staging of the individual as an AIDS patient. Second, misdiagnosis as Kaposi's sarcoma may subject the patient to unnecessary, and potentially detrimental, therapy. Finally, misdiagnosis of EA may result in the progression of a potentially fatal opportunistic infection in an individual who is immunocompromised.

Drug Reactions

A variety of drug reactions have been reported in immunosuppressed individuals, including diffuse ulcerative stomatitis, erythema multiforme, lichenoid drug reactions, and toxic epidermal necrolysis (731–734) (Fig. 5.65). The likelihood of such reactions is increased in HIV-positive individuals due to the use of multiple medications, often including experimental drugs (735–778). All of these conditions have a propensity to involve oral mucosa, often without cutaneous manifestations, and can be confused with oral ulcerations of infectious origin (739,740).

Cutaneous manifestations of these conditions simplify the clinical differential diagnosis; however, when absent, the possibility of vesicular lesions of viral origin, for example, HSV-1,

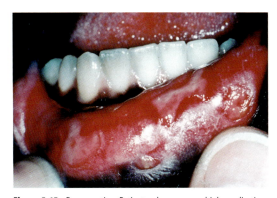

Figure 5.65. Drug reaction. Patients who are on multiple medications are at greater risk for drug reactions and interactions, many of which have oral mucosal manifestations. (*Source*: Courtesy of J. Robert Newland, D.D.S., M.S., University of Texas, Houston Health Science Center, Dental Branch.)

must be ruled out. Corticosteroid therapy and antimicrobial mouth rinses, as well as immunomodulating agents, for example, levamisole, may be effective in some patients (741–744). The determination of the causative agent, or agents, is paramount to preventing recurrent episodes. Unfortunately, the specific etiology can be elusive.

Fungal Infections

A number of opportunistic fungal infections have been reported in the oral cavities of individuals with HIV infection or AIDS (745–747). Causative organisms include *Cryptococcus neoformans, Geotrichum candidum, Histoplasma capsulatum, Mucoraceae (mucormycosis/zygomycosis)*, and *Aspergillus flavus*.

Cryptococcosis is a severe, life-threatening infection in immunosuppressed individuals, which frequently results in meningoencephalitis and the often-rapid demise of the patient, in spite of aggressive antifungal therapy (748,749). It is the fourth most common opportunistic infection in AIDS. The initial route of infection is pulmonary; however, the organism rapidly disseminates to involve a variety of organs, including the skin and oral mucous membranes.

Constitutional symptoms are often vague and may include pulmonary dysfunction, fever of unknown origin and headache. Although the presumptive diagnosis is made on the basis of histopathologic findings or India-ink staining of spinal fluid, confirmatory fungal cultures are

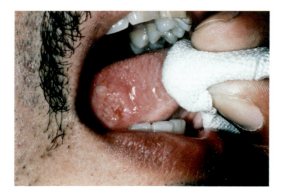

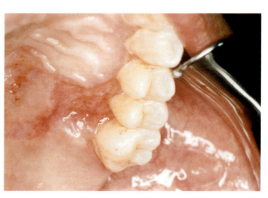

Figure 5.66. Cryptococcosis. Oral mucosal infection with *Cryptococcus neoformans* in AIDS presents as an indurated, nonhealing ulcer with raised, rolled borders, which may mimic other inflammatory and neoplastic mucosal processes. [*Source*: Lynch and Naftolin (755), © 1987 with permission from Elsevier.]

Figure 5.67. Histoplasmosis. *Histoplasma capsulatum* is endemic to the Mississippi valley and indolent oral ulcers due to these infections have been described in AIDS patients.

appropriate. Treatment consists of vigorous antifungal therapy with azole-based antifungal medications, flucytosine, and amphotericin B (750–754). Several cases of oral cryptococcosis involving the tongue and palate have been reported (755–758) (Fig. 5.66). The lesions commonly present as eroded soft tissue masses.

Oral geotrichosis is a rare opportunistic mucosis that has been reported in immunocompromised individuals (759). The major oral manifestation of *Geotrichum candidum* infection in HIV infection consists of nonspecific gingivitis, which resolves with the administration of topical nystatin.

Histoplasmosis, caused by *Histoplasma capsulatum*, is the most common respiratory mycosis in the United States (760). It is endemic to the Mississippi and Ohio river valleys and results in characteristic pulmonary signs and symptoms (761). Oral lesions are uncommon in primary pulmonary histoplasmosis (762). Numerous case reports of oral lesions in HIV-positive and AIDS patients have been published (763–767). Oral lesions have been reported in up to half of all AIDS cases with disseminated histoplasmosis (653).

The most common sites for oral lesions of disseminated histoplasmosis are the tongue, palate, and buccal mucosa. The lesions appear as ill-defined, ulcerated soft tissue swellings (768,769) (Figs. 5.67 and 5.68). The organisms are apparent on examination of histopathologic material; however, the diagnosis should be confirmed by fungal culture (770,771). Specific fluorescent antibody staining has also been used to confirm

the diagnosis. The infection responds fairly well to vigorous antifungal therapy with various azole congeners and amphotericin B (772–776).

Mucormycosis (phycomycosis or *zygomycosis)* is an opportunistic fungal infection caused by various saprophytic fungi found in soil, bread molds, and decaying fruits and vegetables (777). The organisms have a predilection for vascular invasion, resulting in widespread infarction and massive tissue necrosis (778,779). Historically, this condition was seen primarily in poorly controlled diabetics and patients with hematologic malignancies; however, several cases have been reported in HIV-positive and AIDS patients (780,781). Initially, the oral lesions may appear as soft tissue enlargements, but late-stage or poorly controlled lesions are characterized by fulmi-

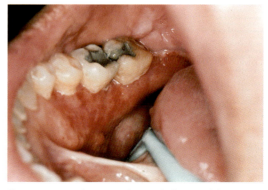

Figure 5.68. Histoplasmosis. Oral mucosal infection with histoplasmosis may also present as nonspecific enlargement and superficial ulceration of the gingiva.

nant tissue necrosis and sloughing of nonvital bone. Amphotericin B is the traditional treatment of choice, although the use of azole-based antifungal agents may also prove to be useful (782).

Aspergillosis is an uncommon fungal infection that has been reported in HIV-positive and AIDS patients (783,784). *Aspergillus flavus*, the most common causative organism, is endemic in the atmosphere and has a predilection for colonizing sinonasal and respiratory systems in immunocompromised individuals (785). The organism has been cultured from the maxillary sinuses in an immunocompromised patient (786). Once established, the hyphal forms of the organism produce exotoxins that destroy epithelial tissues. In addition, *A. flavus* preferentially invades vascular tissues, resulting in thrombosis and vascular necrosis of the surrounding tissues. Amphotericin B and itraconazole are both effective in treating this infection, especially when instituted early in the course of infection (787).

Neurologic Disturbances

Facial palsy, trigeminal neuralgia, and facial paralysis have all been described in HIV infection (788–792). Although peripheral neuropathy is common in HIV-infected and AIDS patients, cranial nerve involvement is infrequent. The fifth and seventh nerves are the most commonly affected; however, it is unclear whether the clinical signs and symptoms result from a direct effect of HIV or an indirect mechanism, for example, vascular spasm (793).

Recurrent Aphthous Stomatitis

Recurrent aphthous stomatitis (RAS), or canker sores, are not unique to HIV infection, as they afflict a substantial percentage of the otherwise healthy population in the United States (794,795). Although many recurrent oral ulcerations represent viral infections of the oral mucosa, oral ulcers seen in HIV-positive individuals often represent nothing more than aphthous ulcers. The clinical appearance, location, lack of other etiologic factors, and response to therapy indicate that RAS occurs with some frequency in HIV-positive individuals, although herpetiform and major aphthae are more common (796–798) (Fig. 5.69). An increasing fre-

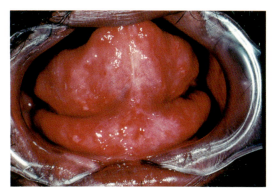

Figure 5.69. Minor aphthous stomatitis. Minor aphthous ulcers are a frequent source of morbidity in HIV-positive and AIDS patients.

quency of major aphthous ulcers may indicate a deteriorating immunologic status (799–801) (Fig. 5.70).

Treatment of minor aphthae with topical corticosteroids is very effective (802). The use of systemic corticosteroids with major aphthae should be approached with caution (803). Antimicrobial mouth rinses have proven to be effective in decreasing the bacterial load in the pseudomembrane covering oral aphthae, as has tetracycline oral suspension (804). Thalidomide has been proposed as a therapeutic agent for aphthous ulcers (805–807), as well as colchicine (808) and interferon (809).

Viral Infections

Cytomegalovirus (CMV) is detectable in the saliva of HIV-positive individuals (810,811) and has been reported as the causative agent for oral

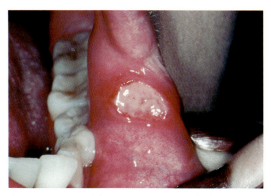

Figure 5.70. Major aphthous stomatitis. Major aphthous ulcers can be a significant source of discomfort for HIV-positive and AIDS patients, often taking weeks to months to heal, despite aggressive therapy.

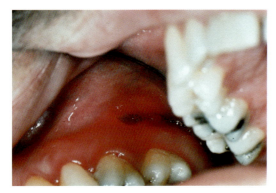

Figure 5.71. Cytomegalovirus. Cytomegalovirus (CMV) ulcers may resemble recurrent aphthous ulcers, with a shallow base covered by pseudomembrane and surrounded by an erythematous halo.

ulcers in some patients (812). Although this aspect of CMV infection is far outshadowed by ocular and other manifestations, it presents a unique oral manifestation of a systemic viral infection in immunocompromised individuals (813). An association has been reported between CMV and necrotizing gingivitis and oral Kaposi's sarcoma (814).

Oral CMV infection normally presents as punched-out, nonindurated ulcers, with variable surrounding erythema (815–822) (Figs. 5.71 and 5.72). Cases of oral CMV infection mimicking HIV-associated periodontal disease (823), as well as some with mandibular involvement (824), have also been reported. The presence of CMV can be confirmed by immunohisto-chemistry utilizing oral smears (825). As a herpesvirus, CMV is responsive to acyclovir; however, interferon, vidarabine, ganciclovir, and foscarnet may also prove to be effective in both

prevention and therapy (826–830). Corticos-teroid therapy increases the risk of CMV infec-tion in HIV-positive individuals and should be avoided, if possible (831).

Molluscum Contagiosum

Molluscum contagiosum is a well-known and well-characterized cutaneous lesion caused by a poxvirus (832–834). Skin involvement has been reported in HIV-positive and AIDS patients (835–843). While lesions of the perioral skin have been described in association with HIV infection (844–847), intraoral lesions are ex-ceedingly rare (848–850). Anti-viral therapy and chemical peeling of infected skin have both been reported as being effective in the treatment of this infection (851,852).

References

1. Laskaris G. Oral manifestations of infectious diseases. Dent Clin North Am 1996;40:395–423.
2. Spruance S, Overall J Jr, Kern E, et al. The natural history of recurrent herpes simplex labialis. N Engl J Med 1977;297:69–74.
3. Spruance SL. The natural history of recurrent oral-facial herpes simplex virus infection. Semin Dermatol 1992;11:200–206.
4. Scully C, Bagg J. Viral infections in dentistry. Curr Opin Dent 1992;2:102–115.
5. Waggoner-Fountain LA, Grossman LB. Herpes simplex virus. Pediatr Rev 2004;25:86–93.
6. Miller CS, Jacob RJ, Hiser DG. UV-enhanced replica-tion of herpes simplex virus in DNA-repair competent and deficient fibroblasts. Oral Surg Oral Med Oral Pathol 1993;75:602–609.
7. Hebert A, Berg J. Oral mucous membrane diseases of childhood. Semin Dermatol 1992;11:80–87.
8. Scully C. Ulcerative stomatitis, gingivitis, and skin lesions. An unusual case of primary herpes simplex infection. Oral Surg Oral Med Oral Pathol 1985;59: 261–263.
9. Fenton SJ, Unkel JH. Viral infections of the oral mucosa in children: a clinical review. Pract Periodon-tics Aesthet Dent 1997;9:683–690.
10. Axell T, Liedholm R. Occurrence of recurrent herpes labialis in an adult Swedish population. Acta Odontol Scand 1990;48:119–123.
11. Griffin JW. Recurrent intraoral herpes simplex virus infection. Oral Surg 1965;19:209–213.
12. Sheridan PJ, Hermann EC. Intraoral lesions of adults associated with herpes simplex virus. Oral Surg 1971;32:390–397.
13. Oranje AP, Folkers E. The Tzanck smear: old, but still of inestimable value. Pediatr Dermatol 1988;5:127–129.
14. Burns JC. Diagnostic methods for herpes simplex infection: a review. Oral Surg 1980;50:346–349.

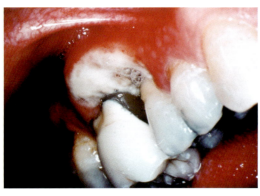

Figure 5.72. Cytomegalovirus. Cytomegalovirus ulcers of the gingiva may be associated with significant periodontal destruction.

15. Kimura H, Shibata M, Kuzushima K, Nishikawa K, Nishiyama Y, Morishima T. Detection and direct typing of herpes simplex virus by polymerase chain reaction. Med Microbiol Immunol 1990;179:177–184.

16. Silverman S Jr, Beumer J. Primary herpetic gingivostomatitis of adult onset. Oral Surg 1973;36:496–503.

17. Stoopler ET, Pinto A, DeRossi SS, et al. Herpes simplex and varicella-zoster infections: clinical and laboratory diagnosis. Gen Dent 2003;51:281–286.

18. Weathers DR, Griffin JW. Intraoral ulcerations of recurrent herpes simplex and recurrent aphthae: two distinct clinical entities. J Am Dent Assoc 1970;81:81–87.

19. Sciubba JJ. Herpes simplex and aphthous ulcerations: presentation, diagnosis and management-an update. Gen Dent 2003;51:510–516.

20. Hirsch M, Schooley R. Treatment of herpesvirus infections. N Engl J Med 1983;309:963–970.

21. Villarreal EC. Current and potential therapies for the treatment of herpes-virus infections. Prog Drug Res 1996;60:263–307.

22. De Clercq E. Antiviral drugs in current clinical use. J Clin Virol 2004;30:115–113.

23. Griffiths PD. Tomorrow's challenges for herpesvirus management: potential applications for valacyclovir. J Infect Dis 2002;186(suppl 1):S131–S137.

24. Wu JJ, Brentjens MH, Torres G, et al. Valacyclovir in the treatment of herpes simplex, herpes zoster, and other viral infections. J Cutan Med Surg 2003;7:372–381.

25. Raborn GW, Grace MG. Recurrent herpes simplex labialis: selected therapeutic options. J Can Dent Assoc 2003;69:498–503.

26. Huber MA. Herpes simplex type-1 virus infection. Quintessence Int 2003;34:453–467.

27. Brady RC, Bernstein DI. Treatment of herpes simplex virus infections. Antiviral Res 2004;61:73–81.

28. Kleymann G. Novel agents and strategies to treat herpes simplex virus infections. Expert Opin Invest Drugs 2003;12:165–183.

29. Habbema L, DeBoulle K, Roders GA, Katz DH. N-Docosanol 10% cream in the treatment of recurrent herpes labialis: a randomized, double-blind, placebo-controlled study. Acta Derm Venerecol 1996;76:479–481.

30. Sasadeusz JJ, Sacks SL. Systemic antivirals in herpesvirus infections. Dermatol Clin 1993;11:171–185.

31. Jensen LA, Hoehns JD, Squires CL. Oral antivirals for the acute treatment of recurrent herpes labialis. Ann Pharmacother 2004;38:705–709.

32. Malmstrom J, Ruokonen H, Konttinen YT, et al. Herpes simplex virus antigens and inflammatory cells in oral lesions in recurrent erythema multiforme. Immunoperoxidase and autoradiographic studies. Acta Derm Venereol 1990;70:405–410.

33. Chilukuri S, Rosen T. Management of acyclovir-resistant herpes simplex virus. Dermatol Clin 2003;21:311–320.

34. Spruance S, Schnipper L, Overall J Jr, et al. Treatment of herpes simplex labialis with topical acyclovir in polyethylene glycol. J Infect Dis 1982;146:85–90.

35. Fiddian A, Ivanyi L. Topical acyclovir in the management of recurrent herpes labialis. Br J Dermatol 1983;109:321–326.

36. Perna JJ, Eskinazi DP. Treatment of oro-facial herpes simplex infections with acyclovir: a review. Oral Surg Oral Med Oral Pathol 1988;65:689–692.

37. Spruance S, Stewart JC, Rowe NH, McKeough MB, Wenerstrom G, Freeman DJ. Treatment of recurrent herpes simplex labialis with oral acyclovir. J Infect Dis 1990;161:185–190.

38. Lavelle CL. Acyclovir: is it an effective virostatic agent for orofacial infections? J Oral Pathol Med 1993;22:391–401.

39. Spruance SL, Rea TL, Thoming C, Tucker R, Saltzman R, Boon R. Penciclovir cream for the treatment of herpes simplex labialis. A randomized, multicenter, double-blind, placebo-controlled trial. Topical Penciclovir Collaborative Study Group. JAMA 1997;277(17):1374–1379.

40. Boon R, Goodman JJ, Martinez J, Marks GL, Gamble M, Welch C. Penciclovir cream for the treatment of sunlight-induced herpes simplex labialis: a randomized, double-blind, placebo-controlled trial. Penciclovir Cream Herpes Labialis Study Group. Clin Ther 2000;22:76–90.

41. Epstein JB, Scully C. Herpes simplex virus in immunocompromised patients: growing evidence of drug resistance. Oral Surg Oral Med Oral Pathol 1991;72:47–50.

42. Bacon TH, Levin MJ, Leary JJ, et al. Herpes simplex virus resistance to acyclovir and penciclovir after two decades of antiviral therapy. Clin Microbiol Rev 2003;16:114–128.

43. Morfin F, Thouvenot D. Herpes simplex virus resistance to antiviral drugs. J Clin Virol 2003;26:29–37.

44. Wachsman M, Aurelian L, Burnett JW. The prophylactic use of cyclooxygenase inhibitors in recurrent herpes simplex infections. Br J Dermatol 1990;123:375–380.

45. Park JB, Park NH. Effect of chlorhexidine on the in vitro and in vivo herpes simplex virus infection. Oral Surg Oral Med Oral Pathol 1989;67:149–153.

46. Spruance S, Stewart J, Freeman D. Early application of topical 15% idoxuridine in dimethyl sulfoxide shortens the course of herpes simplex labialis: a multicenter placebo-controlled trial. J Infect Dis 1990;161:191–197.

47. Rowe NH, Brooks SL, Young SK, et al. A clinical trial of topically applied 3 per cent vidarabine against recurrent herpes labialis. Oral Surg 1979;47:142–147.

48. Kleymann G. New antiviral drugs that target herpesvirus helicase primase enzymes. Herpes 2003;10:46–52.

49. Rooney J, Bryson Y, Mannia O, et al. Prevention of ultraviolet light-induced herpes labialis by sunscreen. Lancet 1991;338:1419–1422.

50. Miller CS, Redding SW. Diagnosis and management of orofacial herpes simplex virus infections. Dent Clin North Am 1992;36:879–895.

51. Birek C, Patterson B, Maximiw WC, Minden MD. EBV and HSV infections in a patient who had undergone bone marrow transplantation: oral manifestations and diagnosis by in situ nucleic acid hybridization. Oral Surg Oral Med Oral Pathol 1989;68:612–617.

52. Barrett AP. Chronic indolent orofacial herpes simplex virus infection in chronic leukemia: a report of three cases. Oral Surg Oral Med Oral Pathol 1988;66:387–390.

53. Epstein JB, Sherlock C, Page JL, Spinelli J, Phillips G. Clinical study of herpes simplex virus infection in leukemia. Oral Surg Oral Med Oral Pathol 1990;70:38–43.

54. Wingard JR. Oral complications of cancer therapies. Infectious and noninfectious systemic consequences. NCI Monogr 1990;9:21–26.

55. Redding SW. Role of herpes simplex virus reactivation in chemotherapy-induced oral mucositis. NCI Monogr 1990;9:103–105.

56. Fleming P. Dental management of the pediatric oncology patient. Curr Opin Dent 1991;1:577–582.

57. Redding SW, Luce EB, Boren MW. Oral herpes simplex virus infection in patients receiving head and neck radiation. Oral Surg Oral Med Oral Pathol 1990;69: 578–580.

58. Heimdahl A, Mattsson T, Dahllof G, Lonnquist B, Ringden O. The oral cavity as a port of entry for early infections in patients treated with bone marrow transplantation. Oral Surg Oral Med Oral Pathol 1989;68: 711–716.

59. Woo SB, Sonis ST, Sonis AL. The role of herpes simplex virus in the development of oral mucositis in bone marrow transplant recipients. Cancer 1990;66:2375–2379.

60. Schubert MM, Peterson DE, Flournoy N, Meyers JC, Truelove EL. Oral and pharyngeal herpes simplex virus infection after allogeneic bone marrow transplantation: analysis of factors associated with infection. Oral Surg Oral Med Oral Pathol 1990;70:286–293.

61. Campton CM. Oral care for the renal transplant patient. ANNA J 1991;18:39–41.

62. Arnow PM. Infections following orthotopic liver transplantation. HPB Surg 1991;3:221–232.

63. Saral R. Oral complications of cancer therapies. Management of acute viral infections. NCI Monogr 1990;9:107–110.

64. Rowe NH, Heine CS, Kowalski CJ. Herpetic whitlow: an occupational disease of practicing dentists. J Am Dent Assoc 1982;105:471–473.

65. Walker LG, Simmons BP, Lovallo JL. Pediatric herpetic hand infections. J Hand Surg 1990;15:176–180.

66. Gunbay T, Gunbay S, Kandemir S. Herpetic whitlow. Quintessence Int 1993;24:363–364.

67. Lewis MA. Herpes simplex virus: an occupational hazard in dentistry. Int Dent J 2004;54:103–111.

68. Schwandt NW, Mjos DP, Lubow RM. Acyclovir and the treatment of herpetic whitlow. Oral Surg Oral Med Oral Pathol 1987;64:255–258.

69. Wander AH. Herpes simplex and recurrent corneal disease. Int Ophthalmol Clin 1984;24:27–38.

70. Mader TH, Stulting RD. Viral keratitis. Infect Dis Clin North Am 1992;6:831–849.

71. Nahmias AJ, Roizman B. Infection with herpes simplex viruses 1 and 2. N Engl J Med 1973;289:781–289.

72. Weller T. Varicella and herpes zoster. N Engl J Med 1983;309:1362–1368.

73. Badger GR. Oral signs of chickenpox (varicella): report of two cases. J Dent Child 1980;47:349–531.

74. Rogers RS 3rd, Tindall JP. Herpes zoster in children. Arch Derm 1972;106:204–207.

75. Beck JD, Watkins C. Epidemiology of nondental oral disease in the elderly. Clin Geriatr Med 1992;8: 461–482.

76. McKenzie CD, Gobetti JP. Diagnosis and treatment of orofacial herpes zoster: report of cases. J Am Dent Assoc 1990;120:679–681.

77. Nally FF, Ross IH. Herpes zoster of the oral and facial structures. Report of five cases and discussion. Oral Surg 1971;32:221–234.

78. Johnson RW, Whitton TL. Management of herpes zoster (shingles) and postherpetic neuralgia. Expert Opin Pharmacother 2004;5:551–559.

79. Johnson RW. Herpes zoster in the immunocompetent patient: management of post-herpetic neuralgia. Herpes 2003;10:38–45.

80. Singh D, Kennedy DH. The use of gabapentin for the treatment of postherpetic neuralgia. Clin Ther 2003; 25:852–889.

81. Muto T, Tsuchiya H, Sato K, Kanazawa M. Tooth exfoliation and necrosis of the mandible—a rare complication following trigeminal herpes zoster: report of a case. J Oral Maxillofac Surg 1990;48:1000–1003.

82. Lin JR, Huang CC. Oral complications following a herpes zoster infection of trigeminal nerve. Chang Keng I Hsueh 1993;16:75–80.

83. Serota F, Starr S, Bryan C, et al. Acyclovir treatment of herpes zoster infections. JAMA 1982;247:2132–2135.

84. Nicholson KG. Antiviral therapy. Varicella-zoster virus infections, herpes labialis and mucocutaneous herpes, and cytomegalovirus infections. Lancet 1984; 2:677–81.

85. Huff J, Bean B, Balfour H, et al. Therapy of herpes zoster with oral acyclovir. Am J Med 1988;85:84–89.

86. Morton P, Thompson A. Oral acyclovir in the treatment of herpes zoster in general practice. N Z Med J 1989;102:93–95.

87. Harding SP, Porter SM. Oral acyclovir in herpes zoster ophthalmicus. Curr Eye Res 1991;10(suppl):177–182.

88. Wallace MR, Bowler WA, Oldfield 3rd EC. Treatment of varicella in the immunocompetent adult. J Med Virol Suppl 1993;1:90–92.

89. Suga S, Yoshikawa T, Ozaki T, Asano Y. Effect of oral acyclovir against primary and secondary viraemia in incubation period of varicella. Arch Dis Child 1993;69:639–642.

90. Wagstaff AJ, Faulds D, Goa KL. Aciclovir. A reappraisal of its antiviral activity, pharmacokinetic properties and therapeutic efficacy. Drugs 1994;47:153–205.

91. Whitley R, Soong S, Dolin R, et al. Early vidarabine therapy to control the complications of herpes zoster in immunosuppressed patients. N Engl J Med 1982; 307:971–975.

92. De Clercq E. Antivirals for the treatment of herpesvirus infections. J Antimicrob Chemother 1993; 32(suppl A):121–132.

93. Porter SR. Infection control in dentistry. Curr Opin Dent 1991;1:429–435.

94. Brentjens MH, Yeung-Yue KA, Lee PC, et al. Vaccines for viral diseases with dermatologic manifestations. Dermatol Clin 2003;21:349–369.

95. Straus SE, Cohen JI, Tosato G, Meier J. NIH conference. Epstein-Barr virus infections: biology, pathogenesis, and management. Ann Intern Med 1993;118:45–58.

96. Mao EJ, Smith CJ. Detection of Epstein-Barr virus (EBV) DNA by the polymerase chain reaction (PCR) in oral smears from health individuals and patients with squamous cell carcinoma. J Oral Pathol Med 1993;22:12–17.

97. Cruchley AT, Williams DM, Niedobitek G, et al. Epstein-Barr virus: biology and disease. Oral Dis 1997;3(suppl 1):S156–S163.

98. Solomides CC, Miller AS, Christman RA, et al. Lymphomas of the oral cavity: histology, immunologic type, and incidence of Epstein-Barr virus infection. Hum Pathol 2002;33:153–157.

99. Ikediobi NI, Tyring SK. Cutaneous manifestations of Epstein-Barr infection. Dermatol Clin 2002;20:283–289.

100. Spano JP, Busson P, Atlan D, et al. Nasopharyngeal carcinomas: an update. Eur J Cancer 2003;39:2121–2135.

101. Ambinder RF. Epstein-Barr virus-associated lymphoproliferative disorders. Rev Clin Exp Hematol 2003; 7:362–374.

102. Shimakage M, Horii K, Tempaku A, et al. Association of Epstein-Barr virus with oral cancers. Hum Pathol 2002;33:608–614.

103. Macsween KF, Crawford DH. Epstein-Barr virus-recent advances. Lancet Infect Dis 2003;3:131–140.

104. Lambore S, McSherry J, Kraus AS. Acute and chronic symptoms of mononucleosis. J Fam Pract 1991;33:33–37.

105. Maddern BR, Werkhaven J, Wessel HB, Yunis E. Infectious mononucleosis with airway obstruction and multiple cranial nerve paresis. Otolaryngol Head Neck Surg 1991;104:529–532.

106. Westmore GA. Cervical abscess: a life-threatening complication of infectious mononucleosis. J Laryngol Otol 1990;104:358–359.

107. Johnson PA, Avery C. Infectious mononucleosis presenting as a parotid mass with associated facial nerve palsy. Int J Oral Maxillofac Surg 1991;20:193–195.

108. Har-El G, Josephson JS. Infectious mononucleosis complicated by lingual tonsillitis. J Laryngol Otol 1990;104:651–653.

109. Magrath IT. African Burkitt's lymphoma. History, biology, clinical features, and treatment. Am J Pediatr Hematol Oncol 1991;13:222–246.

110. Jacobs P. The malignant lymphomas in Africa. Hematol Oncol Clin North Am 1991;5:953–982.

111. Walter PR, Klotz F, Alfy-Gattas T, Minko-Mi-Etoua D, Nguembi-Mbina C. Malignant lymphomas in Gabon (equatorial Africa): a morphologic study of 72 cases. Hum Pathol 1991;22:1040–1043.

112. Okpala IE, Akang EE, Okpala UJ. Lymphomas in University College Hospital, Ibadan, Nigeria. Cancer 1991;68:1356–1360.

113. Syrjanen S, Kallio P, Sainio P, Fuju C, Syrjanen K. Epstein-Barr virus (EBV) genomes and c-myc oncogene in oral Burkitt's lymphomas. Scand J Dent Res 1992;100:176–180.

114. Joske D, Knecht H. Epstein-Barr virus in lymphomas: a review. Blood Rev 1993;7:215–222.

115. Shih LY, Liang DC. Non-Hodgkin's lymphomas in Asia. Hematol Oncol Clin North Am 1991;5:983–1001.

116. Prevot S, Hamilton-Dutoit S, Audouin J, Walter P, Pallesen G, Diebold J. Analysis of African Burkitt's and high-grade B cell non-Burkitt's lymphoma for Epstein-Barr virus genomes using in situ hybridization. Br J Haematol 1992;80:27–32.

117. Soderholdm AL, Lindqvist C, Heikinheimo K, Forssell K, Happonen RP. Non-Hodgkin's lymphomas presenting through oral symptoms. Int J Oral Maxillofac Surg 1990;19:131–134.

118. Anavi Y, Kaplinsky C, Calderon S, Zaizov R. Head, neck, and maxillofacial childhood Burkitt's lymphoma: a retrospective analysis of 31 patients. J Oral Maxillofac Surg 1990;48:708–713.

119. Svoboda WE, Aaron GR, Albano EA. North American Burkitt's lymphoma presenting with intraoral symptoms. Pediatr Dent 1991;13:52–58.

120. Stiller CA, Parkin DM. International variations in the incidence of childhood lymphomas. Paediatr Perinat Epidemiol 1990;4:303–324.

121. Ahmed M, Khan AH, Mansoor A, Khan MA, Saeed S. Burkitt's lymphoma–a study of 50 consecutive cases. J Pak Med Assoc 1993;43:151–153.

122. Loh LE, Chee TS, John AB. The anatomy of the fossa of Rosenmuller—its possible influence on the detection of occult nasopharyngeal carcinoma. Singapore Med J 1991;32:154–155.

123. Cvitkovic E, Bachouchi M, Armand JP. Nasopharyngeal carcinoma. Biology, natural history, and therapeutic implications. Hematol Oncol Clin North Am 1991;5:821–838.

124. Niedobitek G, Herbst H, Young LS. Epstein-Barr virus and carcinomas. Int J Clin Lab Res 1993;23:17–24.

125. Dickens P, Srivastava G, Loke SL, Chan CW, Liu YT. Epstein-Barr virus DNA in nasopharyngeal carcinomas from Chinese patients in Hong Kong. J Clin Pathol 1992;45:396–397.

126. Neel HB 3rd. Nasopharyngeal carcinoma: diagnosis, staging, and management. Oncology 1992;6:87–95.

127. Van Hasselt CA, Skinner DW. Nasopharyngeal carcinoma. An analysis of 100 Chinese patients. S Afr J Surg 1990;28:92–94.

128. Sham JS, Cheung YK, Choy D, Chan FL, Leong L. Cranial nerve involvement and base of the skull erosion in nasopharyngeal carcinoma. Cancer 1991; 68:422–426.

129. Bailey RE. Diagnosis and treatment of infectious mononucleosis. Am Fam Physician 1994;49:879–888.

130. Stein JE, Schwenn MR, Jacir NN, Harris BH. Surgical restraint in Burkitt's lymphoma in children. J Pediatr Surg 1991;26:1273–1275.

131. McGuire LJ, Lee JC. The histopathologic diagnosis of nasopharyngeal carcinoma. Ear Nose Throat J 1990;69:229–236.

132. Hawkins EP, Krischer JP, Smith BE, Hawkins HK, Finegold MJ. Nasopharyngeal carcinoma in children—a retrospective review and demonstration of Epstein-Barr viral genomes in tumor cell cytoplasm: a report of the Pediatric Oncology Group. Hum Pathol 1990;21: 805–810.

133. Akao I, Sato Y, Mukai K, et al. Detection of Epstein-Barr virus DNA in formalin-fixed paraffin-embedded tissue of nasopharyngeal carcinoma using polymerase chain reaction and in situ hybridization. Laryngoscope 1991;101:279–283.

134. Littler E, Baylis SA, Zeng Y, Conway MJ, Mackett M, Arrand JR. Diagnosis of nasopharyngeal carcinoma by means of recombinant Epstein-Barr virus proteins. Lancet 1991;337:685–689.

135. Ohshima K, Kikuchi M, Masuda Y, et al. Epstein-Barr viral genomes in carcinoma metastasis to lymph nodes. Association with nasopharyngeal carcinoma. Acta Pathol Jpn 1991;41:437–443.

136. Pearson GR. Epstein-Barr virus and nasopharyngeal carcinoma. J Cell Biochem Suppl 1993;17F:150–154.

137. Andersson JP. Clinical aspects on Epstein-Barr virus infection. Scand J Infect Dis Suppl 1991;80:94–104.

138. Chetham MM, Roberts KB. Infectious mononucleosis in adolescents. Pediatr Ann 1991;20:206–213.

139. Axelrod P, Finestone AJ. Infectious mononucleosis in older adults. Am Fam Physician 1990;42:1599–1606.

140. Fark AR. Infectious mononucleosis, Epstein-Barr virus, and chronic fatigue syndrome: a prospective case series. J Fam Pract 1991;32:202–209.

141. Daley TD, Lovas LG. Diseases of the salivary glands: a review. J Can Dent Assoc 1991;57:411–414.

142. Lucin P, Pavic I, Polic B, Jonjic S, Koszinowski UH. Gamma interferon-dependent clearance of cytomegalovirus infection in salivary glands. J Virol 1992; 66:1977–1984.

143. Marx JL. Cytomegalovirus: a major cause of birth defects. Science 1975;190:1184–1186.

144. Glick M, Goldman HS. Viral infections in the dental setting: potential effects on pregnant HCWs. J Am Dent Assoc 1993;124:79–86.

145. Landolfo S, Gariglio M, Gribaudo G, et al. The human cytomegalovirus. Pharmacol Ther 2003;98:269–297.

146. Slots J. Update on human cytomegalovirus in destructive periodontal disease. Oral Microbiol Immunol 2004;19:217–223.

147. Peterslund NA. Herpesvirus infection: an overview of the clinical manifestations. Scand J Infect Dis Suppl 1991;80:15–20.

148. Wahren B, Linde A. Virological and clinical characteristics of human herpesvirus 6. Scand J Infect Dis Suppl 1991;80:105–109.

149. Oren I, Sobel JD. Human herpesvirus type 6: review. Clin Infect Dis 1992;14:741–746.

150. Okada K, Ueda K, Kusuhara K, et al. Exanthema subitum and human herpesvirus 6 infection: clinical observations in fifty-seven cases. Pediatr Infect Dis J 1993;12:204–208.

151. Caserta MT, Hall CB. Human herpesvirus-6. Annu Rev Med 1993;44:377–383.

152. Asano Y, Yoshikawa T, Suga S, et al. Clinical features of infants with primary human herpesvirus 6 infection (exanthem subitum, roseola infantum). Pediatrics 1994;93:104–108.

153. Dewhurst S, Skrincosky D, van Loon N. Human Herpesvirus 6. Expert Rev Mol Med 1997;1:1–17.

154. Clark DA. Human herpesvirus 6. Rev Med Virol 2000;10:155–173.

155. Caserta MT, Mock DJ, Dewhurst S. Human herpesvirus 6. Clin Infect Dis 2001;33:829–833.

156. Dockerell DH. Human herpesvirus 6: molecular biology and clinical features. J Med Microbiol 2003; 52(pt 1):5–18.

157. Abdel-Haq NM, Asmar BI. Human herpesvirus 6 (HHV6) infection. Indian J Pediatr 2004;71:89–96.

158. Rathore MH. Human herpesvirus 6. South Med J 1993;86:1197–1201.

159. Lawrence GL, Chee M, Craxton MA, Gompels UA, Honess RW, Barrell BG. Human herpesvirus 6 is closely related to human cytomegalovirus. J Virol 1990;64:287–299.

160. Fox JD, Briggs M, Ward PA, Tedder RS. Human herpesvirus 6 in salivary glands. Lancet 1990;336: 590–593.

161. Bagg J. Human herpesvirus-6: the latest human herpes virus. J Oral Pathol Med 1991;20:465–468.

162. Saito I, Nishimura S, Kudo I, Fox RI, Moro I. Detection of Epstein-Barr virus and human herpes virus type 6 in saliva from patients with lymphoproliferative diseases by the polymerase chain reaction. Arch Oral Biol 1991;36:779–784.

163. Dewhurst S, Skrincosky D, van Loon N. Human herpesvirus 7. Expert Rev Mol Med 1997;1:1–10.

164. De Araujo T, Berman B, Weinstein A. Human herpesviruses 6 and 7. Dermatol Clin 2002;20:301–306.

165. Cannon M, Cesarman E. Kaposi's sarcoma-associated herpes virus and acquired immunodeficiency syndrome-related malignancy. Semin Oncol 2000;27: 409–419.

166. Cathomas G. Human herpes virus 8: a new virus discloses its face. Virchows Arch 2000;436:195–206.

167. Malnati MS, Dagna L, Ponzoni M, et al. Human Herpesvirus 8 (HHV-8/KSHV) and hematologic malignancies. Rev Clin Exp Hematol 2003;7:374–405.

168. Cioc AM, Allen C, Kalmar JR, et al. Oral plasmablastic lymphomas in AIDS patients are associated with human herpesvirus 8. Am J Surg Pathol 2004;28:41–46.

169. Flaitz CM, Nichols CM, Walling DM, et al. Plasmablastic lymphoma: an HIV-associated entity with primary oral manifestations. Oral Oncol 2002;38:96–102.

170. Eversole LR. Papillary lesions of the oral cavity: relationship to human papillomaviruses. J Calif Dent Assoc 2000;28:922–927.

171. Howley PM. The human papillomaviruses. Arch Pathol Lab Med 1982;106:429–432.

172. Scully C, Cox MF, Prime SS, Maitland NJ. Papillomaviruses: the current status in relation to oral disease. Oral Surg Oral Med Oral Pathol 1988;65: 526–532.

173. Lutzner M, Kuffer R, Blanchet-Bardon C, Blanchet-Bardon C, Croissant O. Different papillomaviruses as causes of oral warts. Arch Dermatol 1982;188:393–397.

174. de Villiers EM, Hirsch-Benham A, von-Knebel-Doeberitz C, Neumann C, zur Hausen H. Two newly identified human papillomavirus types (HPV 40 and 57) isolated from oral mucosal lesions. Virology 1989;171:248–253.

175. von Knebel Doeberitz M. Papillomaviruses in human disease: Part I. Pathogenesis and epidemiology of human papillomavirus infections. Eur J Med 1992;1: 415–423.

176. Syrjanen S, Puranen M. Human papillomavirus in fections in children: the potential role of maternal transmission. Crit Rev Oral Biol Med 2000;11:259–274.

177. Steinberg BM, Auborn KJ. Papillomaviruses in head and neck disease: pathophysiology and possible regulation. J Cell Biochem Suppl 1993;17F:155–164.

178. Jalal H, Sanders CM, Prime SS, Scully C, Maitland NJ. Detection of human papilloma virus type 16 DNA in oral squames from normal young adults. J Oral Pathol Med 1992;21:465–470.

179. Kellokoski JK, Syrjanen SM, Chang F, Yliskoski M, Syrjanen KJ. Southern blot hybridization and PCR in detection of oral human papillomavirus (HPV) infections in women with genital HPV infections. J Oral Pathol Med 1992;21:459–464.

180. Lawton G, Thomas S, Schonrock J, Monsour F, Frazer I. Human papillomaviruses in normal oral mucosa: a comparison of methods for sample collection. J Oral Pathol Med 1992;21:265–269.

181. Garlick JA, Taichman LB. Human papillomavirus infection of the oral mucosa. Am J Dermatopathol 1991;13:386–395.

182. Scully C, Epstein J, Porter S, Cox M. Viruses and chronic disorders involving the human oral mucosa. Oral Surg Oral Med Oral Pathol 1991;72:537–544.

183. Premoli-de-Percoco G, Christensen R. Human papillomavirus in oral verrucal-papillary lesions. A comparative histological, clinical and immunohistochemical study. Pathologica 1992;84:383–392.

184. Green TL, Eversole LR, Leider AS. Oral and labial verruca vulgaris: clinical, histological, and immunohistochemical evaluation. Oral Surg Oral Med Oral Pathol 1986;62:410–416.

185. Broich G, Sasaki T. Electron microscopic demonstration of HPV in oral warts. Microbiologica 1990;13:27–34.

186. Eversole LR, Laipis PJ, Green TL. Human papillomavirus type 2 DNA in oral and labial verruca vulgaris. J Cutan Pathol 1987;14:319–325.

187. Eversole LR, Laipis PJ. Oral squamous papillomas: detection of HPV DNA by *in situ* hybridization. Oral Surg Oral Med Oral Pathol 1988;65:545–550.

188. Greer RO Jr, Douglas JM Jr, Breese P, Crosby KL. Evaluation of oral and laryngeal specimens for human papillomavirus (HPV) DNA by dot blot hybridization. J Oral Pathol Med 1990;19:35–38.

189. Zeuss MS, Miller CS, White DK. *In situ* hybridization analysis of human papillomavirus DNA in oral mucosal lesions. Oral Surg Oral Med Oral Pathol 1991;71:714–720.

190. Miller CS, Zeuss MS, White DK. *In situ* detection of HPV DNA in oral mucosal lesions. A comparison of two hybridization kits. J Oral Pathol Med 1991;20:403–408.

191. Miller CS, White DK, Royse DD. *In situ* hybridization analysis of human papillomavirus in orofacial lesions using a consensus biotinylated probe. Am J Dermatopathol 1993;15:256–259.

192. Rodu B. New approaches to the diagnosis of oral soft-tissue disease of viral origin. Adv Dent Res 1993;7:207–212.

193. Premoli-de-Percoco G, Galindo I, Ramirez JL, Perrone M, Rivera H. Detection of human papillomavirus-related oral verruca vulgaris among Venezuelans. J Oral Pathol Med 1993;22:113–116.

194. Welch TB, Barker BF, Williams C. Peroxidase-antiperoxidase evaluation of human oral squamous cell papillomas. Oral Surg Oral Med Oral Pathol 1986;61:603–606.

195. Kellokoski J, Syrjanen S, Syrjanen K, Yliskoski M. Oral mucosal changes in women with genital HPV infection. J Oral Pathol Med 1990;19:142–148.

196. Summers L, Booth DR. Intraoral condyloma acuminatum. Oral Surg 1974;38:273–278.

197. Emmanouil DE, Post AC. Oral condyloma acuminatum in a child: case report. Pediatr Dent 1987;9:232–235.

198. Panici PB, Scambia G, Perrone L, et al. Oral condyloma lesions in patients with extensive genital human papillomavirus infection. Am J Obstet Gynecol 1992;167:451–458.

199. Kui LL, Xiu HZ, Ning LY. Condyloma acuminatum and human papillomavirus infection in the oral mucosa of children. Pediatr Dent 2003;25:149–153.

200. Knapp MJ, Uohara GI. Oral condyloma acuminatum. Oral Surg 1967;23:538–545.

201. Doyle JL, Grodjesk JE, Manhold JH Jr. Condyloma acuminatum occurring in the oral cavity. Oral Surg 1968;26:434–440.

202. Swan RH, McDaniel RK, Dreiman BB, Rome WC. Condyloma acuminatum involving the oral mucosa. Oral Surg 1981;51:503–508.

203. Marquard JV, Racey GL. Combined medical and surgical management of intraoral condyloma acuminatum. J Oral Surg 1981;39:459–461.

204. Eron L, Judson F, Tucker S, et al. Interferon therapy for condyloma acuminata. N Engl J Med 1986;315:1059–1064.

205. Browder JF, Araujo OE, Myer NA, Flowers FP. The interferons and their use in condyloma acuminata. Ann Pharmacother 1992;26:42–45.

206. Archard HO, Heck JW, Stanley HR. Focal epithelial hyperplasia: an unusual and mucosal lesion found in Indian children. Oral Surg Oral Med Oral Pathol 1965;20:201–212.

207. Lamey PJ, Lewis MA, Rennie JS, Beattie AD. Heck's disease. Br Dent J 1990;168:251–252.

208. Ficarra G, Adler-Storthz K, Galeotti F, Shillitoe E. Focal epithelial hyperplasia (Heck's disease): the first reported case from Italy. Tumori 1991;77:83–85.

209. Premoli-de-Percoco G, Cisternas JP, Ramirez JL, Galindo I. Focal epithelial hyperplasia: human-papillomavirus-induced disease with a genetic predisposition in a Venezuelan family. Hum Genet 1993;91:386–388.

210. Cohen PR, Hebert AA, Adler-Storthz K. Focal epithelial hyperplasia: Heck disease. Pediatr Dermatol 1993;10:245–251.

211. Jaramillo F, Rodriquez G. Multiple oral papules in a native South American girl. Focal epithelial hyperplasia (Heck's disease). Arch Dermatol 1991;127:888–892.

212. Morrow DJ, Sandhu HS, Daley TD. Focal epithelial hyperplasia (Heck's disease) with generalized lesions of the gingiva. A case report. J Periodontol 1993;64:63–65.

213. Harris AM, van Wyk CW. Heck's disease (focal epithelial hyperplasia): a longitudinal study. Community Dent Oral Epidemiol 1993;21:82–85.

214. Henke RP, Guerin-Reverchon I, Milde-Langosch K, Koppang HS, Loning T. *In situ* detection of human papillomavirus types 13 and 32 in focal epithelial hyperplasia of the oral mucosa. J Oral Pathol Med 1989;18:419–421.

215. Garlick JA, Calderon S, Buchner A, Mitrani-Rosenbaum S. Detection of human papillomavirus (HPV) DNA in focal epithelial hyperplasia. J Oral Pathol Med 1989;18:172–177.

216. Padayachee A, van Wyk CW. Human papillomavirus (HPV) DNA in focal epithelial hyperplasia by *in situ* hybridization. J Oral Pathol Med 1991;20:210–214.

217. Premoli-de-Percoco G, Galindo I, Ramirez JL. *In situ* hybridization with digoxigenin-labelled DNA probes for the detection of human papillomavirus-induced focal epithelial hyperplasia among Venezuelans. Virchows Arch A Pathol Anat Histopathol 1992;420:295–300.

218. Luomanen M. Oral focal epithelial hyperplasia removed with CO_2 laser. Int J Oral Maxillofac Surg 1990;19:205–207.

219. Obalek S, Janniger C, Jablonska S, Favre M, Orth G. Sporadic cases of Heck disease in two Polish girls: association with human papillomavirus type 13. Pediatr Dermatol 1993;10:240–244.

220. Scully C, Prime S, Maitland N. Papillomaviruses: their possible role in oral disease. Oral Surg Oral Med Oral Pathol 1985;60:166–174.

221. Evans AS, Mueller NE. Viruses and cancer. Causal associations. Ann Epidemiol 1990;1:71–92.

222. Shillitoe EJ. Relationship of viral infection to malignancies. Curr Opin Dent 1991;1:398–403.

223. Chang F, Syrjanen S, Kellokoski J, Syrjanen K. Human papillomavirus (HPV) infections and their associations with oral disease. J Oral Pathol Med 1991;20:305–307.

224. Shroyer KR, Greer RO Jr. Detection of human papillomavirus DNA by *in situ* DNA hybridization and polymerase chain reaction in premalignant and malignant oral lesions. Oral Surg Oral Med Oral Pathol 1991;71:708–713.

225. Shillitoe EJ, Steele C. Inhibition of the transformed phenotype of carcinoma cells that contain human papillomavirus. Ann N Y Acad Sci 1992;660:286–287.

226. Steele C, Sacks PG, Adler-Storthz K, Shillitoe EJ. Effect on cancer cells of plasmids that express antisense RNA of human papillomavirus type 18. Can Res 1992;52:4706–4711.

227. Steele C, Cowsert LM, Shillitoe EJ. Effects of human papillomavirus type 18–specific antisense oligonucleotides on the transformed phenotype of human carcinoma cell lines. Cancer Res 1993;53(suppl 10):2330–2337.

228. Shillitoe EJ, Schantz SP, Spitz MR, Hecht SS. Environmental carcinogenesis and its prevention: the head and neck cancer model. Cancer Res 1993;53:2189–2191.

229. Woods KV, Shillitoe EJ, Spitz MR, Schantz SP, Adler-Storthz K. Analysis of human papillomavirus DNA in oral squamous cell carcinomas. J Oral Pathol Med 1993;22:101–108.

230. Masucci MG. Viral immunology of human tumors. Curr Opin Immunol 1993;5:693–700.

231. Sugerman PB, Shilitoe EJ. The high risk human papillomaviruses and oral cancer: evidence for and against a causal relationship. Oral Dis 1997;3:130–147.

232. Leigh IM, Buchanan JA, Harwood CA, et al. Role of human papillomaviruses in cutaneous and oral manifestations of immunosuppression. J Acquir Immune Defic 1999;21(suppl 1):S49–S57.

233. Hafcamp HC, Manni JJ, Speel EJ. Role of human papillomavirus in the development of head and neck squamous cell carcinomas. Acta Otolaryngol 2004;124:520–526.

234. Syrjanen S. Human papillomavirus infections and oral tumors. Med Microbiol Immunol (Berl) 2003;192:123–128.

235. Herrero R. Human papillomavirus and cancer of the upper aerodigestive tract. J Natl Cancer Inst Monogr 2003;31:47–51.

236. Scully C. Oral squamous cell carcinoma; from an hypothesis about a virus to concern about sexual transmission. Oral Oncol 2002;38:227–234.

237. Watts SL, Brewer EE, Fry TL. Human papillomavirus DNA types in squamous cell carcinomas of the head and neck. Oral Surg Oral Med Oral Pathol 1991;71:701–707.

238. Tsuchiya H, Tomita Y, Shirasawa H, Tanzawa H, Sato K, Simizu B. Detection of human papillomavirus in head and neck tumors with DNA hybridization and immunohistochemical analysis. Oral Surg Oral Med Oral Pathol 1991;71:721–725.

239. Howell RE, Gallant L. Human papillomavirus type 16 in an oral squamous carcinoma and its metastasis. Oral Surg Oral Med Oral Pathol 1992;74:620–626.

240. Cannon CR, Hayne ST. Concurrent verrucous carcinomas of the lip and buccal mucosa. South Med J 1993;86:691–693.

241. Huang ES, Gutsch D, Tzung KW, Lin CT. Detection of low level of human papilloma virus type 16 DNA sequences in cancer cell lines derived from two well-differentiated nasopharyngeal cancers. J Med Virol 1993;40:244–250.

242. Jontell M, Watts S, Wallstrom M, Levin L, Sloberg K. Human papilloma virus in erosive oral lichen planus. J Oral Pathol Med 1990;19:273–277.

243. Young SK, Min KW. *In situ* DNA hybridization analysis of oral papillomas, leukoplakias, and carcinomas for human papillomavirus. Oral Surg Oral Med Oral Pathol 1991;71:726–729.

244. Eversole R. The human papillomaviruses and oral mucosal disease. Oral Surg Oral Med Oral Pathol 1991;71:700.

245. Maden C, Beckmann AM, Thomas DB, et al. Human papillomaviruses, herpes simplex viruses, and the risk of oral cancer in men. Am J Epidemiol 1992;135:1093–1102.

246. McKaig RG, Baric RS, Olshan AF. Human papillomavirus and head and neck cancer: epidemiology and molecular biology. Head Neck 1998;20:250–265.

247. Vasudevan DM, Vijayakumar T. Viruses in human oral cancers. J Exp Clin Cancer Res 1998;17:27–31.

248. Miller CS, Johnstone BM. Human papillomavirus as a risk factor for oral squamous cell carcinoma: a meta-analysis, 1982–1997. Oral Surg Oral Med Oral Pathol Oral Radiol Endod 2001;91:622–635.

249. Grahnen H. Maternal rubella and dental defects. Odontologisk Revy 1958;9:181–192.

250. Perry RT, Halsey NA. The clinical significance of measles: a review. J Infect Dis 2004;189(suppl 1):S4–S16.

251. Cunha BA. Smallpox and measles: historical aspects and clinical differentiation. Infect Dis Clin North Am 2004;18:79–100.

252. Katz J, Guelmann M, Stavropolous F, et al. Gingival and other oral manifestations in measles virus infection. J Clin Periodontol 2003;30:665–668.

253. Perry KR, Brown DW, Parry JV, Panday S, Pipkin C, Richards A. Detection of measles, mumps, and rubella antibodies in saliva using antibody capture radioimmunoassay. J Med Virol 1993;40:235–240.

254. Ito M, Go T, Okuno T, Mikawa H. Chronic mumps virus encephalitis. Pediatr Neurol 1991;7:467–470.

255. Zou ZJ, Wang SL, Zhu JR, Yu SF, Ma DQ, Wu YT. Recurrent parotitis in children. A report of 102 cases. Chin Med J 1990;103:576–582.

256. McQuone SJ. Acute viral and bacterial infections of the salivary glands. Otolaryngol Clin North Am 1999;32:793–811.

257. Waldman HB. Have your young pediatric patients been immuized properly ASDC. J Dent Child 1998;65:107–110.

258. Brook I. Diagnosis and management of parotitis. Arch Otolaryngol Head Neck Surg 1992;118:469–471.

259. Ferson MJ, Bell SM. Outbreak of Coxsackievirus A16 hand, foot, and mouth disease in a child day-care center. Am J Public Health 1991;81:1675–1676.

260. Thomas I, Janniger CK. Hand, foot, and mouth disease. Cutis 1993;52:265–266.

261. Frydenberg A, Starr M. Hand, foot, and mouth disease. Aust Fam Physician 2003;32:594–595.

262. Enterovirus surveillance—United States, 1997–1999. MMWR 2000;49:913–916.

263. Hooi PS, Chua BH, Lee CS, et al. Hand, foot and mouth disease: University of Malaya Medical Center experience. Med J Malaysia 2002;57:88–91.

264. McMinn PC. An overview of the evolution of enterovirus 71 and its clinical and public health significance. FEMS Microbiol Rev 2002;26:91–107.

265. Chan KP, Goh KT, Chong CY, et al. Epidemic hand, foot and mouth disease caused by human enterovirus 71, Singapore. Emerg Infect Dis 2003;9:78–85.

266. Adler JL, Mostow SR, Mellin H, Janney JH, Joseph JM. Epidemiologic investigation of hand, foot, and mouth disease. Am J Dis Child 1970;120:309–313.

267. Chong CY, Chan KP, Shah VA, et al. Hand, foot and mouth disease in Singapore: a comparison of fatal and non-fatal cases. Acta Paediatr 2003;92:1163–1169.

268. Shah VA, Chong CY, Chan KP, et al. Clinical characteristics of an outbreak of hand, foot and mouth disease in Singapore. Ann Acad Med Singapore 2003;32:381–387.

269. McKinney RV. Hand, foot, and mouth disease: a viral disease of importance to dentists. J Am Dent Assoc 1975;91:122–127.

270. Lopez-Sanchez A, Guijarro Guijarro B, Vallejo Hernanez G. Human repercussions of foot and mouth disease and other similar viral diseases. Med Oral 2003;8:26–32.

271. Toida M, Watanabe F, Goto K, et al. Usefulness of low-level laser for control of painful stomatitis in patients with hand-foot-and-mouth disease. J Clin Laser Med Surg 2003;21:363–367.

272. McCourt JW. Hand-foot-and-mouth disease: report of a case in Texas. Tex Dent J 1991;108:13–16.

273. Buchner A. Hand, foot, and mouth disease. Oral Surg 1976;41:333–337.

274. Zahorsky J. Herpetic sore throat. South Med J 1920;13:871–872.

275. Zahorsky J. Herpangina (a specific disease). Arch Pediatr 1924;41:181–184.

276. Robinson CR, Doane FW, Rhodes AJ. Report of an outbreak of febrile illness with pharyngeal lesions and exanthem: Toronto, 1957, isolation of group A Coxsackie virus. Can Med Assoc J 1958;79:615–621.

277. Yamadera S, Yamashita K, Kato N, Akatsuka M, Miyamura K, Yamazaki S. Herpangina surveillance in Japan, 1982–1989. A report of the national epidemiological surveillance of infectious agents in Japan. Jpn J Med Sci Biol 1991;44:29–39.

278. Lang SD, Singh K. The sore throat. When to investigate and when to prescribe. Drugs 1990;40:854–862.

279. Edwards P, Wodak A, Cooper DA, Thompson IL, Penny R. The gastrointestinal manifestations of AIDS. Aust N Z J Med 1990;20:141–148.

280. Royce RA, Luckmann RS, Fusaro RE, Winkelstein W Jr. The natural history of HIV-1 infection: staging classifications of disease. AIDS 1991;5:355–364.

281. Mbopi-Keou FX, Belec L, Teo CG, et al. Synergism between HIV and other viruses in the mouth. Lancet Infect Dis 2002;2:416–424.

282. Greenspan D, Greenspan JS. Oral manifestations of HIV infection. AIDS Clin Care 1997;9:29–33.

283. Grbic JT, Lamster IB. Oral manifestations of HIV infection. AIDS Patient Care STDS 1997;11:18–24.

284. Patton LL, Phelan JA, Ramos-Gomez FJ, et al. Prevalence and classification of HIV-associated oral lesions. Oral Dis 2002;8(suppl 2):98–109.

285. Shirlaw PJ, Chikte U, MacPhail L, et al. Oral and dental care and treatment protocols for the management of HIV-infected patients. Oral Dis 2002;8(suppl 2):136–143.

286. Ramos-Gomez FJ, Flaitz C, Catapano P, et al. Classification, diagnostic criteria, and treatment recommendations for orofacial manifestations in HIV-infected pediatric patients. Collaborative Workgroup on Oral Manifestations of Pediatric HIV Infection. J Clin Pediatr Dent 1999;23:85–96.

287. Ramos-Gomez FJ, Petru A, Hilton JF, et al. Oral manifestations and dental status in paediatric HIV infection. Int J Paediatr Dent 2000;10:3–11.

288. Kozinetz CA, Carter AB, Simon C, et al. Oral manifestations of pediatric vertical HIV infection. AIDS Patient Care STDS 2000;14:89–94.

289. Magalhaes MG, Bueno DF, Serra E, et al. Oral manifestations of HIV positive children. J Clin Pediatr Dent 2001;25:103–106.

290. Classification and diagnostic criteria for oral lesions in HIV infection. EC-Clearinghouse on Oral Problems Related to HIV Infection and WHO Collaborating Centre on Oral Manifestations of the Immunodeficiency Virus. J Oral Pathol Med 1993;22:289–291.

291. Rosenberg RA, Schneider KL, Cohen NL. Head and neck presentations of the acquired immunodeficiency syndrome. Laryngoscope 1984;94:642–646.

292. Wofford DT, Miller RI. Acquired immune deficiency syndrome (AIDS): disease characteristics and oral manifestations. J Am Dent Assoc 1985;111:258–261.

293. Marcusen DC, Sooy CD. Otolaryngologic and head and neck manifestations of acquired immunodeficiency syndrome (AIDS). Laryngoscope 1985;95:401–405.

294. Barr CE, Torosian JP, Quinones-Whitmore GD. Oral manifestations of AIDS: the dentist's responsibility in diagnosis and treatment. Quintessence Int 1986;17:711–717.

295. Schiodt M, Pindborg JJ. AIDS and the oral cavity. Epidemiology and clinical oral manifestations of human immune deficiency virus infection: a review. Int J Oral Maxillofac Surg 1987;16:1–14.

296. Reichart PA, Gelderblom HR, Becker J, Kuntz A. AIDS and the oral cavity. The HIV-infection: virology, etiology, origin, immunology, precautions and clinical observations in 110 patients. Int J Oral Maxillofac Surg 1987;16:129–153.

297. Greenspan D, Greenspan JS. Oral mucosal manifestations of AIDS? Dermatol Clin 1987;5:733–737.

298. Silverman S Jr. AIDS update: oral findings, diagnosis, and precautions. J Am Dent Assoc 1987;115:559–563.

299. Roberts MW, Brahim JS, Rinne NF. Oral manifestations of AIDS: a study of 84 patients. J Am Dent Assoc 1988;116:863–866.

300. Greenspan D, Greenspan JS. The oral clinical features of HIV infection. Gastroenterol Clin North Am 1988;17:535–543.

301. Greenspan JS, Greenspan D, Winkler JR. Diagnosis and management of the oral manifestations of HIV infection and AIDS. Infect Dis Clin North Am 1988;2:373–385.

302. Alessi E, Cusini M, Zerboni R. Mucocutaneous manifestations in patients with human immunodeficiency virus. J Am Acad Dermatol 1990;22:1260–1269.

303. Brahim JS, Roberts MW. Oral manifestations of human immunodeficiency virus infection. Ear Nose Throat J 1990;69:464–474.

304. van der Waal I, Schulten EA, Pindborg JJ. Oral manifestations of AIDS: an overview. Int Dent J 1991;41:3–8.

305. Scully C, Laskaris G, Pindborg J, Porter SR, Reichart P. Oral manifestations of HIV infection and their management. I. More common lesions. Oral Surg Oral Med Oral Pathol 1991;71:158–166.

306. Kelly M, Siegel MA, Balciunas BA, Konzelman JL. Oral manifestations of human immunodeficiency virus infection. Cutis 1991;47:44–49.

307. Silverman S Jr. AIDS update. Oral manifestations and management. Dent Clin North Am 1991;35:259–267.

308. Sciubba JJ. Recognizing the oral manifestations of AIDS. Oncology 1992;6:64–75.

309. Barr CE. Oral diseases in HIV-1 infection. Dysphagia 1992;7:126–137.

310. Greenspan D, Greenspan JS. Oral lesions of HIV infection: features and therapy. AIDS Clin Rev 1992; 225–239.

311. Greenspan JS, Barr CE, Sciubba JJ, Winkler JR. Oral manifestations of HIV infection. Definitions, diagnostic criteria, and principles of therapy. The U.S.A. Oral AIDS Collaborative Group. Oral Surg Oral Med Oral Pathol 1992;73:142–144.

312. Pindborg JJ. Global aspects of the AIDS epidemic. Oral Surg Oral Med Oral Pathol 1992;73:138–141.

313. Ficarra G, Shillitoe EJ. HIV-related infections of the oral cavity. Crit Rev Oral Biol Med 1992;3:207–231.

314. Scully C. Oral infections in the immunocompromised patient. Br Dent J 1992;172:401–407.

315. Greenspan D, Greenspan JS. Oral manifestations of human immunodeficiency virus infection. Dent Clin North Am 1993;37:21–32.

316. Itin PH, Lautenschlager S, Fluckiger R, Rufli T. Oral manifestations in HIV-infected patients: diagnosis and management. J Am Acad Dermatol 1993;29: 749–760.

317. Barzan L, Tavio M, Tirelli U, Comoretto R. Head and neck manifestations during HIV infection. J Laryngol Otol 1993;107:133–136.

318. Oral manifestations of HIV infection. In: JS Greenspan, D Greenspan, eds. Proceedings of the Second International Workshop. Chicago: Quintessence, 1994.

319. Itin PH, Lautenschlager S. Viral lesions of the mouth in HIV-infected patients. Dermatology 1997;194:1–7.

320. Samonis G, Mantadakis E, Maraki S. Orofacial viral infections in the immunocompromised host. Oncol Rep 2000;7:1389–1394.

321. Silverman S Jr, Migliorati CA, Lozada-Nur F, Greenspan D, Conant MA. Oral findings in people with or at high risk for AIDS: a study of 375 homosexual males. J Am Dent Assoc 1986;11:187–192.

322. Feigal DW, Katz MH, Greenspan D, et al. The prevalence of oral lesions in HIV-infected homosexual and bisexual men: three San Francisco epidemiological cohorts. AIDS 1991;5:519–525.

323. Barone R, Ficarra G, Gaglioti D, Orsi A, Mazzotta F. Prevalence of oral lesions among HIV-infected intravenous drug abusers and other risk groups. Oral Surg Oral Med Oral Pathol 1990;69:169–173.

324. Bolski E, Hunt RJ. The prevalence of AIDS-associated oral lesions in a cohort of patients with hemophilia. Oral Surg Oral Med Oral Pathol 1988;65:406–410.

325. Hankins CA, Handley MA. HIV disease and AIDS in women: current knowledge and a research agenda. J Acquir Immune Defic Syndr 1992;5:957–971.

326. Silverman S Jr, Wara D. Oral manifestations of pediatric AIDS. Pediatrician 1989;16:185–187.

327. Ketchem L, Berkowitz RJ, McIlveen L, Forrester D, Rakusan T. Oral findings in HIV-seropositive children. Pediatr Dent 1990;12:143–146.

328. Davis MJ. Oral health care in pediatric AIDS. N Y State Dent J 1990;56:25–27.

329. Moniaci D, Cavallari M, Greco D, et al. Oral lesions in children born to HIV-1 positive women. J Oral Pathol Med 1993;22:8–11.

330. Asher RS, McDowell J, Acs G, Belanger G. Pediatric infection with the human immunodeficiency virus (HIV): head, neck, and oral manifestations. Spec Care Dentist. 1993;13:113–116.

331. Fine DH, Tofsky N, Nelson EM, et al. Clinical implications of the oral manifestations of HIV infection in children. Dent Clin North Am 2003;47:159–174.

332. Ramos-Gomez F. Dental considerations for the paediatric AIDS/HIV patient. Oral Dis 2002;8(suppl 2): 49–54.

333. Tukutuku K, Muyembe-Tamfum L, Kayembe K, Odio W, Kandi K, Ntumba M. Oral manifestations of AIDS in heterosexual population in a Zaire hospital. J Oral Pathol Med 1990;19:232–234.

334. Tukutuku K, Muyembe-Tamfum L, Kayembe K, Mavuemba T, Sangua N, Sekele I. Prevalence of dental caries, gingivitis, and oral hygiene in hospitalized AIDS cases in Kinshasa, Zaire. J Oral Pathol Med 1990;19:271–272.

335. Schulten EA, ten Kate RW, van der Waal I. Oral manifestations of HIV infection in 75 Dutch patients. J Oral Pathol Med 1989;18:42–46.

336. Ramirez V, Gonzalez A, de la Rosa E, et al. Oral lesions in Mexican HIV-infected patients. J Oral Pathol Med 1990;19:482–485.

337. Luangjamekorn L, Silverman S Jr, Gallo J, McKnight M, Migliorati C. Findings in 50 AIDS virus-infected patients with positive oral Candida cultures. J Dent Assoc Thai 1990;40:157–164.

338. Laskaris G, Potouridou I, Laskaris M, Stratigos J. Gingival lesions of HIV infection in 178 Greek patients. Oral Surg Oral Med Oral Pathol 1992;74:168–171.

339. Gillespie GM, Marino R. Oral manifestations of HIV infection: a pan-American perspective. J Oral Pathol Med 1993;22:2–7.

340. Dreizen S. Oral candidiasis. Am J Med 1984;77:28–33.

341. Braun-Falco O. International Workshop on Oral and Gastrointestinal Candidosis: From Pathology to Therapy. Introduction. Mycoses 1989;32(suppl 2): 6–8.

342. Samson J. Oral candidiasis: epidemiology, diagnosis and treatment. Schweiz Monatsschr Zahnmed 1990; 100:548–559.

343. Zegarelli DJ. Fungal infections of the oral cavity. Otolaryngol Clin North Am 1993;26:1069–1089.

344. Lynch DP. Oral candidiasis: History, classification, and clinical presentation. Oral Surg Oral Med Oral Pathol 1994;78:189–193.

345. Chandrasekar PH, Molinari JA. Oral candidiasis: forerunner of acquired immunodeficiency syndrome

(AIDS)? Oral Surg Oral Med Oral Pathol 1985;60:532–534.

346. Scully C, Epstein JB, Porter S, Luker J. Recognition of oral lesions of HIV infection. 1. Candidosis. Br Dent J 1990;169:295–296.

347. Stevens DA. Fungal infections in AIDS patients. Br J Clin Pract Symp Suppl 1990;71:11–22.

348. Imam N, Carpenter CC, Mayer KH, Fisher A, Stein M, Danforth SB. Hierarchical pattern of mucosal Candida infections in HIV-seropositive women. Am J Med 1990;89:142–146.

349. Di Silverio A, Brazzelli V, Brandozzi G, Barbarini G, Maccabruni A, Sacchi S. Prevalence of dermatophytes and yeasts (Candida spp., Malassezia furfur) in HIV patients. A study of former drug addicts. Mycopathologia 1991;114:103–107.

350. Felix DH, Wray D. The prevalence of oral candidiasis in HIV-infected individuals and dental attenders in Edinburgh. J Oral Pathol Med 1993;22:418–420.

351. Franker CK, Lucatorto FM, Johnson BS, Jacobson JJ. Characterization of the mycoflora from oral mucosal surfaces of some HIV-infected patients. Oral Surg Oral Med Oral Pathol 1990;69:683–687.

352. Miyasaki SH, Hicks JB, Greenspan D, et al. The identification and tracking of Candida albicans isolates from oral lesions in HIV-seropositive individuals. J Acquir Immune Defic Syndr 1992;5:1039–1046.

353. Powderly WG. Mucosal candidiasis caused by non-albicans species of Candida in HIV-positive patients. AIDS 1992;6:604–605.

354. Sullivan D, Bennett D, Henman M, et al. Oligonucleotide fingerprinting of isolates of Candida species other than C. albicans and of atypical Candida species from human immunodeficiency virus-positive and AIDS patients. J Clin Microbiol 1993;31:2124–2133.

355. Coleman DC, Bennett DE, Sullivan DJ, et al. Oral Candida in HIV infection and AIDS: new perspectives/new approaches. Crit Rev Microbiol 1993; 19:61–82.

356. Syrjanen S, Valle SL, Antonen J, et al. Oral candidal infection as a sign of HIV infection in homosexual men. Oral Surg Oral Med Oral Pathol 1988;65:36–40.

357. Korting HC. Clinical spectrum of oral candidosis and its role in HIV-infected patients. Mycoses 1989;32 (suppl 2):23–29.

358. Schiodt M, Bakilana PB, Hiza JF, et al. Oral candidiasis and hairy leukoplakia correlate with HIV infection in Tanzania. Oral Surg Oral Med Oral Pathol 1990;69:591–596.

359. Dodd CL, Greenspan D, Katz MH, Westenhouse JL, Feigal DW, Greenspan JS. Oral candidiasis in HIV infection: pseudomembranous and erythematous candidiasis show similar rates of progression to AIDS. AIDS 1991;5:1339–1343.

360. Katz MH, Greenspan D, Westenhouse J, et al. Progression of AIDS in HIV-infected homosexual and bisexual men with hairy leukoplakia and oral candidiasis. AIDS 1992;6:95–100.

361. Greenspan D, Gange SJ, Phelan JA, et al. Incidence of oral lesions in HIV-1–infected women: reduction with HAART. J Dent Res 2004;83:145–150.

362. Torssander J, Morfeldt-Manson L, Biberfeld G, Karlsson A, Putkonen PO, Wasserman J. Oral Candida albicans in HIV infection. Scand J Infect Dis 1987;19:291–295.

363. Hauman CH, Thompson IO, Theunissen F, Wolfaardt P. Oral carriage of Candida in healthy and HIV-seropositive persons. Oral Surg Oral Med Oral Pathol 1993;76:570–572.

364. Fetter A, Partisani M, Koenig H, Kremer M, Lang JM. Asymptomatic oral Candida albicans carriage in HIV-infection: frequency and predisposing factors. J Oral Pathol Med 1993;22:57–59.

365. Tylenda CA, Larsen J, Yeh CK, Lane HC, Fox PC. High levels of oral yeasts in early HIV-1 infection. J Oral Pathol Med 1989;18:520–524.

366. Schmidt-Westhausen A, Schiller RA, Pohle HD, Reichart PA. Oral Candida and Enterobacteriaceae in HIV-1 infection: correlation with clinical candidiasis and antimycotic therapy. J Oral Pathol Med 1991;20: 467–472.

367. McCarthy GM, Mackie ID, Koval J, Sandhu HS, Daley TD. Factors associated with increased frequency of HIV-related oral candidiasis. J Oral Pathol Med 1991; 20:332–336.

368. McCarthy GM. Host factors associated with HIV-related oral candidiasis. A review. Oral Surg Oral Med Oral Pathol 1992;73:181–186.

369. Lode H, Hoffken G. Oral candidosis and its role in immunocompromised patients. Mycoses 1989; 32(suppl 2):30–33.

370. Lynch DP, Gibson DK. The use of calcofluor white in the histopathological diagnosis of oral candidiasis. Oral Surg Oral Med Oral Pathol 1987;63:698–703.

371. Axell T, Simonsson T, Birkhed D, Rosenborg J, Edwardsson S. Evaluation of a simplified diagnostic aid (Oricult-N) for detection of oral candidoses. Scand J Dent Res 1985;93:52–55.

372. Hamilton JN, Thompson SH, Schiedt MJ, McQuade MJ, Van Dyke T, Plowman K. Correlation of subclinical candidal colonization of the dorsal tongue surface with the Walter Reed staging scheme for patients infected with HIV-1. Oral Surg Oral Med Oral Pathol 1992;73:47–51.

373. Samaranayake LP, Holmstrup P. Oral candidiasis and human immunodeficiency virus infection. J Oral Pathol Med 1989;18:554–564.

374. Greenspan D. Treatment of oral candidiasis in HIV infection. Oral Surg Oral Med Oral Pathol 1994;78: 211–215.

375. Epstein JB. Oral and pharyngeal candidiasis. Topical agents for management and prevention. Postgrad Med 1989;85:257–269.

376. Lewis MA, Samaranyake LP, Lamey PJ. Diagnosis and treatment of oral candidosis. J Oral Maxillofac Surg 1991;49:996–1002.

377. Rindum JL, Holmstrup P, Pedersen M, Rassing MR, Stoltze K. Miconazole chewing gum for treatment of chronic oral candidosis. Scand J Dent Res 1993;101: 386–390.

378. Pons V, Greenspan D, Debruin M. Therapy for oropharyngeal candidiasis in HIV-infected patients: a randomized, prospective multicenter study of oral fluconazole versus clotrimazole troches. The Multicenter Study Group. J Acquir Immune Defic Syndr 1993;6:1311–1316.

379. Narani N, Epstein JB. Classification of oral lesions of HIV infection. J Clin Periodontol 2001;28:137–145.

380. Shagase L, Feller L, Blignaut E. Necrotising ulcerative gingivitis/periodontitis as indicators of HIV-infection. SADJ 2004;59:105–108.

381. Dupont B, Drouhet E. Fluconazole in the management of oropharyngeal candidosis in a predominantly HIV antibody-positive group of patients. J Med Vet Mycol 1988;26:67–71.

382. Just-Nubling G, Gentschew G, Dohle M, Bottinger C, Helm EB, Stille W. Fluconazole in the treatment of oropharyngeal candidosis in HIV-positive patients. Mycoses 1990;33:435–440.

383. Blatchford NR. Treatment of oral candidosis with itraconazole: a review. J Am Acad Dermatol 1990;23:565–567.

384. Smith DE, Midgley J, Allan M, Connolly GM, Gazzard BG. Itraconazole versus ketaconazole in the treatment of oral and oesophageal candidosis in patients infected with HIV. AIDS 1991;5:1367–1371.

385. Tschechne B, Brunkhorst U, Ruhnke M, Trautmann M, Dempe S, Deicher H. Fluconazole in therapy of candidiasis of the oropharyngeal space in patients with HIV infection. Results of an open multicenter study of assessing the effectiveness and tolerance of fluconazole. Med Klin 1991;86:508–511.

386. Korting HC, Blecher P, Froschl M, Braun-Falco O. Quantitative assessment of the efficacy of oral ketoconazole for oral candidosis in HIV-infected patients. Mycoses 1992;35:173–176.

387. Bruatto M, Marinuzzi G, Raiteri R, Sinicco A. Susceptibility to ketoconazole of Candida albicans strains from sequentially followed HIV-1 patients with recurrent oral candidosis. Mycoses 1992;35:53–56.

388. Silverman S Jr, McKnight ML, Migliorati C, et al. Chemotherapeutic mouth rinses in immunocompromised patients. Am J Dent 1989;2:303–307.

389. Budtz-Jorgensen E. Etiology, pathogenesis, therapy, and prophylaxis of oral yeast infections. Acta Odontol Scand 1990;48:61–69.

390. Just-Nubling G, Gentschew G, Meissner K, et al. Fluconazole prophylaxis of recurrent oral candidiasis in HIV-positive patients. Eur J Clin Microbiol Infect Dis 1991;10:917–921.

391. Reents S, Goodwin SD, Singh V. Antifungal prophylaxis in immunocompromised hosts. Ann Pharmacother 1993;27:53–60.

392. Korting HC, Ollert M, Georgii A, Froschl M. In vitro susceptibilities and biotypes of Candida albicans isolates from the oral cavities of patients infected with human immunodeficiency virus. J Clin Microbiol 1988;26:2626–2631.

393. Fan-Havard P, Capano D, Smith SM, Mangia A, Eng RH. Development of resistance in Candida isolates from patients receiving prolonged antifungal therapy. Antimicrob Agents Chemother 1991;35:2302–2305.

394. Gallagher PJ, Bennett DE, Henman MC, et al. Reduced azole susceptibility of oral isolates of Candida albicans from HIV-positive patients and a derivative exhibiting colony morphology variation. J Gen Microbiol 1992;138:1901–1911.

395. Heinic GS, Stevens DA, Greenspan D, et al. Fluconazole-resistant Candida in AIDS patients. Report of two cases. Oral Surg Oral Med Oral Pathol 1993;76:711–715.

396. Boken DJ, Swindells S, Rinaldi MG. Fluconazole-resistant Candida albicans. Clin Infect Dis 1993;17:1018–1021.

397. Ng TT, Denning DW. Fluconazole resistance in Candida in patients with AIDS—a therapeutic approach. J Infect 1993;26:1117–1125.

398. Redding S, Smith J, Farinacci G, et al. Resistance of Candida albicans to fluconazole during treatment of oropharyngeal candidiasis in a patient with AIDS: documentation by in vitro susceptibility testing and DNA subtype analysis. Clin Infect Dis 1994;18:240–242.

399. Cameron ML, Schell WA, Bruch S, Bartlett JA, Waskin HA, Perfect JR. Correlation of in vitro fluconazole resistance of Candida isolates in relation to therapy and symptoms of individuals seropositive for human immunodeficiency virus type 1. Antimicrob Agents Chemother 1993;37:2449–2453.

400. Pfaller MA, Rhine-Chalberg J, Redding SW, et al. Variations in fluconazole susceptibility and electrophoretic karyotype among oral isolates of Candida albicans from patients with AIDS and oral candidiasis. J Clin Microbiol 1994;32:59–64.

401. Schmid J, Odds FC, Wiselka MJ, Nicholson KG, Soll DR. Genetic similarity and maintenance of Candida albicans strains from a group of AIDS patients, demonstrated by DNA fingerprinting. J Clin Microbiol 1992;30:935–941.

402. Powderly WG, Robinson K, Keath EJ. Molecular epidemiology of recurrent oral candidiasis in human immunodeficiency virus-positive patients: evidence for two patterns of recurrence. J Infect Dis 1993;168:463–466.

403. Challacombe SJ. Immunologic aspects of oral candidiasis. Oral Surg Oral Med Oral Pathol 1994;78:202–210.

404. Coogan MM, Sweet SP, Challacombe SJ. Immunoglobulin A (IgA), IgA₁, and IgA₂ antibodies to Candida albicans in whole and parotid saliva in human immunodeficiency virus infection and AIDS. Infect Immun 1994;62:892–896.

405. Plettenberg A, Reisinger E, Lenzner U, et al. Oral candidosis in HIV-infected patients. Prognostic value and correlation with immunological parameters. Mycoses 1990;33:421–425.

406. Greenspan D, Greenspan JS, Conant M, Petersen V, Silverman S Jr, de Souza Y Oral "hairy" leucoplakia in male homosexuals: evidence of association with both papillomavirus and a herpes-group virus. Lancet 1984;2:831–834.

407. Greenspan JS, Greenspan D, Lennette ET, et al. Replication of Epstein-Barr virus within the epithelial cells of oral "hairy" leukoplakia, an AIDS-associated lesion. N Engl J Med 1985;313:1564–1571.

408. Greenspan D, Greenspan JS, Hearst NG, et al. Relation of oral hairy leukoplakia to infection with the human immunodeficiency virus and the risk of developing AIDS. J Infect Dis 1987;155:475–481.

409. Reichart PA, Langford A, Gelderblom HR, Pohle HD, Becker J, Wolf H. Oral hairy leukoplakia: observations in 95 cases and review of the literature. J Oral Pathol Med 1989;18:410–415.

410. Sciubba J, Brandsma J, Schwartz M, Barrezueta N. Hairy leukoplakia: an AIDS-associated opportunistic infection. Oral Surg Oral Med Oral Pathol 1989;67:404–410.

411. Itin PH. Oral hairy leukoplakia—10 years on. Dermatology 1993;187:159–163.

412. Greenspan D, Hollander H, Friedman-Kien A, Freese UK, Greenspan JS. Oral hairy leukoplakia in two women, a haemophiliac, and a transfusion recipient. Lancet 1986;2:978–979.

413. Rindum JL, Schiodt M, Pindborg JJ, Scheibel E. Oral hairy leukoplakia in three hemophiliacs with human immunodeficiency virus infection. Oral Surg Med Oral Pathol 1987;63:437–440.

414. Greenspan JS, Mastrucci MT, Leggott PJ, et al. Hairy leukoplakia in a child. AIDS 1988;2:143.

415. Kabani S, Greenspan D, de Souza J, Greenspan JS, Cataldo E. Oral hairy leukoplakia with extensive oral mucosal involvement. Report of two cases. Oral Surg Oral Med Oral Pathol 1989;67:411–415.

416. Syrjanen S, Laine P, Niemela M, Happonen RP. Oral hairy leukoplakia is not a specific sign of HIV-infection but related to immunosuppression in general. J Oral Pathol Med 1989;18:28–31.

417. Ramael M, Colebunders R, Colpaert C, et al. The prevalence of hairy leukoplakia in HIV seropositive and HIV seronegative immunocompromised patients. Int J STD AIDS 1992;3:251–254.

418. Epstein JB, Priddy RW, Sherlock CH. Hairy leukoplakia-like lesions in immunosuppressed patients following bone marrow transplantation. Transplantation 1988;46:462–464.

419. Epstein JB, Sherlock CH, Greenspan JS. Hairy leukoplakia-like lesions following bone-marrow transplantation. AIDS 1991;5:101–102.

420. Epstein JB, Sherlock CH, Wolber RA. Hairy leukoplakia after bone marrow transplantation. Oral Surg Oral Med Oral Pathol 1993;75:690–695.

421. Greenspan D, Greenspan JS, de Souza Y, Levy JA, Ungar AM. Oral hairy leukoplakia in an HIV-negative renal transplant recipient. J Oral Pathol Med 1989;18:32–34.

422. Macleod RI, Logan LQ, Soames JV, Ward MK. Oral hairy leukoplakia in an HIV-negative renal transplant patient. Br Dent J 1990;169:208–209.

423. Euvrard S, Kanitakis J, Pouteil-Nobel C, Chardonnet Y, Touraine JL, Thivolet J. Pseudo oral hairy leukoplakia in a renal allograft recipient. J Am Acad Dermatol 1994;30:300–303.

424. Schmidt-Westhausen A, Gelderblom HR, Neuhaus P, Reichart PA. Epstein-Barr virus in lingual epithelium of liver transplant patients. J Oral Pathol Med 1993;22: 274–276.

425. Schmidt-Westhausen A, Gelderblom HR, Reichart PA. Oral hairy leukoplakia in an HIV-seronegative heart transplant patient. J Oral Pathol Med 1990;19:192–194.

426. Ficarra G, Miliani A, Adler-Storthz K, et al. Recurrent oral condylomata acuminata and hairy leukoplakia: an early sign of myelodysplastic syndrome in an HIV-seronegative patient. J Oral Pathol Med 1991;20: 398–402.

427. Eisenberg E, Krutchkoff D, Yamase H. Incidental oral hairy leukoplakia in immunocompetent persons. A report of two cases. Oral Surg Oral Med Oral Pathol 1992;74:332–333.

428. Felix DH, Watret K, Wray D, Southam JC. Hairy leukoplakia in an HIV-negative, nonimmunosuppressed patient. Oral Surg Oral Med Oral Pathol 1992;74: 563–566.

429. Naher H, Gissmann L, von Knebel Doeberitz C, et al. Detection of Epstein-Barr virus-DNA in tongue epithelium of human immunodeficiency virus-infected patients. J Invest Dermatol 1991;97:421–424.

430. Walling DM, Etienne W, Ray AJ, et al. Persistence and transition of Epstein-Barr virus genotypes in the pathogenesis of oral hairy leukoplakia. J Infect Dis 2004;190:387–395.

431. Hille JJ, Webster-Cyriaque J, Palefski JM, et al. Mechanisms of expression of HHV8, EBV and HPV in selected HIV-associated lesions. Oral Dis 2002;8(suppl 2):161–168.

432. Corso B, Eversole LR, Hutt-Fletcher L. Hairy leukoplakia: Epstein-Barr virus receptors on oral keratinocyte plasma membranes. Oral Surg Oral Med Oral Pathol 1989;67:416–421.

433. Sugihara K, Reupke H, Schmidt-Westhausen A, Pohle HD, Gelderblom HR, Reichart PA. Negative staining EM for the detection of Epstein-Barr virus in oral hairy leukoplakia. J Oral Pathol Med 1990;19:367–370.

434. Madinier I, Doglio A, Cagnon L, Lefebvre JC, Monteil RA. Epstein-Barr virus DNA detection in gingival tissues of patients undergoing surgical extractions. Br J Oral Maxillofac Surg 1992;30:237–243.

435. Adler-Storthz K, Ficarra G, Woods KV, Gaglioti D, Di Pietro M, Shillitoe EJ. Prevalence of Epstein-Barr virus and human papillomavirus in oral mucosa of HIV-infected patients. J Oral Pathol Med 1992;21:164–170.

436. Felix DH, Jalal H, Cubie HA, Southam JC, Wray D, Maitland NJ. Detection of Epstein-Barr virus and human papillomavirus type 16 DNA in hairy leukoplakia by in situ hybridisation and the polymerase chain reaction. J Oral Pathol Med 1993;22:277–281.

437. Green TL, Greenspan JS, Greenspan D, de Souza YG. Oral lesions mimicking hairy leukoplakia: a diagnostic dilemma. Oral Surg Oral Med Oral Pathol 1989; 67:422–426.

438. Schulten EA, Snijders PJ, ten Kate RW, et al. Oral hairy leukoplakia in HIV infection: a diagnostic pitfall. Oral Surg Oral Med Oral Pathol 1991;71:32–37.

439. Fisher DA, Daniels TE, Greenspan JS. Oral hairy leukoplakia unassociated with human immunodeficiency virus: pseudo oral hairy leukoplakia. J Am Acad Dermatol 1992;27:257–258.

440. Kanas RJ, Abrams AM, Recher L, Jensen JL, Handlers JP, Wuerker RB. Oral hairy leukoplakia: a light microscopic and immunohistochemical study. Oral Surg Oral Med Oral Pathol 1988;66:334–340.

441. Cubie HA, Felix DH, Southam JC, Wray D. Application of molecular techniques in the rapid diagnosis of EBV-associated oral hairy leukoplakia. J Oral Pathol Med 1991;20:271–274.

442. Greenspan JS, Rabanus JP, Petersen V, Greenspan D. Fine structure of EBV-infected keratinocytes in oral hairy leukoplakia. J Oral Pathol Med 1989;18:565–572.

443. el-Labban N, Pindborg JJ, Rindum J, Nielsen H. Further ultrastructural findings in epithelial cells of hairy leukoplakia. J Oral Pathol Med 1990;19:24–34.

444. Kratochvil FJ, Riordan GP, Auclair PL, Huber MA, Kragel PJ. Diagnosis of oral hairy leukoplakia by ultrastructural examination of exfoliative cytologic specimens. Oral Surg Oral Med Oral Pathol 1990;70: 613–618.

445. Migliorati CA, Jones AC, Baughman PA. Use of exfoliative cytology in the diagnosis of oral hairy leukoplakia. Oral Surg Oral Med Oral Pathol 1993;76: 704–710.

446. Walling DM, Flaitz CM, Adler-Storthz K, et al. A non-invasive technique for studying oral epithelial Epstein-Barr virus infection and disease. Oral Oncol 2003;39:436–444.

447. Greenspan JS, Greenspan D. Oral hairy leukoplakia: diagnosis and management. Oral Surg Oral Med Oral Pathol. 1989;67:396–403.

448. Resnick L, Herbst JS, Ablashi DV, et al. Regression of oral hairy leukoplakia after orally administered acyclovir therapy. JAMA 1988;259:384–388.

449. Glick M, Pliskin ME. Regression of oral hairy leukoplakia after oral administration of acyclovir. Gen Dent 1990;38:374–375.

450. Greenspan D, de Souza YG, Conant MA, et al. Efficacy of desciclovir in the treatment of Epstein-Barr virus infection in oral hairy leukoplakia. J Acquir Immune Defic Syndr 1990;3:571–578.

451. Newman C, Polk BF. Resolution of oral hairy leukoplakia during therapy with 9-(1,3–dihydroxy-2-propoxymethyl)guanine (DHPG). Ann Intern Med 1987;107:348–350.

452. Phelan JA, Klein RS. Resolution of oral hairy leukoplakia during treatment with azidothymidine. Oral Surg Oral Med Oral Pathol 1988;65:717–720.

453. Kessler HA, Benson CA, Urbanski P. Regression of oral hairy leukoplakia during zidovudine therapy. Arch Intern Med 1988;148:2496–2497.

454. Katz MH, Greenspan D, Heinic GS, et al. Resolution of hairy leukoplakia: an observational trial of zidovudine versus no treatment. J Infect Dis 1991;164: 1240–1241.

455. Walling DM, Flaitz CM, Nichols CM. Epstein-Barr virus replication in oral hairy leukoplakia: response, persistence and resistance to treatment with valacyclovir. J Infect Dis 2003;188:883–890.

456. Schofer H, Ochsendorf FR, Helm EB, Milbradt R. Treatment of oral "hairy" leukoplakia in AIDS patients with vitamin A acid (topically) or acyclovir (systemically). Dermatologica 1987;174:150–151.

457. Lozada-Nur F. Podophyllin resin 25% for treatment of oral hairy leukoplakia: an old treatment for a new lesion. J Acquir Immune Defic Syndr 1991;4:543–546.

458. Lozada-Nur F, Costa C. Retrospective findings of the clinical benefits of podophyllum resin 25% sol on hairy leukoplakia. Clinical results in nine patients. Oral Surg Oral Med Oral Pathol 1992;73:555–558.

459. Herbst JS, Morgan J, Raab-Traub N, Resnick L. Comparison of the efficacy of surgery and acyclovir therapy in oral hairy leukoplakia. J Am Acad Dermatol 1989;21:753–756.

460. Moniaci D, Greco D, Flecchia G, Raiteri R, Sinicco A. Epidemiology, clinical features and prognostic value of HIV-1 related oral lesions. J Oral Pathol Med 1990; 19:477–481.

461. Greenspan D, Greenspan JS, Overby G, et al. Risk factors for rapid progression from hairy leukoplakia to AIDS: a nested case-control study. J Acquir Immune Defic Syndr 1991;4:652–658.

462. Greenspan D, Greenspan JS. Significance of oral hairy leukoplakia. Oral Surg Oral Med Oral Pathol 1992; 73:151–154.

463. Klein RS, Quart AM, Small CB. Periodontal disease in heterosexuals with acquired immunodeficiency syndrome. J Periodontol 1991;62:535–540.

464. Levine RA, Glick M. Rapidly progressive periodontitis as an important clinical marker for HIV disease. Compendium 1991;12:478–482.

465. Riley C, London JP, Burmeister JA. Periodontal health in 200 HIV-positive patients. J Oral Pathol Med 1992; 21:124–127.

466. Masouredis CM, Katz MH, Greenspan D, et al. Prevalence of HIV-associated periodontitis and gingivitis in HIV-infected patients attending an AIDS clinic. J Acquir Immune Defic Syndr 1992;5:479–483.

467. Friedman RB, Gunsolley J, Gentry A, Dinius A, Kaplowitz K, Settle J. Periodontal status of HIV-seropositive and AIDS patients. J Periodontol 1991; 62:623–627.

468. Barr C, Lopez MR, Rua-Dobles A. Periodontal changes by HIV serostatus in a cohort of homosexual and bisexual men. J Clin Periodontol 1992;19:794–801.

469. Yeung SC, Stewart GJ, Cooper DA, Sindhusake D. Progression of periodontal disease in HIV seropositive patients. J Periodontol 1993;64:651–657.

470. Murray PA, Grassi M, Winkler JR. The microbiology of HIV-associated periodontal lesions. J Clin Periodontol 1989;16:636–642.

471. Lucht E, Heimdahl A, Nord CE. Periodontal disease in HIV-infected patients in relation to lymphocyte subsets and specific micro-organisms. J Clin Periodontol 1991;18:252–256.

472. Gornitsky M, Clark DC, Siboo R, et al. Clinical documentation and occurrence of putative periodontopathic bacteria in human immunodeficiency virus-associated periodontal disease. J Periodontol 1991;62:576–585.

473. Moore LV, Moore WE, Riley C, Brooks CN, Burmeister JA, Smibert RM. Periodontal microflora of HIV positive subjects with gingivitis or adult periodontitis. J Periodontol 1993;64:48–56.

474. Rosenstein DI, Riviere GR, Elott KS. HIV-associated periodontal disease: new oral spirochete found. J Am Dent Assoc 1993;124:76–80.

475. Ryder MI. Periodontal considerations in the patient with HIV. Curr Opin Periodontol 1993;43–51.

476. Winkler JR, Murray PA, Grassi M, Hammerle C. Diagnosis and management of HIV-associated periodontal lesions. J Am Dent Assoc. 1989;(suppl);25S–34S.

477. Greenspan D, Greenspan JS. Management of the oral lesions of HIV infection. J Am Dent Assoc 1991; 122:26–32.

478. Hammerle C, Grassi M, Winkler JR. HIV periodontopathies. The diagnosis and therapy of HIV-associated gingivitis/periodontitis. Schweiz Monatsschr Zahnmed 1992;102:940–950.

479. Robinson PG. The significance and management of periodontal lesions in HIV infection. Oral Dis 2002;8(suppl 2):91–97.

480. Dodd CL, Greenspan D, Greenspan JS. Oral Kaposi's sarcoma in a woman as a first indication of HIV infection. J Am Dent Assoc 1991;122:61–63.

481. Epstein JB, Silverman S Jr. Head and neck malignancies associated with HIV infection. Oral Surg Oral Med Oral Pathol 1992;73:193–200.

482. Regezi JA, MacPhail LA, Daniels TE, de Souza YG, Greenspan JS, Greenspan D. Human immunodeficiency virus-associated oral Kaposi's sarcoma. A heterogeneous cell population dominated by spindle-shaped endothelial cells. Am J Pathol 1993;143:240–249.

483. Kaul A, Pearson JD, Petty R, Williams DM, Dalgleish AG. Vascular endothelium: a potential role in HIV infection and the pathogenesis of Kaposi's sarcoma: observations and speculations. Mol Aspects Med 1991;12:297–312.

484. Peterman TA, Jaffe HW, Friedman-Kien AE, Weiss RA. The aetiology of Kaposi's sarcoma. Cancer Surv 1991; 10:23–37.

485. Henke-Gendo C, Schulz TF. Transmission and disease association of Kaposi's sarcoma-associated herpesvirus: recent developments. Curr Opin Infect Dis 2004;17:53–57.

486. Chang Y, Cesarman E, Pessin MS, et al. Identification of herpesvirus-like DNA sequences in AIDS-associated Kaposi's sarcoma. Science 1994;266(5192):1865–1869.

487. Moore PS, Chang Y. Detection of herpesvirus-like DNA sequences in Kaposi's sarcoma in patients with and those without HIV infection. N Engl J Med 1995;332:1181–1185.

488. Koelle DM, Huang ML, Chandran B, Vieira J, Piepkorn M, Corey L. Frequent detection of Kaposi's sarcoma-associated herpesvirus (human herpesvirus 8) DNA in saliva of human immunodeficiency virus-infected men: clinical and immunologic correlates. J Infect Dis 1997;176:94–102.

489. Martin JN. Diagnosis and epidemiology of human herpesvirus 8 infection. Semin Hematol 2003;40:133–142.

490. Pauk J, Huang ML, Brodie SJ, et al. Mucosal shedding of human herpesvirus 8 in men. N Engl J Med 2000;343:1369–1377.

491. Casper C, Redman M, Huang ML, et al. HIV infection and human herpesvirus-8 oral shedding among men who have sex with men. J Acquir Immune Defic Syndr 2004;35:233–238.

492. Flaitz CM, Jin YT, Hicks MJ, Nichols CM, Wang YW, Su IJ. Kaposi's sarcoma-associated herpesvirus-like DNA sequences (KSHV/HHV-8) in oral AIDS-Kaposi's sarcoma: a PCR and clinicopathologic study. Oral Surg Oral Med Oral Pathol Oral Radiol Endod 1997;83:259–264.

493. Di Alberti L, Ngui SL, Porter SR, et al. Presence of human herpesvirus 8 variants in the oral tissues of human immunodeficiency virus-infected individuals. J Infect Dis 1997;175:703–707.

494. Triantos D, Horefti E, Paximadi E, et al. Presence of human herpesvirus-8 in saliva and non-lesional oral mucosa in HIV-infected and oncologic immunocompromised patients. Oral Microbiol Immunol 2004;19:201–204.

495. Teo CG. Viral infections in the mouth. Oral Dis 2002;8(suppl 2):88–90.

496. Bubman D, Cesarman E. Pathogenesis of Kaposi's sarcoma. Hematol Oncol Clin North Am 2003;17:717–745.

497. Aoki Y, Tosato G. Pathogenesis and manifestations of human herpesvirus-8–associated disorders. Semin Hematol 2003;40:143–153.

498. De Paoli P. Human herpesvirus 8: an update. Microbes Infect 2004;6:328–335.

499. Leao JC, Porter S, Scully C. Human herpesvirus 8 and oral health care: an update. Oral Surg Oral Med Oral Pathol Oral Radiol Endod 2000;90:694–704.

500. Porter SP, Di Alberti L, Kumar N. Human herpes virus 8 (Kaposi's sarcoma herpesvirus). Oral Oncol 1998;34:5–14.

501. Goedert JJ. The epidemiology of acquired immunodeficiency syndrome malignancies. Semin Oncol 2002;27:390–401.

502. Reichart PA. Oral manifestations in HIV infection: fungal and bacterial infections, Kaposi's sarcoma. Med Microbiol Immunol (Berl) 2003;192:165–169.

503. Regezi JA, Jordan RC. Oral Kaposi's sarcoma: biopsy accessions as an indication of declining incidence. Oral Surg Oral Med Oral Pathol Oral Radiol Endod 2002;94:399.

504. Langford A, Pohle HD, Reichart P. Primary intraosseous AIDS-associated Kaposi's sarcoma. Report of two cases with initial jaw involvement. Int J Oral Maxillofac Surg 1991;20:366–368.

505. Lausten LL, Ferguson BL, Barker BF, et al. Oral Kaposi sarcoma associated with severe alveolar bone loss: case report and review of the literature. J Periodontol 2003;74:1668–1675.

506. Epstein JB, Scully C. Neoplastic disease in the head and neck of patients with AIDS. Int J Oral Maxillofac Surg 1992;21:219–226.

507. Regezi JA, MacPhail LA, Daniels TE, et al. Oral Kaposi's sarcoma: a 10-year retrospective histopathologic study. J Oral Pathol Med 1993;22:292–297.

508. Newland JR, Adler-Storthz K. Cytomegalovirus in intraoral Kaposi's sarcoma. Oral Surg Oral Med Oral Pathol 1989;67:296–300.

509. Chak LY, Gill PS, Levine AM, Meyer PR, Anselmo JA, Petrovich Z. Radiation therapy for acquired immunodeficiency syndrome-related Kaposi's sarcoma. J Clin Oncol 1988;6:863–867.

510. Schweitzer VG, Visscher D. Photodynamic therapy for treatment of AIDS-related oral Kaposi's sarcoma. Otolaryngol Head Neck Surg 1990;102:639–649.

511. Baumann R, Tauber MG, Opravil M, et al. Combined treatment with zidovudine and lymphoblast interferon-alpha in patients with HIV-related Kaposi's sarcoma. Klin Wochenschr 1991;69:360–367.

512. de Wit R, Danner SA, Bakker PJ, Lange JM, Eeftinck Schattenkerk JK, Veenhof CH. Combined zidovudine and interferon-alpha treatment in patients with AIDS-associated Kaposi's sarcoma. J Intern Med 1991;229:35–40.

513. Epstein JB, Lozada-Nur F, McLeod WA, Spinelli J. Oral Kaposi's sarcoma in acquired immunodeficiency syndrome. Review of management and report of the efficacy of intralesional vinblastine. Cancer 1989;64:2424–2430.

514. Epstein JB. Treatment of oral Kaposi sarcoma with intralesional vinblastine. Cancer 1993;71:1722–1725.

515. Lucatorto FM, Sapp JP. Treatment of oral Kaposi's sarcoma with a sclerosing agent in AIDS patients. A preliminary study. Oral Surg Oral Med Oral Pathol 1993;75:192–198.

516. Shiboski CH, Winkler JR. Gingival Kaposi's sarcoma and periodontitis. A case report and suggested treatment approach to the combined lesions. Oral Surg Oral Med Oral Pathol 1993;76:49–53.

517. Birnbaum W, Hodgson TA, Reichart PA. Prognostic significance of HIV-associated oral lesions with their relation to therapy. Oral Dis 2002;8(suppl 2):110–114.

518. Greenwood I, Zakrzewska JM, Robinson PG. Changes in the prevalence of HIV-associated mucosal disease at a dedicated clinic over 7 years. Oral Dis 2002;8:90–94.

519. Mbopi-Keou FX, Legoff J, Piketty C, et al. Salivary production of IgA and IgG to human herpes virus 8 latent and lytic antigens by patients in whom Kaposi's sarcoma has regressed. AIDS 2004;18:338–340.

520. Ziegler JL, Beckstead JA, Volberding PA, et al. Non-Hodgkin's lymphoma in 90 homosexual men. Relation to generalized lymphadenopathy and the acquired immunodeficiency syndrome. N Engl J Med 1984;311:565–570.

521. Levine AM. Non-Hodgkin's lymphomas and other malignancies in the acquired immune deficiency syndrome. Semin Oncol 1987;14(suppl 3):34–39.

522. Leess FR, Kessler DJ, Mickel RA. Non-Hodgkin's lymphoma of the head and neck in patients with AIDS. Arch Otolaryngol Head Neck Surg 1987;113: 1104–1106.

523. Kaplan LD. AIDS-associated lymphomas. Infect Dis Clin North Am 1988;2:525–532.

524. Jordan RC, Chong L, Dipierdomenico S, et al. Oral lymphoma in HIV infection. Oral Dis 1997;3(suppl 1): S135–S137.

525. Jordan RC, Chong L, Dipierdomenico S, et al. Oral lymphoma in human immunodeficiency virus infection: a report of six cases and review of the literature. Otolaryngol Head Neck Surg 1998;119:672–677.

526. Carbone AIDS-related non-Hodgkin's lymphomas: from pathology and molecular pathogenesis to treatment A. Hum Pathol 2002;33:392–404.

527. Ioachim HL, Dorsett B, Cronnin W, Maya M, Wahl S. Acquired immunodeficiency syndrome-associated lymphomas: clinical, pathologic, immunologic, and viral characteristics of 111 cases. Hum Pathol 1991;22:659–673.

528. Green TL, Eversole LR. Oral lymphomas in HIV-infected patients: association with Epstein-Barr virus DNA. Oral Surg Oral Med Oral Pathol 1989;67: 437–442.

529. Goldschmidts WL, Bhatia K, Johnson JF, et al. Epstein-Barr virus genotypes in AIDS-associated lymphomas are similar to those in endemic Burkitt's lymphomas. Leukemia 1992;6:875–878.

530. Palmer GD, Morgan PR, Challacombe SJ. T-cell lymphoma associated with periodontal disease and HIV infection. A case report. J Clin Periodontol 1993;20: 378–380.

531. Groot RH, van Merkesteyn JP, Bras J. Oral manifestations of non-Hodgkin's lymphoma in HIV-infected patients. Int J Oral Maxillofac Surg 1990;19:194–196.

532. Colmenero C, Gamallo C, Pintado V, Patron M, Sierra I, Valencia E. AIDS-related lymphoma of the oral cavity. Int J Oral Maxillofac Surg 1991;20:2–6.

533. Langford A, Dienemann D, Schurman D, et al. Oral manifestations of AIDS-associated non-Hodgkin's lymphomas. Int J Oral Maxillofac Surg 1991;20:136–141.

534. Rubin MM, Gatta CA, Cozzi GM. Non-Hodgkin's lymphoma of the buccal gingiva as the initial manifestation of AIDS. J Oral Maxillofac Surg 1989;47: 1311–1313.

535. Kaugars GE, Burns JC. Non-Hodgkin's lymphoma of the oral cavity associated with AIDS. Oral Surg Oral Med Oral Pathol 1989;67:433–436.

536. Dodd CL, Greenspan D, Schiodt M, et al. Unusual oral presentation of non-Hodgkin's lymphoma in association with HIV infection. Oral Surg Oral Med Oral Pathol 1992;73:603–608.

537. Hicks MJ, Flaitz CM, Nichols CM, Luna MA, Gresik MV. Intraoral presentation of anaplastic large-cell Ki-1 lymphoma in association with HIV infection. Oral Surg Oral Med Oral Pathol 1993;76:73–81.

538. Dodd CL, Greenspan D, Heinic GS, Rabanus JP, Greenspan JS. Multi-focal oral non-Hodgkin's lymphoma in a AIDS patient. Br Dent J 1993;175:373–377.

539. De Weese TL, Hazuka MB, Hommel DJ, Kinzie JJ, Daniel WE. AIDS-related non-Hodgkin's lymphoma: the outcome and efficacy of radiation therapy. Int J Radiat Oncol Biol Phys 1991;20:803–808.

540. Beck K. Mycobacterial disease associated with HIV infection. J Gen Intern Med 1991;6(suppl 1):S19–S23.

541. Young LS. Mycobacterial diseases and the compromised host. Clin Infect Dis 1993;17:436–441.

542. Joshi VV, Oleske JM, Saad S, Connor EM, Rapkin RH, Minnefor AB. Pathology of opportunistic infections in children with acquired immune deficiency syndrome. Pediatr Pathol 1986;6:145–150.

543. Rutstein RM, Cobb P, McGowan KL, Pinto-Martin J, Starr SE. Mycobacterium avium intracellulare complex infection in HIV-infected children. AIDS 1993;7:507–512.

544. Collins FM. Mycobacterial disease, immunosuppression, and acquired immunodeficiency syndrome. Clin Microbiol Rev 1989;2:360–367.

545. Horsburgh CR Jr, Selik RM. The epidemiology of disseminated nontuberculous mycobacterial infection in the acquired immunodeficiency syndrome (AIDS). Am Rev Respir Dis 1990;139:4–7.

546. Berlin OG, Zakowski P, Bruckner DA, Clancy MN, Johnson BL Jr. Mycobacterium avium: a pathogen of patients with acquired immunodeficiency syndrome. Diagn Microbiol Infect Dis 1984;2:213–218.

547. Hawkins CC, Gold JW, Whimbey E, et al. Mycobacterium avium complex infections in patients with the acquired immunodeficiency syndrome. Ann Intern Med 1986;105:184–188.

548. Klatt EC, Jensen DF, Meyer PR. Pathology of Mycobacterium avium-intracellulare infection in acquired immunodeficiency syndrome. Hum Pathol 1987;18: 709–714.

549. Jacobson MA, Hopewell PC, Yajko DM, et al. Natural history of disseminated Mycobacterium avium complex infection in AIDS. J Infect Dis 1991;164: 994–998.

550. Horsburgh CR Jr, Mason 3rd UG, Farhi DC, Iseman MD. Disseminated infection with Mycobacterium avium-intracellulare. A report of 13 cases and a review of the literature. Medicine 1985;64:36–48.

551. Young LS, Inderlied CB, Berlin OG, Gottlieb MS. Mycobacterial infections in AIDS patients, with an emphasis on the Mycobacterium avium complex. Rev Infect Dis 1986;8:1024–1033.

552. Tenholder MF, Moser 3rd RJ, Tellis CJ. Mycobacteria other than tuberculosis. Pulmonary involvement in patients with acquired immunodeficiency syndrome. Arch Intern Med 1988;148:953–955.

553. Wallace JM, Hannah JB. Mycobacterium avium complex infection in patients with the acquired immunodeficiency syndrome. A clinicopathologic study. Chest 1988;93:926–932.

554. Horsburgh CR Jr. Mycobacterium avium complex infection in the acquired immunodeficiency syndrome. N Engl J Med 1991;324:1332–1338.

555. Masur H, Tuazon C, Gill V, et al. Effect of combined clofazimine and ansamycin therapy on Mycobacterium avium-Mycobacterium intracellulare bacteremia in patients with AIDS. J Infect Dis 1987;155:127–129.

556. Levin RH, Bolinger AM. Treatment of nontuberculous mycobacterial infections in pediatric patients. Clin Pharm 1988;7:545–551.

557. Guthertz LS, Damsker B, Bottone EJ, Ford EG, Midura TF, Janda JM. Mycobacterium avium and Mycobac-

terium intracellulare infections in patients with and without AIDS. J Infect Dis 1989;160:1037–1041.

558. Hoy J, Mijch A, Sandland M, Grayson K, Lucas R, Dwyer B. Quadruple-drug therapy for Mycobacterium avium-intracellulare bacteremia in AIDS patients. J Infect Dis 1990;161:801–805.

559. Benson CA, Kessler HA, Pottage JC Jr, Trenholme GM. Successful treatment of acquired immunodeficiency syndrome-related Mycobacterium avium complex disease with a multiple drug regimen including amikacin. Arch Intern Med 1991;151:582–585.

560. Garrelts JC. Clofazimine: a review of its use in leprosy and Mycobacterium avium complex infection. Drug Intell Clin Pharm 1991;25:525–531.

561. Kemper CA, Meng TC, Nussbaum J, et al. Treatment of Mycobacterium avium complex bacteremia in AIDS with a four-drug oral regimen. Rifampin, ethambutol, clofazimine, and ciprofloxacin. The California Collaborative Treatment Group. Ann Intern Med 1992;116: 466–472.

562. Abrams DI, Mitchell TF, Child CC, Shiboski SC, Brosgart CL, Mass MM. Clofazimine as prophylaxis for disseminated Mycobacterium avium complex infection in AIDS. J Infect Dis 1993;167:1459–1463.

563. Benson CA, Ellner JJ. Mycobacterium avium complex infection and AIDS: advances in theory and practice. Clin Infect Dis 1993;17:7–20.

564. Centers for Disease Control US. Department of Health and Human Services. Diagnosis and management of mycobacterial infection and disease in persons with human immunodeficiency virus infection. Ann Intern Med 1987;106:254–256.

565. Goodman PC. Pulmonary tuberculosis in patients with acquired immunodeficiency syndrome. J Thorac Imaging 1990;5:38–45.

566. Sunderam G, McDonald RJ, Maniatis T, Oleske J, Kapila R, Reichman LB. Tuberculosis as a manifestation of the acquired immunodeficiency syndrome (AIDS). JAMA 1986;256:362–366.

567. Braun MM, Byers RH, Heyward WL, et al. Acquired immunodeficiency syndrome and extrapulmonary tuberculosis in the United States. Arch Intern Med 1990;150:1913–1916.

568. Horsburgh CR Jr, Pozniak A. Epidemiology of tuberculosis in the era of HIV. AIDS 1993;7(suppl 1): S109–S114.

569. Yoder KM. Tuberculosis: a reemerging hazard for oral healthcare workers. J Dent Hyg 1993;67:208–213.

570. Pitchenik AE, Fertel D. Medical management of AIDS patients. Tuberculosis and nontuberculous mycobacterial disease. Med Clin North Am 1992;76: 121–171.

571. Barnes PF, Le HQ, Davidson PT. Tuberculosis in patients with HIV infection. Med Clin North Am 1993;77:1369–1390.

572. Miller B. Preventive therapy for tuberculosis. Med Clin North Am 1993;77:1263–1275.

573. Cohn DL, Dobkin JF. Treatment and prevention of tuberculosis in HIV infection. AIDS 1993;7(suppl 1): S195–S202.

574. Johnson MP, Chaisson RE. Tuberculosis and HIV disease. AIDS Clin Rev 1993;94;73–93.

575. Brudney K, Dobkin J. Resurgent tuberculosis in New York City. Human immunodeficiency virus, homelessness, and the decline of tuberculosis control programs. Am Rev Respir Dis 1991;144:745–749.

576. Liang GS, Daikos GL, Serfling U, et al. An evaluation of oral ulcers in patients with AIDS and AIDS-related complex. J Am Acad Dermatol 1993;29:563–568.

577. Langford A, Pohle HD, Gelderblom H, Zhang X, Reichart PA. Oral hyperpigmentation in HIV-infected patients. Oral Surg Oral Med Oral Pathol 1989;67:301–307.

578. Langford AA, Gelderblom H, Kunze RO, Pohle HD, Reichart PA. Hyperpigmentation of the oral mucosa in HIV infection. Schweiz Monatsschr Zahnmed 1990; 100:1037–1041.

579. Ficarra G, Shillitoe EJ, Adler-Storthz K, et al. Oral melanotic macules in patients infected with human immunodeficiency virus. Oral Surg Oral Med Oral Pathol 1990;70:748–755.

580. Greenberg RG, Berger TG. Nail and mucocutaneous hyperpigmentation with azidothymidine therapy. J Am Acad Dermatol 1990;22:327–330.

581. Tadini G, D'Orso M, Cusini M, Alessi E. Oral mucosa pigmentation: a new side effect of azidothymidine therapy in patients with acquired immunodeficiency syndrome. Arch Dermatol 1991;127:267–268.

582. Porter SR, Glover S, Scully C. Oral hyperpigmentation and adrenocortical hypofunction in a patient with acquired immunodeficiency syndrome. Oral Surg Oral Med Oral Pathol 1990;70:59–60.

583. Zhang X, Langford A, Gelderblom H, Reichart P. Ultrastructural findings in oral hyperpigmentation of HIV-infected patients. J Oral Pathol Med 1989;18:471–474.

584. Langford A, Ruf B. Diagnosis and differential diagnosis of oral hyperpigmentation. Quintessenz 1990;41: 1989–2001.

585. Winkler JR, Robertson PB. Periodontal disease associated with HIV infection. Oral Surg Oral Med Oral Pathol 1992;73:145–150.

586. Williams CA, Winkler JR, Grassi M, Murray PA. HIV-associated periodontitis complicated by necrotizing stomatitis. Oral Surg Oral Med Oral Pathol 1990;69: 351–355.

587. Melnick SL, Engel D, Truelove E, et al. Oral mucosal lesions: association with the presence of antibodies to the human immunodeficiency virus. Oral Surg Oral Med Oral Pathol 1989;68:37–43.

588. Felix DH, Wray D, Smith GL, Jones GA. Oro-antral fistula: an unusual complication of HIV-associated periodontal disease. Br Dent J 1991;171:61–62.

589. Scully C, McCarthy G. Management of oral health in persons with HIV infection. Oral Surg Oral Med Oral Pathol 1992;73:215–225.

590. Anneroth G, Anneroth I, Lynch DP. Acquired immune deficiency syndrome (AIDS) in the United States in 1986: etiology, epidemiology, clinical manifestations, and dental implications. J Oral Maxillofac Surg 1986; 44:956–964.

591. Schiodt M, Greenspan D, Daniels TE, et al. Parotid gland enlargement and xerostomia associated with labial sialadenitis in HIV-infected patients. J Autoimmun 1989;2:415–425.

592. Schiodt M, Greenspan D, Levy JA, et al. Does HIV cause salivary gland disease? AIDS 1989;3:819–822.

593. Zeitlen S, Shaha A. Parotid manifestations of HIV infection. J Surg Oncol 1991;47:230–232.

594. Schiodt M. HIV-associated salivary gland disease: a review. Oral Surg Oral Med Oral Pathol 1992;73:164–167.

595. Fox PC, van der Ven PF, Sonies BC, Weiffenbach JM, Baum BJ. Xerostomia: evaluation of a symptom with increasing significance. J Am Dent Assoc 1985;110: 519–525.

596. Couderc LJ, D'Agay MF, Danon F, Harzic M, Brocheriou C, Clauvel JP. Sicca complex and infection with human immunodeficiency virus. Arch Intern Med 1987;147:898–901.

597. Kaye BR. Rheumatologic manifestations of infection with human immunodeficiency virus (HIV). Ann Intern Med 1989;111:158–167.

598. Chapnik JS, Noyek AM, Berris B, et al. Parotid gland enlargement in HIV infection: clinical/imaging findings. J Otolaryngol 1990;19:189–194.

599. Knox WF, McWilliam LJ, Banerjee SS. Benign lymphoepithelial lesion of salivary gland in a patient with AIDS. J Clin Pathol 1990;43:780–781.

600. Terry JH, Loree TR, Thomas MD, Marti JR. Major salivary gland lymphoepithelial lesions and the acquired immunodeficiency syndrome. Am J Surg 1991;162: 324–329.

601. Fox PC. Saliva and salivary gland alterations in HIV infection. J Am Dent Assoc 1991;122:46–48.

602. Schiodt M, Dodd CL, Greenspan D, et al. Natural history of HIV-associated salivary gland disease. Oral Surg Oral Med Oral Pathol 1992;74:326–331.

603. Rosenberg ZS, Joffe SA, Itescu S. Spectrum of salivary gland disease in HIV-infected patients: characterization with Ga-67 citrate imaging. Radiology 1992;184: 761–764.

604. Scully C, Davies R, Porter S, Eveson J, Luker J. HIV-salivary gland disease. Salivary scintiscanning with technetium pertechnetate. Oral Surg Oral Med Oral Pathol 1993;76:120–123.

605. Ryan JR, Ioachim HL, Marmer J, Loubeau JM. Acquired immune deficiency syndrome-related lymphadenopathies presenting in the salivary gland lymph nodes. Arch Otolaryngol 1985;111:554–556.

606. Shaha A, Thelmo W, Jaffee BM. Is parotid lymphadenopathy a new disease or part of AIDS? Am J Surg 1988;156:297–300.

607. Vargas PA, Mauad T, Boehm GM, et al. Parotid gland involvement in advanced AIDS. Oral Dis 2003;9:55–61.

608. Lecatsas G, Houff S, Macher A, et al. Retrovirus-like particles in salivary glands, prostate and testes of AIDS patients. Proc Soc Exp Biol Med 1985;178:653–655.

609. Schiodt M, Atkinson JC, Greenspan D, et al. Sialochemistry in human immunodeficiency virus associated salivary gland disease. J Rheumatol 1992;19:26–29.

610. Fox PC. Salivary gland involvement in HIV-1 infection. Oral Surg Oral Med Oral Pathol 1992;73:168–170.

611. Yeung SC, Kazazi F, Randle CG, et al. Patients infected with human immunodeficiency virus type 1 have low levels of virus in saliva even in the presence of periodontal disease. J Infect Dis 1993;167:803–809.

612. Barr CE, Miller KL, Lopez MR, et al. Recovery of infectious HIV-1 from whole saliva. J Am Dent Assoc 1992; 123:36–48.

613. Barr CE, Lopez MR, Rua-Dobles A, Miller LK, Mathur-Wagh U, Turgeon LR. HIV-associated oral lesions; immunologic, virologic and salivary parameters. J Oral Pathol Med 1992;21:295–298.

614. Mandel ID, Barr CE, Turgeon L. Longitudinal study of parotid saliva in HIV-1 infection. J Oral Pathol Med 1992;21:209–213.

615. Yeh CK, Fox PC, Goto Y, Austin HA, Brahim JS, Fox CH. Human immunodeficiency virus (HIV) and HIV infected cells in saliva and salivary glands of a patient with systemic lupus erythematosus. J Rheumatol 1992; 19:1810–1812.

616. Epstein JB, Scully C, Porter SR. The risk of transmission of human immunodeficiency virus in dental practice. Oral Health 1992;82:33–38.

617. Moore BE, Flaitz CM, Coppenhaver DH, et al. HIV recovery from saliva before and after dental treatment. Inhibitors may have a critical role in viral inactivation. J Am Dent Assoc 1993;124:67–74.

618. Branson BM. FDA approves OraQuick for use in saliva. AIDS Clin Care 2004;16:39.

619. Oral HIV test approved by FDA. Lancet 2004;363:1125.

620. Pinheiro A, Marcenes W, Zakrzewska JM, et al. Dental and oral lesions in HIV infected patients: a study in Brazil. Int Dent J 2004;54:131–137.

621. Mulligan R, Phelan JA, Brunelle J, et al. Baseline characteristics of participants in the oral health component of the Women's Interagency HIV Study. Community Dent Oral Epidemiol 2004;32:86–98.

622. Abelson DC, Barton J, Mandel ID. The effect of chewing sorbitol-sweetened gum on salivary flow and cemental plaque pH in subjects with low salivary flow. J Clin Dent 1990;2:3–5.

623. Fox RI, Michelson P. Approaches to the treatment of Sjögren's syndrome. J Rheumatol 2000;suppl 61:15–21.

624. Fox RI, Stern M, Michelson P. Update in Sjögren syndrome. Curr Opin Rheumatol 2000;12:391–398.

625. Karpatkin S, Nardi MA. Immunologic thrombocytopenic purpura in human immunodeficiency virus–seropositive patients with hemophilia. Comparison with patients with classic autoimmune thrombocytopenic purpura, homosexuals with thrombocytopenia, and narcotic addicts with thrombocytopenia. J Lab Clin Med 1988;111:441–448.

626. Rarick MU, Espina B, Mocharnuk R, Trilling Y, Levine AM. Thrombotic thrombocytopenic purpura in patients with human immunodeficiency virus infection: a report of three cases and review of the literature. Am J Hematol 1992;40:103–109.

627. Najean Y, Rain JD. The mechanism of thrombocytopenia in patients with HIV infection. J Lab Clin Med 1994;123:415–420.

628. Ficarra G. Oral lesions of iatrogenic and undefined etiology and neurologic disorders associated with HIV infection. Oral Surg Oral Med Oral Pathol 1992;73:201–211.

629. van der Waal I. Organ-specific manifestations of HIV infection. IV. Oral manifestations of HIV infection: comments on the present classification. AIDS 1993;7(suppl 1):S223–224.

630. Cabrera VP, Rodu B. Differential diagnosis of oral mucosal petechial hemorrhages. Compendium 1991; 12:418–422.

631. Speight PM, Zakrzewska J, Fletcher CD. Epithelioid angiomatosis affecting the oral cavity as a first sign of HIV infection. Br Dent J 1991;171:367–370.

632. Nishioka GJ, Chilcoat CC, Aufdemorte TB, Clare N. The gingival biopsy in the diagnosis of thrombotic thrombocytopenic purpura. Oral Surg Oral Med Oral Pathol 1988;65:580–585.

633. Costello C, Treacy M, Lai L. Treatment of immune thrombocytopenic purpura in homosexual men. Scand J Haematol 1986;36:507–510.

634. Reichart PA. Oral ulceration and iatrogenic disease in HIV infection. Oral Surg Oral Med Oral Pathol 1992; 73:212–214.

635. Silverman Jr S, Gallo J, Stites DP. Prednisone management of HIV-associated recurrent oral aphthous ulcerations. J Acquir Immune Defic Syndr 1992;5:952–953.

636. Youle M, Clarbour J, Farthing C, et al. Treatment of resistant aphthous ulceration with thalidomide in patients positive for HIV antibody. BMJ 1989;298:432.

637. Erlich KS, Mills J. Other virus infections in AIDS. II. Herpes simplex virus. Immunol Ser 1989;44:534–554.

638. Corey JP, Seligman I. Otolaryngology problems in the immune compromised patient—an evolving natural history. Otolaryngol Head Neck Surg 1991;104:196–203.

639. Eversole LR. Viral infections of the head and neck among HIV-seropositive patients. Oral Surg Oral Med Oral Pathol 1992;73:155–163.

640. Katz MH, Mastrucci MT, Leggott PJ, Westenhouse J, Greenspan JS, Scott GB. Prognostic significance of oral lesions in children with perinatally acquired human immunodeficiency virus infection. Am J Dis Child 1993;147:45–48.

641. Jones AC, Migliorati CA, Baughman RA. The simultaneous occurrence of oral herpes simplex virus, cytomegalovirus, and histoplasmosis in an HIV-infected patient. Oral Surg Oral Med Oral Pathol 1992;74:334–339.

642. Heinic GS, Northfelt DW, Greenspan JS, MacPhail LA, Greenspan D. Concurrent oral cytomegalovirus and herpes simplex virus infection in association with HIV infection. A case report. Oral Surg Oral Med Oral Pathol 1993;75:488–494.

643. Miller RG, Whittington WL, Coleman RM, Nigida Jr SM. Acquisition of concomitant oral and genital infection with herpes simplex virus type 2. Sex Transm Dis 1987;14:41–43.

644. Cohen SG, Greenberg MS. Chronic oral herpes simplex virus infection in immunocompromised patients. Oral Surg Oral Med Oral Pathol 1985;59:465–471.

645. Reichart PA. Oral manifestations of recently described viral infections, including AIDS. Curr Opin Dent 1991;1:377–383.

646. Barrett AP, Buckley DJ, Greenberg ML, Earl MJ. The value of exfoliative cytology in the diagnosis of oral herpes simplex infection in immunosuppressed patients. Oral Surg Oral Med Oral Pathol 1986;62:175–178.

647. Bagg J, Mannings A, Munro J, Walker DM. Rapid diagnosis of oral herpes simplex or zoster virus infections by immunofluorescence: comparison with Tzanck cell preparations and viral culture. Br Dent J 1989;167:235–238.

648. Epstein JB, Page JL, Anderson GH, Spinelli J. The role of an immunoperoxidase technique in the diagnosis of oral herpes simplex virus infection in patients with leukemia. Diagn Cytopathol 1987;3:205–209.

649. Laga Jr EA, Toth BB, Rolston KV, Tarrand JJ. Evaluation of a rapid enzyme-linked immunoassay for the diagnosis of herpes simplex virus in cancer patients with oral lesions. Oral Surg Oral Med Oral Pathol 1993;75:168–172.

650. Mintz GA, Rose SL. Diagnosis of oral herpes simplex virus infections: practical aspects of viral culture. Oral Surg Oral Med Oral Pathol 1984;58:486–492.

651. Declerq E. Antivirals for the treatment of herpesvirus infections. J Antimicrob Chemother 1993;32(suppl A):121–132.

652. Perry CM, Wagstaff AJ. Famciclovir. A review of its pharmacological properties and therapeutic efficacy in herpesvirus infections. Drugs 1995;50:396–415.

653. Cirelli R, Herne K, McCrary M, Lee P, Tyring SK. Famciclovir: review of clinical efficacy and safety. Antiviral Res 1995;29:141–151.

654. Alrabiah FA, Sachs SL. New antiherpesvirus agents. Their targets and therapeutic potential. Drugs 1996;52:17–32.

655. Perry CM, Faulds D. Valaciclovir. A review of its antiviral activity, pharmacokinetic properties and therapeutic efficacy in herpesvirus infections. Drugs 1996;52:754–772.

656. Stein GE. Pharmacology of new antiherpes agents: famciclovir and valacyclovir. J Am Pharm Assoc 1997;37:157–163.

657. Acosta EP, Fletcher CV. Valacyclovir. Ann Pharmacother 1997;31:185–191.

658. Hamuy R, Berman B. Treatment of herpes simplex virus infections with topical antiviral agents. Eur J Dermatol 1998;8:310–319.

659. Snoeck R. Antiviral therapy of herpes simplex. Int J Antimicrob Agents 2000;16:157–169.

660. Emmert DH. Treatment of common cutaneous herpes simplex virus infections. Am Fam Physician 2000;61:1697–1708.

661. Safrin S. Treatment of acyclovir-resistant herpes simplex virus infections in patients with AIDS. J Acquir Immune Defic Syndr 1992;5(suppl 1):S29–S32.

662. Safrin S, Kemmerly S, Plotkin B, et al. Foscarnet-resistant herpes simplex virus infection in patients with AIDS. J Infect Dis 1994;169:193–196.

663. Marks GL, Nolan PE, Erlich KS, Ellis MN. Mucocutaneous dissemination of acyclovir-resistant herpes simplex virus in a patient with AIDS. Rev Infect Dis 1989;11:474–476.

664. Butler S, Molinari JA, Plezia RA, Chandrasekar P, Venkat H. Condyloma acuminatum in the oral cavity: four cases and a review. Rev Infect Dis 1988;10:544–550.

665. Zunt SL, Tomich CE. Oral condyloma acuminatum. J Dermatol Surg Oncol 1989;15:591–594.

666. Syrjanen SM, Syrjanen KJ, Lamberg MA. Detection of human papillomavirus DNA in oral mucosal lesions using in situ DNA-hybridization applied on paraffin sections. Oral Surg Oral Med Oral Pathol 1986;62:660–667.

667. Eversole LR, Laipis PJ, Merrell P, Choi E. Demonstration of human papillomavirus DNA in oral condyloma acuminatum. J Oral Pathol 1987;16:266–272.

668. Luomanen M. Experience with a carbon dioxide laser for removal of benign oral soft-tissue lesions. Proc Finn Dent Soc 1992;88:49–55.

669. Vilmer C, Cavelier-Balloy B, Pinquier L, Blanc F, Dubertret L. Focal epithelial hyperplasia and multifocal human papillomavirus infection in an HIV-seropositive man. J Am Acad Dermatol 1994;30:497–498.

670. Syrjanen S, von Krogh G, Kellokoski J, Syrjanen K. Two different human papillomavirus (HPV) types associated with oral mucosal lesions in an HIV-seropositive man. J Oral Pathol Med 1989;18:366–370.

671. Schulten EA, ten Kate RW, van der Waal I. Oral findings in HIV-infected patients attending a department of internal medicine: the contribution of intraoral examination towards the clinical management of HIV disease. Q J Med 1990;76:741–745.

672. Laskaris G, Hadjivassiliou M, Stratigos J. Oral signs and symptoms in 160 Greek HIV-infected patients. J Oral Pathol Med 1992;21:120–123.

673. de Villiers EM. Prevalence of HPV 7 papillomas in the oral mucosa and facial skin of patients with human immunodeficiency virus. Arch Dermatol 1989;125:1590.

674. Greenspan D, de Villiers EM, Greenspan JS, de Souza YG, zur Hausen H. Unusual HPV types in oral warts in association with HIV infection. J Oral Pathol 1988;17:482–488.

675. DeRossi SS, Laudenbach J. The management of oral human papillomavirus with topical cidofir: a case report. Cutis 2004;73:191–193.

676. Quinnan Jr GV, Masur H, Rook AH, et al. Herpesvirus infections in the acquired immune deficiency syndrome. JAMA 1984;252:72–77.

677. Perronne C, Lazanas M, Leport C, et al. Varicella in patients infected with the human immunodeficiency virus. Arch Dermatol 1990;126:1033–1036.

678. Kelley R, Mancao M, Lee F, Sawyer M, Nahmias A, Nesheim S. Varicella in children with perinatally acquired human immunodeficiency virus infection. J Pediatr 1994;124:271–273.

679. Jura E, Chadwick EG, Josephs SH, et al. Varicella-zoster virus infections in children infected with human immunodeficiency virus. Pediatr Infect Dis J 1989;8:586–590.

680. Patterson LE, Butler KM, Edwards MS. Clinical herpes zoster shortly following primary varicella in two HIV-infected children. Clin Pediatr 1989;28:354.

681. Gulick RM, Heath-Chiozzi M, Crumpacker CS. Varicella-zoster virus disease in patients with human immunodeficiency virus infection. Arch Dermatol 1990;126:1086–1088.

682. Hoppenjans WB, Bibler MR, Orme RL, Solinger AM. Prolonged cutaneous herpes zoster in acquired immunodeficiency syndrome. Arch Dermatol 1990;126:1048–1050.

683. Gnann JW, Whitley RJ. Natural history and treatment of varicella-zoster in high-risk populations. J Hosp Infect 1991;18(suppl A):317–329.

684. Leibovitz E, Kaul A, Rigaud M, Bebenroth D, Krasinski K, Borkowsky W. Chronic varicella zoster in a child infected with human immunodeficiency virus: case report and review of the literature. Cutis 1992;49:27–31.

685. Srugo I, Israele V, Wittek AE, Courville T, Viman VM, Brunell PA. Clinical manifestations of varicella-zoster virus infections in human immunodeficiency virus-infected children. Am J Dis Child 1993;147:742–745.

686. Wallace MR, Hooper DG, Pyne JM, Graves SJ, Malone JL. Varicella immunity and clinical disease in HIV-infected adults. South Med J 1994;87:74–76.

687. Sindrup JH, Weismann K, Sand Petersen C, et al. Skin and oral mucosal changes in patients infected with human immunodeficiency virus. Acta Derm Venereol 1988;68:440–443.

688. Jensen JL, Kanas RJ, De Boom GW. Multiple oral and labial ulcers in an immunocompromised patient. J Am Dent Assoc 1987;114:235–236.

689. Van de Perre P, Bakkers E, Batungwanayo J, et al. Herpes zoster in African patients: an early manifestation of HIV infection. Scand J Infect Dis 1988;20:277–282.

690. Moskow BS, Hernandez G. Aggressive periodontal destruction and herpes zoster in a suspected AIDS patient. J Parodontol 1991;10:359–369.

691. Balfour Jr HH. Acyclovir and other chemotherapy for herpes group viral infections. Annu Rev Med 1984;35:279–291.

692. Straus SE. The management of varicella and zoster infections. Infect Dis Clin North Am 1987;1:367–382.

693. Hermans PE, Cockerill 3rd FR. Antiviral agents. Mayo Clin Proc 1987;62:1108–1115.

694. Straus SE, Ostrove JM, Inchauspe G, et al. NIH conference. Varicella-zoster virus infections. Biology, natural history, treatment, and prevention. Ann Intern Med 1988;108:221–237.

695. Huff JC. Antiviral treatment in chickenpox and herpes zoster. J Am Acad Dermatol 1988;18:204–206.

696. Balfour Jr HH. Varicella zoster virus infections in immunocompromised hosts. A review of the natural history and management. Am J Med 1988;85:68–73.

697. Sellitti TP, Huang AJ, Schiffman J, Davis JL. Association of herpes zoster ophthalmicus with acquired immunodeficiency syndrome and acute retinal necrosis. Am J Ophthalmol 1993;116:297–301.

698. Tricot G, De Clercq E, Boogaerts MA, Verwilghen RL. Oral bromovinyldeoxyuridine therapy for herpes simplex and varicella-zoster virus infections in severely immunosuppressed patients: a preliminary clinical trial. J Med Virol 1986;18:11–20.

699. Safrin S, Berger TG, Gilson I, et al. Foscarnet therapy in five patients with AIDS and acyclovir-resistant varicella-zoster virus infection. Ann Intern Med 1991;115:19–21.

700. Cole NL, Balfour Jr HH. Varicella-zoster virus does not become more resistant to acyclovir during therapy. J Infect Dis 1986;153:605–608.

701. Jacobson MA, Berger TG, Fikrig S, et al. Acyclovir-resistant varicella zoster virus infection after chronic oral acyclovir therapy in patients with the acquired immunodeficiency syndrome (AIDS). Ann Intern Med 1990;112:187–191.

702. Balfour Jr HH, Benson C, Braun J, et al. Management of acyclovir-resistant herpes simplex and varicella-zoster virus infections. J Acquir Immune Defic Syndr 1994;7:254–260.

703. Lyall EG, Ogilvie MM, Smith NM, Burns S. Acyclovir resistant varicella zoster and HIV infection. Arch Dis Child 1994;70:133–135.

704. Zambon JJ, Reynolds HS, Genco RJ. Studies of the subgingival microflora in patients with acquired immunodeficiency syndrome. J Periodontol 1990;61:699–704.

705. Watkins KV, Richmond AS, Langstein IM. Nonhealing extraction site due to Actinomyces naeslundii in a patient with AIDS. Oral Surg Oral Med Oral Pathol 1991;71:675–677.

706. Cockerell CJ, Le Boit PE. Bacillary angiomatosis: a newly characterized, pseudoneoplastic, infectious, cutaneous vascular disorder. J Am Acad Dermatol 1990;22:501–512.

707. Cockerell CJ. The clinicopathologic spectrum of bacillary (epithelioid) angiomatosis. Prog AIDS Pathol 1990;2:111–226.

708. Cotell SL, Noskin GA. Bacillary angiomatosis. Clinical and histologic features, diagnosis, and treatment. Arch Intern Med 1994;154:524–528.

709. Koehler JE, Le Boit PE, Egbert BM, Berger TG. Cutaneous vascular lesions and disseminated cat-scratch disease in patients with the acquired immunodeficiency syndrome (AIDS) and AIDS-related complex. Ann Intern Med 1988;109:449–455.

710. Le Boit PE, Berger TG, Egbert BM, et al. Epithelioid haemangioma-like vascular proliferation in AIDS: manifestation of cat scratch disease bacillus infection? Lancet 1988;1:960–963.

711. Pilon VA, Echols RM. Cat-scratch disease in a patient with AIDS. Am J Clin Pathol 1989;92:236–240.

712. Szaniawski WK, Don PC, Bitterman SR, Schachner JR. Epithelioid angiomatosis in patients with AIDS. Report of seven cases and review of the literature. J Am Acad Dermatol 1990;23:41–48.

713. Schwartzman WA, Marchevsky A, Meyer RD. Epithelioid angiomatosis or cat scratch disease with splenic and hepatic abnormalities in AIDS: case report and review of the literature. Scand J Infect Dis 1990;22:121–133.

714. Kemper CA, Lombard CM, Deresinski SC, Tompkins LS. Visceral bacillary epithelioid angiomatosis: possible manifestations of disseminated cat scratch disease in the immunocompromised host: a report of two cases. Am J Med 1990;89:216–122.

715. Relman DA, Loutit JS, Schmidt TM, Falkow S, Tompkins LS. The agent of bacillary angiomatosis. An approach to the identification of uncultured pathogens. N Engl J Med 1990;323:1573–1580.

716. Cockerell CJ, Tierno PM, Friedman-Kien AE, Kim KS. Clinical, histologic, microbiologic, and biochemical characterization of the causative agent of bacillary (epithelioid) angiomatosis: a rickettsial illness with features of bartonellosis. J Invest Dermatol 1991;97:812–817.

717. McDonald G. Cat-scratch disease. Postgrad Med 1992;92:47.

718. Birtles RJ, Harrison TG, Taylor AG. Cat scratch disease and bacillary angiomatosis: aetiological agents and the link with AIDS. Commun Dis Rep CDR Rev 1993;3:R107–110.

719. Tappero JW, Mohle-Boetani J, Koehler JE, et al. The epidemiology of bacillary angiomatosis and bacillary peliosis. JAMA 1993;269:770–775.

720. Marasco WA, Lester S, Parsonnet J. Unusual presentation of cat scratch disease in a patient positive for antibody to the human immunodeficiency virus. Rev Infect Dis 1989;11:793–803.

721. Abrams J, Farhood AI. Infection-associated vascular lesions in acquired immunodeficiency syndrome patients. Hum Pathol 1989;20:1025–1026.

722. Berger TG, Tappero JW, Kaymen A, Le Boit PE. Bacillary (epithelioid) angiomatosis and concurrent Kaposi's sarcoma in acquired immunodeficiency syndrome. Arch Dermatol 1989;125:1543–1547.

723. Le Boit PE, Berger TG, Egbert BM, Beckstead JH, Yen TS, Stoler MH. Bacillary angiomatosis. The histopathology and differential diagnosis of a pseudoneoplastic infection in patients with human immunodeficiency virus disease. Am J Surg Pathol 1989;13:909–920.

724. Walford N, Van der Wouw PA, Das PK, Ten Velden JJ, Hulsebosch HJ. Epithelioid angiomatosis in the acquired immunodeficiency syndrome: morphology and differential diagnosis. Histopathology 1990;16:83–88.

725. Glick M, Cleveland DB. Oral mucosal bacillary epithelioid angiomatosis in a patient with AIDS associated with rapid alveolar bone loss: case report. J Oral Pathol Med 1993;22:235–239.

726. Rudikoff D, Phelps RG, Gordon RE, Battone EJ. Acquired immunodeficiency syndrome-related bacillary vascular proliferation (epithelioid angiomatosis): rapid response to erythromycin therapy. Arch Dermatol 1989;125:706–707.

727. van der Wouw PA, Hadderingh RJ, Reiss P, Hulsebosch HJ, Walford N, Lange JM. Disseminated cat-scratch disease in a patient with AIDS. AIDS 1989;3:751–753.

728. Mui BS, Mulligan ME, George WL. Response of HIV-associated disseminated cat scratch disease to treatment with doxycycline. Am J Med 1990;89:229–231.

729. Holley Jr HP. Successful treatment of cat-scratch disease with ciprofloxacin. JAMA 1991;265:1563–1565.

730. Innocenzi D, Cerio R, Barduagni O, Bosman C, Carlesimo OA. Bacillary epithelioid angiomatosis in acquired immunodeficiency syndrome (AIDS)—clinicopathological and ultrastructural study of a case with a review of the literature. Clin Exp Dermatol 1993;18:133–137.

731. Salomon D, Saurat JH. Erythema multiforme major in a 2-month-old child with human immunodeficiency virus (HIV) infection. Br J Dermatol 1990;123:797–800.

732. Belfort Jr R, de Smet M, Whitcup SM, et al. Ocular complications of Stevens-Johnson syndrome and toxic epidermal necrolysis in patients with AIDS. Cornea 1991;10:536–538.

733. Lewis DA, Brook MG. Erythema multiforme as a presentation of human immunodeficiency virus seroconversion illness. Int J STD AIDS 1992;3:56–57.

734. Schuval SJ, Bonagura VR, Ilowite NT. Rheumatologic manifestations of pediatric human immunodeficiency virus infection. J Rheumatol 1993;20:1578–1582.

735. Porteous DM, Berger TG. Severe cutaneous drug reactions (Stevens-Johnson syndrome and toxic epidermal necrolysis) in human immunodeficiency virus infection. Arch Dermatol 1991;127:740–741.

736. Saiag P, Caumes E, Chosidow O, Revuz J, Roujeau JC. Drug-induced toxic epidermal necrolysis (Lyell syndrome) in patients infected with the human immunodeficiency virus. J Am Acad Dermatol 1992;26:567–574.

737. Azon-Masoliver A, Vilaplana J. Fluconazole-induced toxic epidermal necrolysis in a patient with human immunodeficiency virus infection. Dermatology 1993;187:268–269.

738. Schlienger RG, Haefeli WE, Bircher A, Leib SL, Luscher TF. Drug-induced Stevens-Johnson syndrome in a patient with AIDS. Schweiz Rundsch Med Prax 1993;82:888–892.

739. Krippaehne JA, Montgomery MT. Erythema multiforme: a literature review and case report. Spec Care Dentist 1992;12:125–130.

740. Williams DM. Non-infectious diseases of the oral soft tissue: a new approach. Adv Dent Res 1993;7:213–219.

741. Corticosteroids for erythema multiforme? Pediatr Dermatol 1989;6:229–250.

742. Lozada-Nur F, Gorsky M, Silverman Jr S. Oral erythema multiforme: clinical observations and

treatment of 95 patients. Oral Surg Oral Med Oral Pathol 1989;67:36–40.

743. Lozada-Nur F, Cram D, Gorsky M. Clinical response to levamisole in thirty-nine patients with erythema multiforme. An open prospective study. Oral Surg Oral Med Oral Pathol 1992;74:294–298.

744. Schofield JK, Tatnall FM, Leigh IM. Recurrent erythema multiforme: clinical features and treatment in a large series of patients. Br J Dermatol 1993;128: 542–545.

745. Stein DK, Sugar AM. Fungal infections in the immunocompromised host. Diagn Microbiol Infect Dis 1989; 12(suppl 4):221S–228S.

746. de Almeida OP, Scully C. Oral lesions in the systemic mycoses. Curr Opin Dent 1991;1:423–428.

747. Samaranayake LP. Oral mycoses in HIV infection. Oral Surg Oral Med Oral Pathol 1992;73:171–180.

748. Grant IH, Armstrong D. Fungal infections in AIDS. Cryptococcosis. Infect Dis Clin North Am 1988;2: 457–464.

749. Sugar AM. Overview: cryptococcosis in the patient with AIDS. Mycopathologia 1991;114:153–157.

750. Stern JJ, Hartman BJ, Sharkey P, et al. Oral fluconazole therapy for patients with acquired immunodeficiency syndrome and cryptococcosis: experience with 22 patients. Am J Med 1988;85:477–480.

751. Hostetler JS, Denning DW, Stevens DA. US experience with itraconazole in aspergillus, cryptococcus and histoplasma infections in the immunocompromised host. Chemotherapy 1992;8(suppl 1):12–22.

752. Laroche R, Dupond B, Touze JE, et al. Cryptococcal meningitis associated with acquired immunodeficiency syndrome (AIDS) in African patients: treatment with fluconazole. J Med Vet Mycol 1992;30: 71–78.

753. Dismukes WE. Management of cryptococcosis. Clin Infect Dis 1993;17(suppl 2):S507–512.

754. Como JA, Dismukes WE. Oral azole drugs as systemic antifungal therapy. N Engl J Med 1994;330:263–272.

755. Lynch DP, Naftolin LZ. Oral Cryptococcus neoformans infection in AIDS. Oral Surg Oral Med Oral Pathol 1987;64:449–453.

756. Glick M, Cohen SG, Cheney RT, Crooks GW, Greenberg MS. Oral manifestations of disseminated Cryptococcus neoformans in a patient with acquired immunodeficiency syndrome. Oral Surg Oral Med Oral Pathol 1987;64:454–459.

757. Heimdahl A, Nord CE. Oral yeast infections in immunocompromised and seriously diseased patients. Acta Odontol Scand 1990;48:77–84.

758. Tzerbos F, Kabani S, Booth D. Cryptococcosis as an exclusive oral presentation. J Oral Maxillofac Surg 1992;50:759–760.

759. Heinic GS, Greenspan D, MacPhail LA, Greenspan JS. Oral Geotrichum candidum infection associated with HIV infection. A case report. Oral Surg Oral Med Oral Pathol 1992;73:726–728.

760. Hay RJ. Histoplasmosis. Semin Dermatol 1993;12:310–314.

761. Cobb CM, Shultz RE, Brewer JH, Dunlap CL. Chronic pulmonary histoplasmosis with an oral lesion. Oral Surg Oral Med Oral Pathol 1989;67:73–76.

762. Cohen PR, Bank DE, Silvers DN, Grossman ME. Cutaneous lesions of disseminated histoplasmosis in human immunodeficiency virus-infected patients. J Am Acad Dermatol 1990;23:422–428.

763. Oda D, MacDougall L, Fritsche T, Worthington P, MacDougall L. Oral histoplasmosis as a presenting disease in acquired immunodeficiency syndrome. Oral Surg Oral Med Oral Pathol 1990;70:631–636.

764. Eisig S, Boguslaw B, Cooperband B, Phelan J. Oral manifestations of disseminated histoplasmosis in acquired immunodeficiency syndrome: report of two cases and review of the literature. J Oral Maxillofac Surg 1991;49:310–313.

765. Heinic GS, Greenspan D, MacPhail LA, et al. Oral Histoplasma capsulatum infection in association with HIV infection: a case report. J Oral Pathol Med 1992; 21:85–89.

766. Swindells S, Durham T, Johansson SL, Kaufman L. Oral histoplasmosis in a patient infected with HIV. A case report. Oral Surg Oral Med Oral Pathol 1994;77: 126–130.

767. Mandell W, Goldberg DM, Neu HC. Histoplasmosis in patients with the acquired immune deficiency syndrome. Am J Med 1986;81:974–978.

768. Dobleman TJ, Scher N, Goldman M, Doot S. Invasive histoplasmosis of the mandible. Head Neck 1989;11: 81–84.

769. Scully C, de Almeida OP. Orofacial manifestations of the systemic mycoses. J Oral Pathol Med 1992;21:289–294.

770. Stein DK, Sugar AM. Fungal infections in the immunocompromised host. Diagn Microbiol Infect Dis 1989; 12(suppl 4):221S–228S.

771. Wheat LJ. Diagnosis and management of histoplasmosis. Eur J Clin Microbiol Infect Dis 1989;8:480–490.

772. Saag MS, Dismukes WE. Treatment of histoplasmosis and blastomycosis. Chest 1988;93:848–851.

773. Quinones CA, Reuben AG, Hamill RJ, Musher DM, Gorin AB, Sarosi GA. Chronic cavitary histoplasmosis. Failure of oral treatment with ketoconazole. Chest 1989;95:914–916.

774. Hay RJ. Antifungal therapy and the new azole compounds. J Antimicrob Chemother 1991;28(suppl A): 35–46.

775. Negroni R, Taborda A, Robies AM, Archevala A. Itraconazole in the treatment of histoplasmosis associated with AIDS. Mycoses 1992;35:281–287.

776. Berger TG. Treatment of bacterial, fungal, and parasitic infections in the HIV-infected host. Semin Dermatol 1993;12:296–300.

777. Rinaldi MG. Zygomycosis. Infect Dis Clin North Am 1989;3:19–41.

778. Ng KH, Chin CS, Jalleh RD, Siar CH, Ngui CH, Singaram SP. Nasofacial zygomycosis. Oral Surg Oral Med Oral Pathol 1991;72:685–688.

779. Jones AC, Bentsen TY, Freedman PD. Mucormycosis of the oral cavity. Oral Surg Oral Med Oral Pathol 1993;75:455–460.

780. Clark R, Greer DL, Carlisle T, Carroll B. Cutaneous zygomycosis in a diabetic HTLV-I-seropositive man. J Am Acad Dermatol 1990;22:956–959.

781. Hopwood V, Hicks DA, Thomas S, Evans EG. Primary cutaneous zygomycosis due to Absidia corymbifera in a patient with AIDS. J Med Vet Mycol 1992;30:399–402.

782. Graybill JR. Treatment of systemic mycoses in patients with AIDS. Arch Med Res 1993;24:403–412.

783. Klapholz A, Salomon N, Perlman DC, Talavera W. Aspergillosis in the acquired immunodeficiency syndrome. Chest 1991;100:1614–1618.

784. Denning DW, Follansbee SE, Scolaro M, Norris S, Edelstein H, Stevens DA. Pulmonary aspergillosis in the acquired immunodeficiency syndrome. N Engl J Med 1991;324:654–662.

785. Morace G, Tamburrini E, Manzara S, Antinori A, Maiuro G, Dettori G. Epidemiological and clinical aspects of mycoses in patients with AIDS-related pathologies. Eur J Epidemiol 1990;6:398–403.

786. Shannon MT, Sclaroff A, Colm SJ. Invasive aspergillosis of the maxilla in an immunocompromised patient. Oral Surg Oral Med Oral Pathol 1990;70:425–427.

787. Hostetler JS, Stevens DA. The treatment of aspergillosis, cryptococcosis and histoplasmosis in immunocompromised patients. Report of experience in the United States. Med Klin 1991;86(suppl 1):8–10.

788. Brown MM, Thompson A, Goh, Forster GE, Swash M. Bell's palsy and HIV infection. J Neurol Neurosurg Psychiatry 1988;51:425–426.

789. Belec L, Gherardi R, Georges AJ, et al. Peripheral facial paralysis and HIV infection: report of four African cases and review of the literature. J Neurol 1989; 236:411–414.

790. Murr AH, Benecke Jr JE. Association of facial paralysis with HIV positivity. Am J Otol 1991;12:450–451.

791. Belec L, Georges AJ, Bouree P, et al. Peripheral facial nerve palsy related to HIV infection: relationship with the immunological status and the HIV staging in Central Africa. Cent Afr J Med 1991;37:88–93.

792. Linstrom CJ, Pincus RL, Leavitt EB, Urbina MC. Otologic neurotologic manifestations of HIV-related disease. Otolaryngol Head Neck Surg 1993;108:680–687.

793. Uldry PA, Regli F. Isolated and recurrent peripheral facial paralysis in human infection with human immunodeficiency virus (HIV). Schweiz Med Wochenschr 1988;118:1029–1031.

794. Hutton KP, Rogers 3rd RS. Recurrent aphthous stomatitis. Dermatol Clin 1987;5:761–768.

795. Scully C, Porter S. Recurrent aphthous stomatitis: current concepts of etiology, pathogenesis and management. J Oral Pathol Med 1989;18:21–27.

796. MacPhail LA, Greenspan D, Feigal DW, Lennette ET, Greenspan JS. Recurrent aphthous ulcers in association with HIV infection. Description of ulcer types and analysis of T-lymphocyte subsets. Oral Surg Oral Med Oral Pathol 1991;71:678–683.

797. MacPhail LA, Greenspan D, Greenspan JS. Recurrent aphthous ulcers in association with HIV infection. Diagnosis and treatment. Oral Surg Oral Med Oral Pathol 1992;73:283–288.

798. Siegel RD, Granich R. Recurrent aphthous ulcers in association with HIV infection. Oral Surg Oral Med Oral Pathol 1993;76:406–407.

799. Phelan JA, Eisig S, Freedman PD, Newsome N, Klein RS. Major aphthous-like ulcers in patients with AIDS. Oral Surg Oral Med Oral Pathol 1991;71:68–72.

800. Reyes-Teran G, Ramirez-Amador V, De la Rosa E, Gonzalez-Guevara M, Ponce de Leon S. Major recurrent oral ulcers in AIDS: report of three cases. J Oral Pathol Med 1992;21:409–411.

801. Muzyka BC, Glick M. Major aphthous ulcers in patients with HIV disease. Oral Surg Oral Med Oral Pathol 1994;77:116–120.

802. Thompson AC, Nolan A, Lamey PJ. Minor aphthous oral ulceration: a double-blind cross-over study of

beclomethasone dipropionate aerosol spray. Scott Med J 1989;34:531–532.

803. Vincent SC, Lilly GE. Clinical, historic, and therapeutic features of aphthous stomatitis. Literature review and open clinical trial employing steroids. Oral Surg Oral Med Oral Pathol 1992;74:79–86.

804. Chadwick B, Addy M, Walker DM. Hexetidine mouthrinse in the management of minor aphthous ulceration and as an adjunct to oral hygiene. Br Dent J 1991;171:83–87.

805. Grinspan D. Significant response of oral aphthosis to thalidomide treatment. J Am Acad Dermatol 1985;12: 85–90.

806. Strazzi S, Lebbe C, Geoffray C, et al. Aphthous ulcers in HIV-infected patients: treatment with thalidomide. Genitourin Med 1992;68:424–425.

807. Ghigliotti G, Repetto T, Farris A, Roy MT, De Marchi R. Thalidomide: treatment of choice for aphthous ulcers in patients seropositive for human immunodeficiency virus. J Am Acad Dermatol 1993;28:271–272.

808. Ruah CB, Stram JR, Chasin WD. Treatment of severe recurrent aphthous stomatitis with colchicine. Arch Otolaryngol Head Neck Surg 1988;114:671–675.

809. Hutchinson VA, Angenend JL, Mok WL, Cummins JM, Richards AB. Chronic recurrent aphthous stomatitis: oral treatment with low-dose interferon alpha. Mol Biother 1990;2:160–164.

810. Marder MZ, Barr CE, Mandel ID. Cytomegalovirus presence and salivary composition in acquired immunodeficiency syndrome. Oral Surg Oral Med Oral Pathol 1985;60:372–376.

811. Lucht E, Albert J, Linde A, et al. Human immunodeficiency virus type 1 and cytomegalovirus in saliva. J Med Virol 1993;39:156–162.

812. Schubert MM. Oral manifestations of viral infections in immunocompromised patients. Curr Opin Dent 1991;1:384–397.

813. Epstein J, Scully C. Cytomegalovirus: a virus of increasing relevance to oral medicine and pathology. J Oral Pathol Med 1993;22:348–353.

814. Glick M, Cleveland DB, Salkin LM, Alfaro-Miranda M, Fielding AF. Intraoral cytomegalovirus lesion and HIV-associated periodontitis in a patient with acquired immunodeficiency syndrome. Oral Surg Oral Med Oral Pathol 1991;72:716–720.

815. Kanas RJ, Jensen JL, Abrams AM, Wuerker RB. Oral mucosal cytomegalovirus as a manifestation of the acquired immune deficiency syndrome. Oral Surg Oral Med Oral Pathol 1987;64:183–189.

816. Langford A, Kunze R, Timm H, Ruf B, Reichart P. Cytomegalovirus associated oral ulcerations in HIV-infected patients. J Oral Pathol Med 1990;19:71–76.

817. Jones AC, Freedman PD, Phelan JA, Baughman RA, Kerpel SM. Cytomegalovirus infections of the oral cavity. A report of six cases and review of the literature. Oral Surg Oral Med Oral Pathol 1993;75:76–85.

818. Heinic GS, Greenspan D, Greenspan JS. Oral CMV lesions and the HIV infected. Early recognition can help prevent morbidity. J Am Dent Assoc 1993;124: 99–105.

819. Epstein JB, Sherlock CH, Wolber RA. Oral manifestations of cytomegalovirus infection. Oral Surg Oral Med Oral Pathol 1993;75:443–451.

820. Schubert MM, Epstein JB, Lloid ME, Cooney E. Oral infections due to cytomegalovirus in immunocompromised patients. J Oral Pathol Med 1993;22:268–273.

821. Pedersen A, Hornsleth A. Recurrent aphthous ulceration: a possible clinical manifestation of reactivation of varicella zoster or cytomegalovirus infection. J Oral Pathol Med 1993;22:64–68.

822. Syrjanen S, Leimola-Virtanen R, Schmidt-Westerhausen A, et al. Oral ulcers in AIDS patients frequently associated with cytomegalovirus (CMV) and Epstein-Barr virus (EBV) infection. J Oral Pathol Med 1999;28:204–209.

823. Dodd CL, Winkler JR, Heinic GS, Daniels TE, Yee K, Greenspan D. Cytomegalovirus infection presenting as acute periodontal infection in a patient infected with the human immunodeficiency virus. J Clin Periodontol 1993;20:282–285.

824. Flaitz CM, Hicks MJ, Nichols CM. Cytomegaloviral infection of the mandible in acquired immunodeficiency syndrome. J Oral Maxillofac Surg 1994;52: 305–308.

825. Langford A, Kunze R, Schmelzer S, Wolf H, Pohle HD, Reichart P. Immunocytochemical detection of herpes viruses in oral smears of HIV-infected patients. J Oral Pathol Med 1992;21:49–57.

826. Meyers JD. Treatment of herpesvirus infections in the immunocompromised host. Scand J Infect Dis Suppl 1985;47:128–236.

827. Meyers JD. Chemoprophylaxis of viral infection in immunocompromised patients. Eur J Cancer Clin Oncol 1989;25:1369–1374.

828. Balfour Jr HH, Fletcher CV, Dunn D. Cytomegalovirus infections in the immunocompromised transplant patient. Prevention of cytomegalovirus disease with oral acyclovir. Transplant Proc 1991;23(suppl 1): 17–19.

829. De Armond B. Future directions in the management of cytomegalovirus infections. J Acquir Immune Defic Syndr 1991;4(suppl 1):S53–56.

830. Griffiths PD. Current management of cytomegalovirus disease. J Med Virol 1993;suppl 1;106–111.

831. Nelson MR, Erskine D, Hawkins DA, Gazzard BG. Treatment with corticosteroids—a risk factor for the development of clinical cytomegalovirus disease in AIDS. AIDS 1993;7:375–378.

832. Gellis SE. Warts and molluscum contagiosum in children. Pediatr Ann 1987;16:69–76.

833. Epstein WL. Molluscum contagiosum. Semin Dermatol. 1992;11:184–189.

834. Porter CD, Blake NW, Cream JJ, Archard LC. Molluscum contagiosum virus. Mol Cell Biol Hum Dis Ser 1992;1:233–257.

835. Katzman M, Elmets CA, Lederman MM. Molluscum contagiosum and the acquired immunodeficiency syndrome. Ann Intern Med 1985;102:413–414.

836. Sarma DP, Weilbaecher TG. Molluscum contagiosum in the acquired immunodeficiency syndrome. J Am Acad Dermatol 1985;13:682–683.

837. Delescluse J, Goens J. Molluscum contagiosum disclosing HTLV III infection. Dermatologica 1986;172: 283–285.

838. Katzman M, Carey JT, Elmets CA, Jacobs GH, Lederman MM. Molluscum contagiosum and the acquired immunodeficiency syndrome: clinical and immunological details of two cases. Br J Dermatol 1987;116:131–138.

839. Cotton DW, Cooper C, Barrett DF, Leppard BJ. Severe atypical molluscum contagiosum infection in an immunocompromised host. Br J Dermatol 1987;116: 871–876.

840. Prose NS, Mendez H, Menikoff H, Miller HJ. Pediatric human immunodeficiency virus infection and its cutaneous manifestations. Pediatr Dermatol 1987; 4:67–74.

841. Matis WL, Triana A, Shapiro R, Eldred L, Polk BF, Hood AF. Dermatologic findings associated with human immunodeficiency virus infection. J Am Acad Dermatol 1987;17:746–751.

842. Petersen CS, Gerstoft J. Molluscum contagiosum in HIV-infected patients. Dermatology 1992;184:19–21.

843. Schwartz JJ, Myskowski PL. Molluscum contagiosum in patients with human immunodeficiency virus infection. A review of twenty-seven patients. J Am Acad Dermatol 1992;27:583–588.

844. Barsh LI. Molluscum contagiosum of the oral mucosa. Report of a case. Oral Surg Oral Med Oral Pathol 1966;22:42–46.

845. Phelan JA, Saltzman BR, Friedland GH, Klein RS. Oral findings in patients with acquired immunodeficiency syndrome. Oral Surg Oral Med Oral Pathol 1987;64: 50–56.

846. Ficarra G, Gaglioti D. Facial molluscum contagiosum in HIV-infected patients. Int J Oral Maxillofac Surg 1989;18:200–201.

847. Sugihara K, Reichart PA, Gelderblom HR. Molluscum contagiosum associated with AIDS: a case report with ultrastructural study. J Oral Pathol Med 1990;19:235–239.

848. Laskaris G, Sklavounou A. Molluscum contagiosum of the oral mucosa. Oral Surg Oral Med Oral Pathol 1984;58:688–691.

849. Svirsky JA, Sawyer DR, Page DG. Molluscum contagiosum of the lower lip. Int J Dermatol 1985;24:668–669.

850. Whitaker SB, Wiegand SE, Budnick SD. Intraoral molluscum contagiosum. Oral Surg Oral Med Oral Pathol 1991;72:334–336.

851. Betlloch I, Pinazo I, Mestre F, Altes J, Villalonga C. Molluscum contagiosum in human immunodeficiency virus infection: response to zidovudine. Int J Dermatol 1989;28:351–352.

852. Garrett SJ, Robinson JK, Roenigk Jr HH. Trichloroacetic acid peel of molluscum contagiosum in immunocompromised patients. J Dermatol Surg Oncol 1992;18:855–858.

6

Common Respiratory Viruses and Pulmonary Mucosal Immunology

David B. Huang

The lungs play a critical role in supplying the body's cells with oxygen and in removal of waste products. The lungs are continually exposed to inhaled gases, particulates, and airborne pathogens as a result of daily inhalation of tens of thousands of liters of air. Despite these exposures, the lower airways are able to remain sterile due to the remarkable and efficient host defense systems of the pulmonary mucosal surface. This surface consists of epithelial cells and other cells, such as T cells and dendritic cells, that are able to respond to microbial exposures by activation of humoral and cell-mediated immune responses and the production of inflammatory mediators (i.e., cytokines, chemokines, and antimicrobial peptides). A breakdown in these host defenses, especially when the lungs are exposed to highly virulent organisms or an overwhelming inoculum, can result in an infection of the respiratory tract. Infections of the lung typically occur by aspiration of upper airway resident flora, inhalation of aerosolized material, and metastatic seeding of the lung by infectious agents systemically. This chapter describes the common viral causes of respiratory tract infections and the pulmonary mucosal immunologic response to common respiratory viruses.

Common Viral Causes of Respiratory Tract Infections

Many viruses can infect the respiratory tract. Respiratory viruses target the ciliated respiratory epithelium. In most cases, viral infection is limited to the upper airways; however, infections of the lower airway regions occur in a significant number of infected individuals. Worldwide, approximately 90% of the cases of the "common cold" are caused by viruses and most are seen in the winter months. These viruses are spread from person to person and commonly during hand-to-hand contact. This chapter focuses on the common viral causes of respiratory tract infections: adenovirus, coronavirus including severe acute respiratory syndrome–associated coronavirus (SARS-CoV), influenza virus, rhinovirus, and respiratory syncytial virus (RSV) (Table 6.1). This chapter describes the pathogen, epidemiology, clinical syndromes, pathogenesis, diagnosis, and treatment of each virus.

Adenovirus

In 1953, adenovirus was discovered after the removal of adenoids and tonsils from children yielded a transmissible cytopathic agent after it was cultured for several weeks (1). Serially passed epithelial cells produced enlarged rounded cells with strands connecting each other from 2 to 5 days after infection. Adenovirus has also been isolated from adenoidal tissues and pulmonary secretions of adults with respiratory tract symptoms of military recruits, and from the eyes of shipyard workers with conjunctivitis. Adenovirus causes approximately 5% of all upper respiratory infections in children younger than 5 years and 10% of pneumonias of childhood.

Table 6.1. Common viral causes of respiratory tract infections

Virus	Epidemiology	Clinical syndrome	Pathogenesis	Diagnosis	Treatment
Adenovirus	Primary infection takes place in the first few years of life. Seasonal patterns occur in winter, spring, and summer.	Bronchiolitis, pneumonia, pharyngoconjunctival fever, hemorrhagic cystitis, diarrhea, central nervous system disease	Lytic, latent, or chronic infection or oncogenic transformation	Viral isolation, antigen detection, PCR, or serology	Self-limited
Coronavirus including SARS-associated coronavirus (SARS-CoV)	Seasonal pattern occurs in winter and spring. SARS-CoV has first described in November 2002 in Guangdong Province of China causing >8000 cases and 800 deaths.	Upper and lower respiratory disease, fever, headache, chills, mucopurulent nasal discharge, sore throat, cough, diarrhea, and neurologic syndromes; SARS-CoV causes similar symptoms. SARS-CoV has an overall case-fatality of 7–17% and up to 50% in persons with underlying medical condition or age over 65 years.	SARS-CoV can cause pulmonary hyaline membrane formation, interstitial infiltration, and desquamation of pneumocytes.	Viral isolation, antigen detection, RT-PCR, or serology	Self-limited; steroids may be of benefit in patients infected with SARS-CoV.
Influenza virus	An average attack rate is 10–20% but may be as high as 50% in the young and the elderly; 31 pandemics have occurred. The 1918–1919 pandemic resulted in 21 million deaths worldwide.	Fever, chills, headaches, dry cough, pharyngeal pain, nasal obstruction, hoarseness, myalgia, malaise, anorexia, and ocular symptoms; myositis, cardiac complications, toxic shock syndrome, and central nervous complications can occur. A secondary bacterial infection can complicate influenza infection.	Hemagglutinin and neuraminidase are surface antigens. M1 and M2 are integral membrane proteins. The incubation period is 18 to 72 hours. Diffuse inflammation of the upper and lower respiratory mucosa.	Viral isolation, culture in embryonated eggs, antigen detection, PCR (used in research settings), or serology	M2-inhibitors (amantadine, rimantadine) and neuraminidase inhibitors (zanamavir, oseltamivir) if started within 48 hours of symptoms
Rhinovirus	Primary infection takes place in the first few years of life. Seasonal pattern occurs in fall, spring, and summer.	Nasal, pharyngeal, or lower respiratory tract involvement; median duration of symptoms is 7 days and up to 2 weeks in 25% of infected persons.	Incubation period is 8–10 hours. Only slight damage to the mucosal epithelium occurs.	Viral isolation using cell culture systems (WI-38, MRC-5 strains, M-HeLA cells), PCR, and serology	Self-limited
Respiratory syncytial virus	Primary infection takes place in the first few years of life. Seasonal pattern occurs in the winter and spring. Risk factors include young age, male sex, and lower socioeconomic status.	Bronchiolitis, croup, tracheobronchitis, and pneumonia; central nervous system and cardiovascular symptoms, and rashes are uncommon manifestations.	Surface proteins (F, G) are integral to pathogenesis. Infection involves lymphocytic peribronchiolar infiltration, edema of the bronchiole epithelium.	Viral isolation, EIA (sensitivity 60–70%, specificity 90–95%), RT-PCR (used in research settings), or serology	Supportive treatment; infants with severe disease may benefit from ribavirin, bronchodilators, corticosteroids, and RSV-IVIG or palivizumab.

EIA, enzyme immunoassay; RT-PCR, reverse-transcriptase polymerase chain reaction; SARS, severe acute respiratory syndrome.

Adenovirus has the unique characteristic of possessing high oncogenic potential. This was the first human virus to demonstrate oncogenic potential in rodents. Modified adenovirus has been used as vectors for the insertion of genetic material into different types of cells for gene therapy and for immunization against other pathogens. Since the initial description of adenovirus, there have been at least 51 serotypes of adenovirus described based on their hemagglutination pattern of red blood cells, the ability to cause tumors in rodents, and the percentage of guanine plus cytosine content of their DNA (2). However, less than half of these serotypes play a role in human disease. All serotypes have similar morphology and nucleic acid composition, and they produce characteristic cytopathic effect.

Adenoviruses are medium sized, nonenveloped, icosahedron with fiber-like projections, double-stranded DNA viruses with a molecular weight of approximately 23×10^6. The fiber-like projections from the capsomeres are rod-like structures with knobs, and they function as an attachment apparatus for the virus (Fig. 6.1). The adenovirus attaches to a cellular receptor named CAR and are internalized in endosomes, where it undergoes a process of initial uncoating. The virus has an outer protein coat called a capsid. The capsid contains 252 subunits, referred to as capsomeres, which are arranged in a structure with 20 sides and 12 vertices comprising hexons. Neutralizing antibody is directed at the major type-specific neutralizing epitopes on both the fiber and the hexon.

Adenovirus synthesizes several proteins, which include a family of transforming proteins from the E1A and E1B regions, three proteins from the E2A and E2B regions (responsible for replication of the viral genome), proteins from the E3 region (which control the host immune and cytokine response to infection), and proteins from the E4 regions [which facilitate viral messenger RNA (mRNA) transcription].

Most people have experienced a primary infection with adenovirus during their first few years of life (3). Infected individuals with adenovirus may be asymptomatic or the infection may result in a respiratory illness, such as pneumonia, croup, or bronchitis (1). Acute respiratory disease caused by adenovirus was described during conditions of crowding and stress among military recruits during World War II (4). Seasonal patterns of adenovirus most commonly occur in the late winter, spring, and early summer. Adenovirus can cause nonrespiratory tract infections such as gastroenteritis, conjunctivitis, cystitis, and rash. The incubation period of adenovirus is 4 to 5 days. The most common symptoms associated with adenovirus infection are cough, fever, sore throat, and rhinorrhea. These symptoms typically last 3 to 5 days. The classification system of adenovirus serotypes has clinical significance, as there is some association of the serotype, age, and the clinical spectrum of disease. In infants, serotype 7 causes fulminant bronchiolitis and pneumonia. In children, adenovirus serotypes 1, 2, and 4 to 6 are associated with upper respiratory disease. In young adults, particularly military recruits, serotypes 3, 4, and 7 are associated with acute respiratory disease, tracheobronchitis, and pneumonia, and in immunocompromised patients, serotypes 5, 31, 34, 35, and 39 are associated with pneumonia and dissemination. Transmission of adenovirus can occur by direct contact, fecal–oral transmission, and occasionally waterborne transmission. Adenoviruses are able to survive outside of the body for a prolonged period of time due to their ability to remain stable to chemical and physical agents and adverse pH conditions.

Adenovirus is capable of a lytic infection (5), a latent or chronic infection (1), or oncogenic

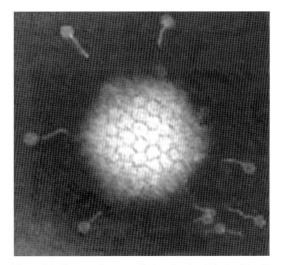

Figure 6.1. Electron microscopy of adenovirus in which six of the 12 vertices are visible. (From www.clinical-virology.org/gallery/images/em/adenovirus2.gif.)

transformation (6) during their interaction with epithelial cells. A lytic interaction with pulmonary epithelial cells results in cell death by inhibition of both host macromolecular synthesis and transport of cellular mRNA to the cytoplasm. During the lytic interaction, up to one million virions per cell can be released. Natural killer cell and lymphocyte recognition of infected cells elicits a cytokine response and induction of cytotoxic T cells and neutralizing and nonneutralizing antibodies to adenovirus. Latent or chronic infection with adenovirus involves lymphoid cells. Adenovirus sequences have been found in human lymphocytes and tonsils where a small number of viruses are released over a period of time even in the presence of a neutralizing antibody response. Oncogenic transformation occurs when adenoviral DNA is integrated into and replicated with the host cell's DNA; however, no virions are produced. All three types of interactions with epithelial cells result in virus-specific proteins (T antigens), which indicate the presence of adenovirus.

A definitive diagnosis of adenovirus infections can be made by antigen detection, polymerase chain reaction (PCR) assay, virus isolation, and serology (7). Because adenovirus can be excreted for prolonged periods of time, isolation of adenovirus does not imply disease. Adenovirus typing is accomplished by hemagglutination-inhibition or neutralization with type-specific antisera. The virus can be cultured from sputum, the nasopharynx, stool, urine, or conjunctival scrapings in monolayers of human epithelial cells. Characteristic cytopathic changes can be visualized after 2 to 5 days. The presence of adenovirus antigens can be detected in samples and tested by immunofluorescence or enzyme-linked immunosorbent assay (ELISA). Serologic diagnosis of adenovirus requires a fourfold rise in antibodies that fix complement, neutralize the virus, or prevent adenoviral hemagglutination by ELISA or by radioimmunoassay.

Most adenovirus infections are self-limited in immunocompetent patients. Symptomatic treatment can be offered. In severe cases among immunocompromised patients, cidofovir, ribavirin, vidarabine, or human immune globulin alone or in combination have been administered with variable success (8–10). However, the overall efficacy of these treatment agents has not been established or thoroughly studied.

Coronavirus Including Severe Acute Respiratory Syndrome–Associated Coronavirus

Coronavirus was first isolated from chickens in 1937. Tyrrell and Bynoe (11) passaged coronavirus, from nasal wash fluids of patients with common colds, in human ciliated embryonal trachea and nasal epithelium cells in 1965. The medium from these cultures caused respiratory symptoms in volunteers. At about the same time of Tyrell and Bynoe's descriptions of coronavirus, Hamre and Procknow (12,13) described the cytopathic effect of coronavirus 229E isolated from medical students who developed acute respiratory illnesses. A number of animal coronaviruses causing disease have since been described.

Severe acute respiratory syndrome (SARS)-associated coronavirus (SARS-CoV) was described in November 2002 (14). This virus was first identified in the Guangdong Province of China, where it subsequently spread to Hong Kong and countries in Southeast Asia, Europe, North America, and eventually throughout the world with more than 8000 cases and 800 deaths by June 2003. The SARS-CoV genome was quickly sequenced and found to be related to previously characterized human and animal coronaviruses (14).

Electron microscopy of a coronavirus shows particles that are medium sized (80 to 150 nm), pleomorphic with an outer envelope covered with crown-like surface proteins; hence, the name coronavirus (Figs. 6.2 and 6.3). The family Coronaviridae has two genera: *Coronavirus* and *Torovirus*. Coronaviruses have a nonsegmented, positive sense, single-stranded, 5′ methylated cap and a 3′ polyadenylated RNA. The RNA codes for a large polyprotein, which forms several nonstructural and structural proteins after being cleaved by virus-encoded proteases. The structural proteins include a surface hemagglutinin-esterase (HE) protein; a surface spike glycoprotein on the virion envelope, which is responsible for receptor binding and cell fusion (S protein), a small envelope protein; a membrane glycoprotein, which is responsible for budding and envelope formation (M); and a nucleocapsid protein complexed with RNA. The S protein mediates attachment to sialic acid, which resides in the plasma membrane of host cells. Antibody to the S protein neutralizes viral

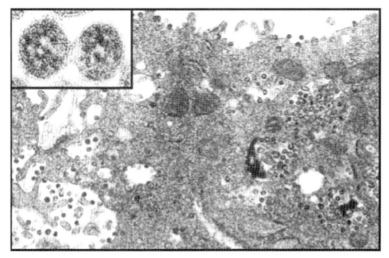

Figure 6.2. Electron microscopy of an infected Vero E6 cell showing coronaviruses within cytoplasmic membrane-bound vacuoles and accumulating on the lining of the surface of the plasma membrane. A higher magnification of the coronaviruses is shown in the inset. (From www.cdc.gov/mmwr/preview/mmwrhtml/mm5212a1.html.)

infectivity. Two strains of coronavirus are able to grow in cultured cells: 229E and OC43. Coronaviruses have the ability to undergo genetic recombination if two viruses simultaneously infect the same cell. The 5′-positive sense of coronavirus is translated to produce a viral polymerase, which results in a full-length negative-sense strand. This strand is used as a template to produce mRNA as a nested set of transcripts that possess a 3′ polyadenated end. Assembly of the virion occurs by budding from cytoplasmic vesicles from the membranes of endoplasmic reticulum. Particles are then transferred to the surface of the cell and released from the cell when the cell dies (15).

Over 85% of adults have antibody to coronavirus OC43 and 229E, the two most studied strains of coronavirus. Respiratory coronaviruses are transmitted from person to person and occur mostly in the winter and spring in coun-

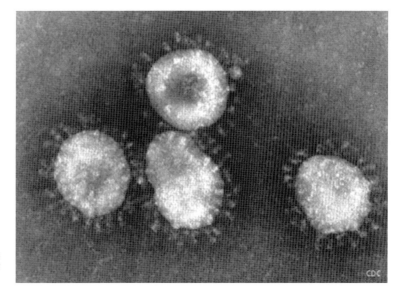

Figure 6.3. Electron microscopy of a coronavirus. (From www.fda.gov/fdac/features/ 2003/403_sars.html.)

tries with temperate climates. This virus is responsible for 15% of upper respiratory tract infections in adults. During peak viral activity, respiratory illness due to coronaviruses may be as high as 35%. In the United States, strains OC43 and 229E are commonly found as causes of large epidemics (16,17). Half of individuals infected with respiratory coronaviruses become ill as evidenced by increased antibody titers. Reinfection likely occurs when there is a rapid decrease in antibody levels after infection.

SARS-CoV causes severe, acute atypical pneumonia in persons who have a clear history of exposure either to a SARS patient or to a setting in which SARS-CoV transmission is occurring (i.e., China, Hong Kong, Hanoi, Taiwan, or Singapore). Epidemiologic studies identified numerous cases occurring in hospitals that involved health care workers, visitors, patients, and family members who had direct contact with infected persons with the SARS-CoV. At the end of the epidemic, in June 2003, the overall case-fatality rate ranged from 7% to 17%. Among elderly individuals (over the age of 65 years) and persons who had underlying medical conditions, the mortality rate was as high as 50%.

Persons infected with coronavirus develop upper respiratory tract illness and cold symptoms such as fever, headache, malaise, cough, sore throat, mucopurulent nasal discharge, and chills (11,18). Enteric infections, neurologic syndromes, and lower respiratory tract involvement (i.e., pneumonia and pleural effusion) with coronavirus have also been described. The mean duration of symptoms is 7 days with a range of 2 to 18 days. Reinfection occurs because of the antigenic heterogeneity of coronaviruses, and immunity is serotype specific. The elderly are more susceptible to severe respiratory infection compared to younger adults. In persons infected with SARS-CoV, the most common symptoms include fever, headache, malaise, myalgia, nonproductive cough, dyspnea, and diarrhea. Rhinorrhea and sore throat are not commonly reported symptoms among patients with SARS. In 25% of patients, especially patients over 50 years of age or have some underlying medical condition, pulmonary disease progresses to acute respiratory distress syndrome (ARDS) (19), which has an estimated mortality rate of 10%. Patients infected with the SARS-CoV often have abnormal laboratory values such as lymphopenia, as well as elevated creatine kinase,

lactic dehydrogenase, and aspartate aminotransferase levels.

Respiratory coronavirus including SARS-CoV infect a variety of mammals and birds. The number of serotypes of antigenic variation is unknown because most coronavirus isolates cannot be grown in culture. Aerosols of respiratory tract secretions result in coronavirus attachment to the epithelium of the nasopharynx by a virus–receptor interaction. Coronavirus replicates in epithelial cells, resulting in a cytolytic effect on ciliated epithelial cells and release of cytokines and chemokines such as CXCL10/interferon-γ (IFN-γ)-inducible protein 10 and CCL2/monocyte chemotactic protein 1 (20). These inflammatory mediators are responsible for many of the respiratory symptoms that occur with infection. The incubation period and viral shedding of coronavirus ranges from 3 to 5 days (18). The incubation period of SARS-CoV ranges from 4 to 7 days, with viral shedding reported in some cases over several weeks (21). The pulmonary histology of SARS-CoV infection has been described by the presence of hyaline membrane formation, interstitial infiltration with lymphocytes and mononuclear cells, and desquamation of pneumocytes in the alveolar spaces (22).

Respiratory coronaviruses are isolated from clinical specimens from tracheal or nasopharyngeal epithelium. These viruses can be rapidly detected by antigen detection methods that utilize immunofluorescence of respiratory cells, or enzyme immunoassays of respiratory secretions, or reverse transcriptase PCR (RT-PCR) (23,24). Some strains, such as 229E and OC43, can be grown in human diploid fibroblast cells lines. SARS-CoV can be isolated from the upper and lower respiratory tract, blood, and stool and urine specimens by RT-PCR and can be grown in the respiratory tract specimens in Vero E6 and fetal rhesus monkey kidney cells. Serum antibodies to SARS-CoV in a single serum specimen or a fourfold or greater increase in SARS-CoV antibody titer between acute- and convalescent-phase serum specimen tests has been used to detect infection with SARS-CoV (25). Immunoglobulin M (IgM) can be detected in cases for a limited period of time and IgG can be detected after the first week of infection. IgA antibody is probably the primary mediator of resistance to coronavirus infections since these infections are initiated in the nasopharynx.

The treatment of coronaviruses is self-limited and supportive. For the SARS-CoV, anecdotal reports suggest that steroid treatment may be of benefit (26). Interferons have shown in vitro activity against the SARS-CoV but ribavirin has not. SARS-CoV is currently being tested against various antiviral drugs to determine if effective treatment can be found.

Influenza Virus

One of the first descriptions of an influenza virus outbreak was that of Sydenham in 1679; however, it was not identified until 1933 (27). Influenza attack rates have been reported as high as 40%. A recurrent epidemic of febrile respiratory disease due to influenza virus occurs every 1 to 3 years, and a worldwide pandemic every 10 to 20 years (28). Epidemics typically occur during the winter months and have a characteristic pattern where outbreaks in children are usually followed by influenza-like illness among adults, which is then followed by increased hospital admissions for patients with pneumonia, exacerbation of chronic obstructive pulmonary diseases, and congestive heart failure. An average attack rate of an influenza epidemic is approximately 10% to 20%, but in certain populations the attack rates may be as high as 50%. Rates of infection are highest among children, and serious illness and death are highest among persons ≥65 years and persons with underlying medical conditions. Influenza is responsible for approximately 36,000 deaths annually in the United States. Pandemic influenza has occurred 31 times thus far, with the greatest pandemic occurring in 1918–1919, causing between 20 and 40 million deaths worldwide (29).

Influenza viruses belong to the family of Orthomyxoviridae and are classified as influenza A, B, and C based on their antigenic, structural, genetic, and epidemiologic differences. Influenza A has been most studied and is further characterized into subtypes based on its two surface antigens—hemagglutinin (HA) and neuraminidase (NA). The nomenclature for influenza strains includes the influenza type, place of initial isolation, strain designation, and year of isolation. Influenza virus contains eight separate segments of linear negative-sense single-stranded RNA, and has a ribonucleoprotein core arranged as a helical nucleocapsid that is surrounded by a lipid-containing envelope

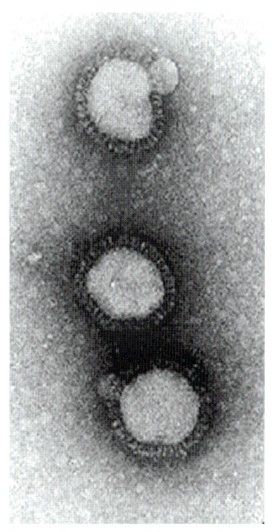

Figure 6.4. Electron microscopy of influenza A virus. (From www.virology.net/Big_Virology/BVRNAortho.html.)

covered with about 500 surface projections that possess HA or NA activity (Fig. 6.4). The HAs extend out from the lipid envelope as a globular rod–shaped head and are involved with attachment of the virus to neuraminic acid–containing mucopolysaccharide receptors on the cell surface membrane. There are at least 15 described antigenically distinct HAs and at least nine distinct NAs. The NA is mushroom shaped and functions as an enzyme that catalyzes the removal of terminal sialic acids from sialic acid–containing glycoproteins during the early stages of penetration of the virus. Other integral membrane proteins include the M2 protein present in the viral envelope and the M1 protein present in the matrix of the virus. Eight discrete nucleocapsid segments exist within the envelope that are associated with viral nucleoprotein and three polymerase proteins (PB1, PB2, and PB3), which are important targets for cross-reactive, viral-specific cytotoxic T lymphocytes.

Influenza causes respiratory-related illnesses that follow a U-shaped epidemic curve with attack rates greatest among the young and the elderly who have lower antibody levels (30). Children are commonly infected due to crowding of children in schoolrooms where aerosol spread is efficient. Influenza epidemics occur mostly in October to April in the Northern Hemisphere and May to September in the Southern Hemisphere. Influenza pandemics are quite variable and occur when there is an emergence of a new virus to which the overall population does not have immunity. Underlying medical disease, such as cardiovascular and pulmonary conditions, diabetes, renal dysfunction, hemoglobinopathies, and immunodeficiencies, are risk factors for severe influenza (31).

A reason for the continued threat of influenza epidemics and pandemics is the unique ability of this virus to alter its antigenic structure, referred to as antigenic variation, of two external glycoproteins HA and NA. The genes that code for HA and NA of influenza A are relatively unstable and continuously undergo mutations that alter their antigenic structure. New influenza variants develop from small antigenic variations known as antigenic drift and large antigenic variations known as antigenic shift. Antigenic drift occurs when relatively minor antigenic changes or an accumulation of point mutations during viral replication occurs at the major antigenic sites on the HA or NA molecule (32). Influenza A virus undergoes antigenic drift more rapidly than influenza B viruses. Antigenic shift results in a new virus to which the population has no immunity. Major antigenic shifts can lead to pandemic influenza.

Infection with the influenza virus occurs by inhalation of small particle aerosols (<10 µm) that contain virus. The aerosols are created by sneezing, coughing, and talking. In experimental conditions, nasal drops are infectious at doses of 137 to 300 times the median tissue-culture infective dose ($TCID_{50}$) and 0.6 to 3.0 $TCID_{50}$ by the aerosol route (33,34). The incubation period for influenza is 1 to 4 days, although viral multiplication is detectable within 24 hours. Virus shedding in respiratory secretions disappears after 5 to 10 days in most individuals but can occur up to several months in severely immunocompromised persons. These particles are relatively stable at different temperatures and humidity, although its survival is favored by lower relative humidity and lower temperatures (35). The histopathology of individuals with typical, uncomplicated acute inflammation is characterized by diffuse inflammation (hyperemia and edema) of the mucosa of the larynx, trachea, and bronchi with vacuolization of columnar cells and desquamation of the ciliated columnar epithelium down to the basal layer of cells. Individual cells may show shrinkage, pyknotic nuclei, and a loss of cilia. Viral antigens are found in the epithelial cells but not the basal layers. Lymphocytes and histiocytes are also found in areas with epithelial damage.

In individuals with severe disease, histologic findings include extensive necrotizing tracheobronchitis, with ulceration and sloughing of the bronchial mucosa, extensive hemorrhage, hyaline membrane formation, and a paucity of polymorphonuclear cell infiltrations. Viral replication occurs intracellularly and leads to death of the host cell through decreased host–cell protein synthesis, degradation, and blockage of translation of cellular mRNAs, degradation of coexpressed proteins, and apoptosis (36,37). Cell death due to apoptosis is related to the induction of the Fas antigen by double-stranded RNA during viral replication or poisoning of the mitochondria by a protein PB1-F2, which is encoded by a second reading frame in the *PB1* gene (38). The incubation period of influenza virus is 18 to 72 hours. Influenza virus is released several hours prior to cell death and infects nearby epithelial cells and peripheral blood mononuclear cells. Infection of epithelial cells and peripheral blood mononuclear cells elicits release of cytokines that are responsible for systemic symptoms. Infection of peripheral blood mononuclear cells, such as polymorphonuclear leukocytes, lymphocytes, and monocytes, results in defects in these cells' chemotaxis and phagocytosis and a decreased proliferation and co-stimulation by mononuclear cells (39,40). Defects in peripheral blood mononuclear cells are related to virus replication and to the direct toxic effects of certain virus proteins such as the HA, NA, and nucleoprotein.

Infection with influenza virus elicits a humoral and cell-mediated immune response, which is vital in recovery from infection and resistance to reinfection (41). There are variable degrees of protection within subtypes but not protection across subtypes. The production of interferon and the generation of cytotoxic lymphocytes best correlate with the recovery from acute influenza. Immunity to reinfection and reduction of the severity of disease with influenza viruses is mediated by antibodies in the serum and the respiratory tract secretions. A systemic antibody response of IgM, IgG, and IgA to influenza virus results in the development of antibody to the glycoproteins HA, NA, and matrix and nucleoproteins (42). Serum IgG neutralizing antibody is the primary mediator of resistance to influenza virus infection. However, antibody to one influenza virus type confers limited or no protection against another influenza virus type. Peak antibody responses are found at 4 to 7 weeks after infection. Anti-neuraminidase antibody parallels that of hemagglutinin-inhibiting antibodies (HAI) (43) and is the primary method of detecting antigenic relatedness among hemagglutinins of influenza virus. Some studies suggest that a serum HAI titer of ≥1:40 or a serum neutralizing titer of ≥1:8 is associated with protection against infection. Antibody to NA can be measured by NA inhibition or ELISA. Anti-NA antibody can provide protection against influenza infection by reducing efficient release of virus from infected cells and decreased severity of illness (44,45). Antibodies to internal proteins (matrix and nucleoproteins) are cross-reactive but not protective against infection. Immunoglobulins G and A are found in nasal secretions. The IgG to HA in the nasal secretions is the IgG_1 subtype, which is the same subtype found in the serum, suggesting that nasal IgG_1 originates from diffusion from the serum (46). Nasal HA-specific IgA is the polymeric and IgA_1 subtype. The IgA_1 subtype suggests local production and derivation from peripheral lymphoid tissue by memory cells derived from the mucosa.

Cell-mediated responses are important in recovery and resistance to reinfection of influenza. Both CD4 and CD8 T cells effect clearance of influenza A virus. Virus-specific CD8 cytotoxic T lymphocytes recognize class I human leukocyte antigen (HLA) and mediate immunity through lysis of infected cells and expression of antiviral cytokines. CD4 cells recognize epitopes on HA, matrix proteins, and nucleocapsid proteins and stimulate B cells to produce antibody to HA and NA and antiviral cytokine expression (47,48). Both animal and human studies have found T helper (Th)1 and Th2 responses to infection with influenza virus. Infected epithelial cells can be lysed by antibody in the presence of complement, antibody-dependent cellular cytotoxicity or by cytotoxic T lymphocytes (49). Cytotoxic lymphocytes peak at day 14 of infection. Class I–restricted cytotoxic lymphocytes are associated with reduced duration and level of influenza A virus replication in epithelial cells (50).

Infection with influenza can present different complications. The first signs and symptoms of influenza infection are an abrupt onset of fever (37.7°C) and a dry cough. Uncomplicated influenza infection is characterized by fever, chills, headaches, dry cough, pharyngeal pain, nasal obstruction, hoarseness, myalgia, malaise, anorexia, and ocular symptoms. Otitis media, nausea, and vomiting are commonly reported among children with influenza illness. Influenza illness and symptoms typically resolve over several days for most individuals. Cough and malaise may persist for 2 weeks or more. Primary influenza viral pneumonia and secondary bacterial infection are possible pulmonary complications of influenza infection. Primary influenza viral pneumonia occurs among susceptible persons with underlying cardiovascular and pulmonary disease. These patients have a relentless progression from classic 3-day influenza and have bilateral findings on physical examination and a normal flora sputum bacteriology. Secondary bacterial pneumonia occurs among older patients (>65 years) with underlying pulmonary, cardiac, metabolic, or other disease, followed by a period of improvement within approximately 1 to 2 weeks and then a recrudescence of symptoms of fever with signs and symptoms of bacterial pneumonia. Secondary bacterial infection likely occurs from direct physical damage to bronchial epithelium and impairment of normal ciliary activity in physical clearance of bacteria from the lung. Gram stain and culture of sputum of these individuals most commonly reveal *Staphylococcus aureus, Streptococcus pneumoniae,* and *Haemophilus influenzae*. Nonpulmonary complications have also been described with influenza, such as myositis, cardiac complication (myocarditis and pericarditis), toxic shock syndrome, central nervous system complications (Guillain-Barré syndrome, transverse myelitis, and encephalopathy) and Reye's syndrome.

The sensitivity and specificity of the clinical diagnosis of influenza and influenza-like illness compared to viral culture range from 63% to 78% and 55% to 71%, respectively. Influenza infection is definitively diagnosed by isolation of virus or detection of viral antigen in respiratory secretions, nasal swab specimens, throat swab specimens, nasal washes, or combined nose and throat swab specimens. Incubation of the virus onto rhesus monkey kidney, cynomolgus monkey kidney, or Madin-Darby canine kidney cell line can produce a characteristic cytopathic effect. Influenza virus can be cultured on embryonated eggs and detected within 3 to 7 days (51). Rapid detection of viral antigens in respiratory secretions can be performed with enzyme immunoassay and direct immunofluorescence—Directigen Flu A+B (Becton-Dickenson, Cockeysville, MO), Flu OIA (Biostar, Boulder, CO), QuickVue Influenza A+B test (Quide Corp, San Diego, CA), and ZstatFlu (ZymeTX, Oklahoma City, OK). These tests have varied sensitivities of 40% to 80% and specificities of 85% to 100% compared to cell culture (52–55). These tests vary in complexity and in the skill and time required for performance and interpretation. Sensitivity appears to be improved with nasopharyngeal swabs and aspirates compared with throat swabs and gargles. Nucleic acid hybridization and PCR amplification have been used in the research setting and offer a higher sensitivity at the cost of being more labor intensive and technically demanding. Serologic tests by using complement fixation and hemagglutination inhibition, consisting of both acute and convalescent sera, are available for making a retrospective diagnosis of influenza.

The treatment of persons infected with susceptible influenza isolates includes M2 inhibitors (amantadine and rimantadine) and neuraminidase inhibitors (zanamivir and oseltamivir). The Centers for Disease Control and Prevention recommends that any person experiencing a potentially life-threatening influenza-related illness or any person at high risk for serious complication of influenza and who is within the first 2 days of illness onset should be treated with antiviral medications (56). Both M2 inhibitors, amantadine and rimantadine, have been shown to be effective in experimentally induced and naturally occurring influenza A virus infection. Resistant viruses have been seen in less than 1% of unexposed individuals (57–59). The M2 inhibitors inhibit the M2 ion channel activity of susceptible viruses. The M2 ion channel is responsible for acidifying the interior of the virus, disrupting the interaction between the matrix and nucleoproteins, and allowing the ribonucleoproteins to be transported to the nucleus where replication occurs (60). The most common side effects of amantadine are central nervous system related (insomnia, dizziness, and difficulty in concentration). Neuraminidase inhibitors have been shown to be effective in human experimental challenge models and clinical trials. Viruses with reduced susceptibility to oseltamivir have been isolated in 1% of adults and 6% of pediatric children. Neuraminidase cleaves terminal sialic acid from sialic acid–containing glycoprotein that serves as host receptors for attachment of influenza virus and facilitates the penetration of virus through secretions in the respiratory tract, which are rich in sialic acid-containing macromolecules (61,62). Neuraminidase inhibitors are generally well tolerated. The most common side effects of the neuraminidase inhibitors are related to gastrointestinal symptoms.

Rhinovirus

The infectious nature of colds was described in 1914 by Kruse. Volunteer studies with intranasal instillation of bacteria-free filtrates of nasal secretions from cold sufferers into healthy subjects resulted in cold-like symptoms in these individuals (63). It was not until the 1940s that rhinovirus was isolated from the nasal secretions of cold sufferers (64). Rhinovirus is derived from the Greek root *rhin* meaning nose. Since the 1950s, research on isolation and characterization of rhinovirus, development of a highly sensitive human embryonic lung cell line (65), a classification system for known rhinovirus and immunotypes (66), and epidemiologic studies have demonstrated rhinovirus as a cause of the common cold (67). Rhinoviruses are estimated to cause up to 35% of all adult colds and occur mostly in early fall, spring, and summer.

Rhinoviruses are single-stranded, positive-sense RNA viruses, belong to the Picornaviridae family, and contain four structural proteins that have exterior projections that interact with neutralizing antibodies VP1, VP2, VP3, and VP4. These structural proteins from a nonenveloped capsid with icosahedral symmetry. Rhinovirus also contains nonstructural proteins including two proteases with specific viral cleavage sites, an RNA-dependent RNA polymerase, and a small protein covalently bound to the 5′ end of the viral RNA designated as VPg. There are 12 capsomers per nucleocapsid. A deep conserved hydrophobic cleft on the viral surface functions in maintaining the structural integrity of the viral capsid and in facilitating the conformational changes for uncoating of viral RNA. The 5′ end of the genome has a genome-linked protein and the 3′ end has a polyadenylated tract. Rhinoviruses are similar to enteroviruses with 40% to 60% homology between their genomes. There are more than 110 distinct rhinovirus types reflecting the chronology of isolation of the prototypic strains of each serotype (66).

Rhinoviruses are susceptible to inactivation by acid (pH < 5) and a higher density in cesium chloride gradients, which distinguishes them from enteroviruses. These viruses are resistant to organic solvents such as ether and chloroform, and other chemicals such as trichlorofluoroethane, ethanol, and weak phenol. Rhinovirus grows best at temperatures of 33° to 35°C, the temperature inside the human nose and large airways (68). Rhinovirus serotypes are based on their receptor specificity. The major group (80%) of rhinovirus utilizes the leukocyte attachment protein known as intercellular adhesion molecule-1 (ICAM-1, CD54) receptor, a member of the immunoglobulin superfamily (69). It is found on most cells of human origin including HeLa cells, fibroblasts, and cells in the respiratory epithelium. A minority of rhinoviruses bind to low-density lipoprotein receptor.

Worldwide, infections with rhinovirus occur in early childhood and continue throughout life. During childhood and adolescence, antibody levels to rhinovirus are rapidly acquired; they peak in young adulthood and begin to decline and then remain constant throughout adulthood (70,71). Military recruits often encounter respiratory disease in the form of rhinovirus due to the close contact with others in crowded quarters, which allows exposure to infections secretions over a short distance (72). Volunteer studies have shown that effective transmission occurs with infected nasal secretions spread from hand to hand and exposure to fomites followed by autoinoculation of the nasal and conjunctiva mucosa. The aerosol route may also be a means for rhinovirus transmission; however,

studies have shown that under ordinary indoor conditions of 70°F and 40% relative humidity, rhinovirus in aerosol is rapidly inactivated (73). The annual seasonal pattern of rhinovirus infection peaks in the fall, with a smaller peak in March, April, and May (70,71). These seasonal patterns can be partially explained by changes in living conditions, and because rhinoviruses survive better under conditions of high relative humidity. In some studies, rhinovirus is recovered in cell cultures in 25% of patients with colds and in up to 50% by the combination of cultures and PCR methods (74). Infection rates of rhinovirus range from 1.2 infections per person-year in children up to 1 year of age to 0.7 in young adults (12,13). Transmission of this virus occurs mostly in the home setting, schools, and day-care centers by school-aged children. Secondary transmission occurs in young siblings and mothers, and attack rates have ranged from 25% to 70% (75). Other conditions that may lead to increased susceptibility include fatigue, emotional stress, poor nutrition, smoking, or living or working in crowded conditions.

The clinical manifestations of rhinovirus infection can be broadly classified into nasal, pharyngeal, or lower respiratory involvement, and include watery eyes, malaise, anorexia, rhinorrhea, nasal obstructions, sore throat, cough, sneezing, and hoarseness, but fever is uncommon. The incubation period is 24 to 72 hours following deposition of virus in the nasopharynx. The median duration of these symptoms is 7 days and up to 2 weeks in 25% of infected persons. Usually virus can be shed for 7 to 10 days, but there is documentation of shedding for several weeks.

Rhinovirus has a 95% infection rate in volunteer challenge studies. According to these studies, rhinovirus replicates in nasal passages and can be detected in the nasal secretions of volunteers as early as 8 to 10 hours, with viral shedding peaking on the second and third day (76). Replication of rhinovirus occurs in ciliated epithelial cells, and cell death results in large quantities of protein including fibrinogen released from the mucous membrane of the nose. Histologic examination of the nasal mucous membranes reveals only slight damage to the epithelium. This damage elicits release of inflammatory mediators such as interleukin-1 (IL-1), IL-6, IL-8, and IL-16 and other mediators such as bradykinin, lysyl-bradykinin, prostaglandin, histamine, and RANTES (regulated on activation, normal T-cell expressed and secreted) (77). Histamine increases the blood flow to the infected cells and causes swelling, congestion, and increased mucus production. Serum neutralizing antibodies IgM, IgA, and IgG are produced in nasal passages during infection with rhinovirus (78,79). These antibodies provide protection against reinfection with the same serotype. IgA antibody is however the primary mediator of resistance to rhinovirus.

Rhinovirus is usually diagnosed by clinical suspicion. Rhinovirus can be isolated from nasal secretions, and the yield can be improved when the secretions are collected directly or are washed from the mucosal surface with a physiologic salt solution. Definitive diagnosis of a rhinovirus infection requires viral identification in cell culture systems such as human embryonic lung cell line (WI-38 and MRC-5 strains), HEp-2, and M-HeLA cells at 33° to 34°C. The cytopathic effect usually occurs within 2 to 6 days. Polymerase chain reaction with nucleic acid probes is being increasingly used to identify rhinovirus (80). Serodiagnosis of rhinovirus infection has been done with a neutralization test; however, this technique cannot be used on a routine basis due to the multiplicity of rhinovirus types.

The treatment of rhinovirus is supportive, with a combination of a first-generation antihistamine and a nonsteroidal antiinflammatory drug (81). Supportive management also includes rest, hydration, decongestants, saline gargles, and cough suppressants. These supportive treatments may relieve the symptoms of sneezing, runny nose, itchy eyes, and congestion. There are currently no antiviral therapies available for rhinovirus infection.

Respiratory Syncytial Virus

Respiratory syncytial virus (RSV) was first described in 1956 by Morris and colleagues (82). They identified a chimpanzee coryza agent (CCA) from a chimpanzee suffering from a cold. Later, an indistinguishable strain was isolated from humans with respiratory illness and increases in the specific neutralizing antibody to CCA were found. Epidemiologic studies have found that 95% of children have antibodies to CCA by age 2, and the CCA virus was subsequently renamed respiratory syncytial virus based on its clinical and laboratory manifesta-

tions (83). It is the most common cause of lower respiratory tract infection in infants. In older children and adults, it is a common cause of upper respiratory tract infection. Most persons have experienced infection with this virus within the first few years of life as evidenced by serology. In children under 1 year of age, the yearly attack rate for RSV lower respiratory tract disease has been estimated to be 23 per 100 (84).

Respiratory syncytial virus belongs to the Paramyxoviridae family. This virus does not possess hemagglutinin and neuraminidase activity. It is divided into two major groups, A and B (85). The antigenic relatedness between the two groups is 25%. The major difference between the two groups is the variability in the G, F, SH, and NS1 proteins. Both major groups, A and B, are found to be circulating simultaneously during outbreaks. Respiratory syncytial virus is an enveloped, nonsegmented, single-stranded, negative-sense RNA virus. The RNA encodes for several proteins, of which three are associated with the nucleocapsid, five with the envelope, three with the glycosylated transmembrane surface proteins (F, G, and SH), M and M2 with nonglycosylated matrix proteins, and two glycosylated surface proteins (F and G) that are integral in the infectivity and pathogenesis of RSV. The F protein is responsible for viral penetration by fusing viral and cellular membranes. The G protein is responsible for mediating the attachment of the virus to host cells. The viral envelope has a membrane derived from the plasma membrane of host cells and a transmembrane surface with several glycoprotein spikes. Respiratory syncytial virus is unstable in the environment and does not withstand temperature and pH changes (86). It is stable on nonporous surfaces for 3 to 30 hours and porous surfaces for less than 1 hour, and its optimal pH is 7.5. It is readily inactivated with soap, water, and disinfectants.

Respiratory syncytial virus is found in all geographic areas of differing climates every year (87). Seasonal outbreaks tend to occur in the winter or spring in the United States. In northern tropical areas, RSV infection is associated with an increase in rainfall and a decrease in temperature. In the southern tropical areas, RSV is associated with a decrease in rainfall and in temperature. Respiratory syncytial virus infection in children causes bronchiolitis, croup, tracheobronchitis, and pneumonia. Respiratory syncytial virus is rarely found (0.3%) in children without respiratory disease (88). It is transmitted by respiratory secretions through close contact with infected persons or contaminated objects. Infection occurs when infectious material contacts the mucous membranes of the eyes, mouth, or nose, or possibly by inhalation of infectious droplets. Risk factors for RSV disease are age (severest disease occurs among infants), sex (males are affected more than females), and socioeconomic factors such as crowding, lower income, day-care attendance, multiple siblings, and exposure to passive smoke within 6 months of onset of the RSV infection (89). Immunity to RSV infection is incomplete, variable, and not durable. In a day-care study, 98% of children were infected with a first RSV exposure (88), 74% became infected or reinfected after a second RSV exposure, and 65% became infected or reinfected after a third RSV exposure. Transmission of RSV occurs with inoculation of the nose or eyes, and less efficiently through the mouth. The incubation period is 2 to 8 days (90–92). Respiratory syncytial virus infection is confined to the respiratory tract—the upper respiratory tract in early disease and the lower respiratory tract in late disease.

The pathogenesis of RSV infection involves the spread and destruction of ciliated epithelial cells of the respiratory tract (Fig. 6.5), and lymphocytic peribronchiolar infiltration with edema of the walls and surrounding tissue, which leads to proliferation and necrosis of the bronchiole epithelium (93). The pulmonary epithelium responds to injury by the production of opsonins, collectins, and multiple chemokines and cytokines such as IL-1β, IL-6, IL-8, IL-10, IL-11, RANTES, and macrophage inflammatory protein-1α (MIP-1α) (94). The production of chemokines and cytokines results in recruitment of effector molecules, neutrophils, macrophages, natural killer cells, and eosinophils. The bronchiole lumina become obstructed from inflammation, necrotic epithelium, and the secretion of mucus. Hyperinflation results from lumen narrowing with positive expiratory pressure. Atelectasis develops with complete obstruction and absorbed trapped air. Bronchiolitis results when an increase in lung volume and expiratory resistance occurs (95). Pneumonia results from an interstitial infiltration of mononuclear cells with an accompanied edema, and necrosis leads to alveolar filling (93).

Although antibodies to RSV does not prevent viral replication in the nasal passages, serum

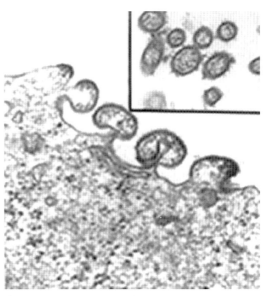

Figure 6.5. Electron microscopy of respiratory syncytial virus budding at the host cell membrane. (From www.epidemic.org/theFACTS/viruses/viralReplication.html.)

antibody to specific RSV proteins has been associated with protection against RSV infection and provides the rationale for administration of RSV hyperimmune globulin and monoclonal antibody to infected persons with more severe RSV disease (94). The qualitative and quantitative antibody responses to the F and G proteins of RSV are not well defined but appear to correlate with resistance to reinfection and are influenced by the presence of preexisting antibody and age of the host (95,96). Immunoglobulin M, IgG, IgE, and IgA are found in nasal secretions of persons infected with RSV (97). Immunoglobulin M appears early in RSV infection. An IgE and histamine response in nasopharynx secretions is associated with wheezing during acute illness and episodes of airway bronchospasm. IgA appears to be correlated with diminished titers of the virus.

Cell-mediated immunity is important in the clearance of RSV and recovery of the host. Immunosuppressed individuals with deficiencies of cellular immunity have more severe disease and prolonged shedding of virus (98). Patients have depressed lymphocyte function in T-cell subsets, depressed IL-12 levels and elevated IL-8 levels (99). Cytotoxic T lymphocyte (CTL) responses in persons infected with RSV are complex. In some studies, CTL and helper T lymphocyte responses are correlated with clinical response and viral clearance. The response to RSV infection is likely a result of both a helper T cell with both a Th1 response and a Th2 response (100). The quantity, timing, specificity, and types of T cells determine the immunologic and clinical outcomes of RSV infection.

The clinical manifestations of RSV infection vary by age and the presence of underlying diseases of the host. The major clinical manifestations associated with RSV infection include bronchiolitis, pneumonia, croup, otitis media, apnea, and sudden infant death syndrome. In infants and children under 1 year of age, infection with RSV is the most common cause of lower respiratory tract infection and upper respiratory tract illness (101,102). Up to 2% of young children with first-time RSV infection require hospitalization. The majority of children hospitalized for RSV infection are under 6 months of age (102). Symptoms include wheezing, rhonchi, rales, rhinorrhea, nasal congestion, cough, low-grade fever, dyspnea, and hypoxemia (103). Hypoxemia represents lower respiratory tract involvement and diffuse viral involvement of the lung parenchyma (104). The mean arterial oxygen saturation on admission of one group of hospitalized infants was 87%. Otitis media is a common complication of RSV infection in young children. In infants who are preterm or have low birth weight (<2500 g) or children with underlying diseases such as chronic lung disease, congenital heart disease, immunosuppressive conditions (i.e., those undergoing transplantation of bone marrow and solid organs, HIV), or other chronic diseases, RSV infection can result in complications with prolonged morbidity and mortality. In older children and adults, the clinical manifestation of RSV infection depends on the immune status of the host. In this older population, an infection with RSV represents a reinfection. In healthy individuals, reinfections with RSV are milder and range from being asymptomatic to causing upper respiratory tract illness such as nasal congestion, cough, hoarseness, sore throat, low-grade fever, and conjunctivitis. The average duration of clinical illness is 9.5 days and for viral shedding it is 1 to 6 days. In persons with medical conditions, especially those with underlying cardiac and pulmonary disease, the disease may be quite severe (105). Respiratory syncytial virus is a cause of cardiovascular and chronic obstructive pulmonary disease exacerbations, especially among individuals who are institutionalized.

Diagnosis of RSV depends on clinical suspicion and is confirmed by viral isolation (Fig. 6.6), detection of viral antigens or viral RNA, or serology. Isolation of RSV is time-consuming and expensive. Respiratory syncytial virus is isolated from nasopharyngeal washes, tracheal secretions, or nasal swabs. Specific cytopathic change is usually seen within 3 to 7 days. The shell vial technique hastens the identification of RSV (106). Antigen detection tests are rapid and less expensive and are used by most clinical laboratories. Available antigen detection tests include direct and indirect immunofluorescent assays, and an enzyme immunoassay (EIA) method (sensitivity 60–70% and specificity 90–95%) (107). Reverse transcriptase PCR consistently has higher sensitivity and specificity rates than other diagnostic tools but is used mainly in research laboratories (107,108). Serologic diagnosis of RSV is usually done using enzyme immunoassays and neutralizing assays. Serologic tests for RSV infection have been used in epidemiologic studies, but are not always practical in patient management due to the delay in acquiring convalescent sera. Also, production of a significant rise in antibody titer does not always occur especially in young infants and individuals with underlying medical conditions.

Treatment of most patients with mild disease is supportive, including adequate fluid intake, acetaminophen to reduce fever, and rest. Hospitalized infants may require oxygen therapy and sometimes mechanical ventilation and aerosolized ribavirin, bronchodilating agents, and corticosteroids (109). Aerosolized

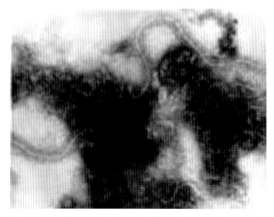

Figure 6.6. Electron microscopy of respiratory syncytial virus. (From www.cdc.gov/ncidod/aip/images/rsv_germ.jpg.)

ribavirin is an approved antiviral therapy for RSV lower respiratory tract disease in hospitalized infants. Studies have shown that this broad-spectrum antiviral agent may reduce long-term pulmonary sequelae and recurrent wheezing and result in rapid improvement in clinical illness and oxygenation (110,111). Bronchodilators and corticosteroids have been of benefit in infants with bronchiolitis. The use of RSV-neutralizing antibody [RSV–intravenous immunoglobulin (IVIG) polyclonal antibody] or intramuscular monoclonal antibody (palivizumab—humanized mouse IgG monoclonal antibody that binds the F protein of RSV) has been shown to reduce the risk of hospitalization in 41% to 50% of high-risk children with prematurity with or without chronic lung disease. RSV-IVIG may also be beneficial in the treatment of patients with compromised immune systems (112).

Pulmonary Mucosal Immunity

The pulmonary defense system includes the anatomic and mechanical barriers of the respiratory tract and mucosal immunity. These barriers form a first line of defense against mucosal transmitted pathogens. Most particles are filtered out of the inhaled air by the mouth and nose. Small particles (<4 μm in diameter) can travel to the lower respiratory tract. Mechanical barriers and reflex mechanisms (coughing and sneezing) prevent and reduce the amount of inhaled pathogens in the respiratory tract. The mucosal defense system consists of both innate barriers such as mucus, epithelium, and innate immune mechanisms (phagocytic cells), and adaptive host immunity, which consists of humoral and cell-mediated immunity such as secretory immunoglobulin A, CD4 T cells, and antigen-specific cytotoxic T lymphocytes (113).

The conducting airways, made up of the trachea and bronchi, are lined with ciliated columnar epithelial cells, which secrete antimicrobial factors and clear mucus, and generate inflammatory chemokines and cytokines that attract inflammatory and phagocytic cells into the lung. There are three types of epithelial cells: type I, type II and Clara cells. Type I cells are alveolar epithelial cells responsible for gas exchange. Type II cells produce a surface airway liquid that contains lysozyme, lactoferrin, secretory leukocyte proteinase inhibitor, cathelicidins, and β-defensins, nitric oxide and extracellular superoxide dismutase all of which posses microbicidal activity (114,115). Clara cells are localized in the bronchiole and involved in lung remodeling. The β-defensins may function as nonspecific immune lung host response and in communicating with memory T cells and dendritic cells. In the lower respiratory tract, humoral and cell-mediated host defenses play an important role in pulmonary mucosal immunity. Like the conducting airways, a surface airway liquid is present that lines the alveoli, and this microbicidal liquid contains surfactant, fibronectin, immunoglobulin, complement factors, free fatty acid, and iron-binding proteins. From an immunology point of view, surfactant, which has several components, serves to increase the microbicidal capacity of macrophages, affect free-radical production and lymphocyte activity, and bind various microorganisms including viruses (113). Binding microorganisms with surfactant may result in reduced microbial virulence or enhanced phagocytosis by neutrophils and alveolar macrophages (116).

The respiratory tract is made up of phagocytic cells, which include the alveolar macrophage, interstitial macrophage, dendritic cells, and intravascular macrophage (117). The alveolar macrophages are located in the alveolar lining fluid and defend the lower airways from inspired foreign materials (118). Alveolar macrophages are drawn into the lung when epithelial cells and resident macrophages produce chemokines such as macrophage inflammatory protein-1 (MIP-1), MIP-2, and macrophage chemoattractant protein-1 (MCP-1). The alveolar macrophages function by phagocytosing and eliminating organisms, and becoming a mediator of an inflammatory response recruiting neutrophils through the production of cytokines and chemokines (119). The interstitial macrophages are located in the connective tissue of the lung and serve as a phagocytic cell and a class II antigen presenting cell. Compared to alveolar macrophages, interstitial macrophages do not demonstrate Fc-receptor–independent phagocytosis, cytokine production, or oxygen radical production. Dendritic cells are located within the epithelium of the trachea, conducting airways, terminal airways, alveolar septa, pulmonary vasculature, and visceral pleura, and also serve as phagocytic cells and a class II antigen presenting cell. These cells are derived from monocytes and migrate to lymphoid tissue and stimulate T-cell immune responses. Dendritic cells are also capable of the production of a variety of chemokines and cytokines that stimulate both T-cell and B-cell immune responses (119). Intravascular macrophages are located in the capillary endothelial cells and they serve as phagocytic cells and remove foreign or damaged material entering the lung through the bloodstream.

The recruitment of neutrophils, occurring by chemokine and cytokine mediation, is vital to the inflammatory response in the lung. Complement factors, specifically the fifth component of complement, leukotriene B_4, and peptides of bacterial cell walls are also responsible for eliciting mediators of inflammation and recruitment of neutrophils into areas of inflammation. These mediators are responsible for attracting and maintaining neutrophils to areas of lung injury as well as regulating the inflammatory process involved. Key chemokine and cytokines involved in this process include IL-1, tumor necrosis factor-α (TNF-α), IFN-γ, IL-8, IL-10, IL-12, and α-chemokines (120).

The lung epithelium produces antibody through humoral immunity to prevent bacterial adherence or growth. The humoral immunity of the respiratory tract involves secretory immunoglobulin (sIgA), IgG, and IgM. Secretory IgA is a major immunoglobulin involved in the humoral immunity of the respiratory tract, and it provides an important first line of defense against the invasion of deeper tissues by respiratory tract viruses. It prevents or reduces the attachment of respiratory viruses and thus prevents internalization of host cells. Polymeric IgA has also been shown to neutralize virus intracellularly, and the transport of polymeric IgA (pIgA) across epithelial cells allows active elimination of immune complexes at mucosal sites and even virus inside epithelial cells. Many live attenuated viral vaccines optimize the production of sIgA in forming mucosal immune protection to respiratory tract viruses. Secretory IgA accounts for 10% of the total protein of nasal secretions and provides antibacterial and antiviral activity (121). Other immunoglobulins, IgG and IgM, also provide antibacterial and antiviral activity and aid in bacterial opsonization, complement activation, agglutination, and neutralization activity. They enter upper and lower airways by the route of transudation from the blood. Deficiencies in either IgG_2 or IgG_4 are

associated with bronchiectasis, a progressive dilation of the bronchi or bronchioles, as a consequence of inflammation.

The cell-mediated immunity of the respiratory tract, as manifested by $CD4^+$ Th1 cells and $CD8^+$ cytotoxic T lymphocytes, is especially important against viruses and intracellular parasites because these organisms can survive within pulmonary macrophages. The CD4 to CD8 ratio is approximately 2:1. There is an estimated 4×10^8 lymphocytes on the epithelial surface of the human lung and more than 95% are T cells. T cells outnumber B cells by 10:1 in the lung. The lung has lymphoid tissue, located in follicles along the bronchial tree in bronchus-associated lymphoid tissue (BALT) and nasal-associated lymphoid tissue (NALT) commonly in the upper respiratory tract, tonsils, and adenoids, where uncommitted cells to memory T and B cells exist and differentiate (122). The pulmonary mucosal immune system allows for production of specific immune effectors that promote the removal or elimination of any viral pathogen that challenges the respiratory mucosa without damaging the mucosal surface or impairing gas exchange. The BALT, similar to the gut-associated lymphoid tissue of the intestinal tract, is intimately involved with the pulmonary mucosa and is considered the site where antigen presentation to T cells occurs before clonal expansion. Inhaled antigens that cross the respiratory epithelia surface encounter the antigen processing cells where B and T lymphocytes, in the BALT, are stimulated to become memory and effector cells. The increased memory T-cell numbers result in local proliferation or migration of these cells from the BALT. The memory lymphocytes are located in the submucosa and lamina propria. The lymphocytes are made up mostly of T cells of which 40% are CD4 cells (helper) and up to 32% represent CD8 cells (suppressor) (123). Pulmonary lymphocytes migrate between lymphoid tissue and lung parenchyma. The effector cells are located between epithelial cells and in the interstitium of the lung. The cytotoxic T lymphocytes are present in the pulmonary mucosa.

Inhaled antigens that reach the alveolus encounter antigen-presenting cells, which activate alveolar lymphoid cells. Activated alveolar lymphoid cells stimulate memory lymphocytes to migrate to areas of inflammation, which results in a localized accumulation of antigen-specific T and B lymphocytes. The critical step in this inflammatory process is the binding of T cells to endothelium via the interaction of leukocyte function–associated antigen-1 integrins on the lymphocyte cell surface with endothelium ligands such as ICAM-1 and -2 and vascular cell adhesion molecule-1. Ligands on the endothelium are expressed via the upregulation of inflammatory mediators IL-1, IFN-γ, and TNF-α. Activated lymphocytes function by producing antibody, having cytotoxic activity, or producing inflammatory mediators. Cytotoxic cells within the lung, including natural killer cells, antibody-dependent cytotoxic cells, and antigen-restricted cytotoxic cells, can produce cytokines when they interact with pathogens. Unstimulated T lymphocytes produce IL-2 and memory T lymphocytes are able to produce Th1 and Th2 pattern of cytokines. A Th1 immune response involves cell-mediated inflammatory reactions with the production of cytokines such as IFN-γ, IL-2, IL-6, and IL-10. A Th2 immune response involves humoral immunity with cytokine production such as IL-4, IL-5, and IL-10 and antibody production. Other cytokines may be secreted by both a Th1 and Th2 immune response such as TNF-α, IL-3, and granulocyte-macrophage colony-stimulating factor. Stimulated alveolar epithelial cells express a variety of integrins such as ICAM-1 that allow the infiltration and retention of lymphocytes, leukocytes, and monocytes.

Conclusion

Respiratory viruses, such as adenovirus, SARS-CoV, influenzavirus, rhinovirus, and respiratory syncytial virus, are common causes of pulmonary and nonpulmonary clinical syndromes. These viruses are a significant cause of morbidity and mortality worldwide. The pulmonary defense has the difficult task of balancing the ability to destroy or remove the constant barrage of foreign antigens that assault our mucosal tissues without causing undue and injurious inflammation. Pulmonary mucosal immunity consists of both innate and acquired immunity, which are essential in protecting the host from developing a respiratory tract illness. Specifically, humoral immunity (B cells and immunoglobulin) and cell-mediated immunity (antigen-presenting cells and T cells) are vital to protecting the host against respiratory viruses. It is these protective mechanisms that defend the pulmonary mucosa from colonization and invasion by viral pathogens and keep most people healthy despite a continuous barrage of inhaled pathogens and particulates.

References

1. Rowe WP, Huebner RJ, Gilmore LK, Parrott RH, Ward TG. Isolation of a cytopathogenic agent from human adenoids undergoing spontaneous degeneration in tissue culture. Proc Soc Exp Biol Med 1953;84(3): 570–573.
2. De Jong JC, Wermenbol AG, Verweij-Uijterwaal MW, et al. Adenoviruses from human immunodeficiency virus-infected individuals, including two strains that represent new candidate serotypes Ad50 and Ad51 of species B1 and D, respectively. J Clin Microbiol 1999;37(12):3940–3945.
3. Jennings LC, Anderson TP, Werno AM, Beynon KA and Murdoch DR. Viral etiology of acute respiratory tract infections in children presenting to hospital: role of polymerase chain reaction and demonstration of multiple infections. Pediatr Infect Dis J 2004;23(11): 1003–1007.
4. Hilleman MR, Werner JH. Recovery of new agent from patients with acute respiratory illness. Proc Soc Exp Biol Med 1954;85(1):183–188.
5. Shenk T. Adenoviridae: The viruses and their replication. In: Fields BN, Knipe DM, Howley PM, eds. Virology, 3rd ed. Philadelphia: Lippincott-Raven, 1996: 2111–2148.
6. Huebner RJ, Rowe WP, Lane WT. Oncogenic effects in hamsters of human adenovirus types 12 and 18. Proc Natl Acad Sci USA 1962;48:2051–2058.
7. Krafft AE, Russell KL, Hawksworth AW, et al. Evaluation of PCR testing of ethanol-fixed nasal swab specimens as an augmented surveillance strategy for influenza virus and adenovirus identification. J Clin Microbiol 2005;43(4):1768–1775.
8. Schievning M, Buxbavm-Conradi H, Jager G, Kolb HJ. Intravenous ribavirin for eradication of respiratory syncytial virus (RSV) and adenovirus isolates from respiratory and/or gastrointestinal tract in recipients of allogeneic hematopoietic stem cell transplants. Hematol J 2004;5(2):135–144.
9. Dagan R, Schwartz RH, Insel RA, Menegus MA. Severe diffuse adenovirus 7a pneumonia in a child with combined immunodeficiency: possible therapeutic effect of human immune serum globulin containing specific neutralizing antibody. Pediatr Infect Dis 1984;3(3): 246–251.
10. Legrand F, Berrebi D, Houhou N, et al. Early diagnosis of adenovirus infection and treatment with cidofovir after bone marrow transplantation in children. Bone Marrow Transplant 2001;27(6):621–626.
11. Tyrrell DA, Bynoe ML. Cultivation of a novel type of common-cold virus in organ cultures. Br Med J 1965; 5448:1467–1470.
12. Hamre D, Connelly AP Jr, Procknow JJ. Virologic studies of acute respiratory disease in young adults. IV. Virus isolations during four years of surveillance. Am J Epidemiol 1966;83(2):238–249.

13. Hamre D, Procknow JJ. A new virus isolated from the human respiratory tract. Proc Soc Exp Biol Med 1966; 121(1):190–193.

14. Ksiazek TG, Erdman D, Goldsmith CS, et al. A novel coronavirus associated with severe acute respiratory syndrome. N Engl J Med 2003;348(20):1953–1966.

15. Heath RB. The pathogenesis of respiratory viral infection. Postgrad Med J 1979;55(640):122–127.

16. Sakai K, Kawaguchi Y, Kishino Y, Kido H. Electron immunohistochemical localization in rat bronchiolar epithelial cells of tryptase Clara, which determines the pneumotropism and pathogenicity of Sendai virus and influenza virus. J Histochem Cytochem 1993; 41(1):89–93.

17. Sakai K, Kohri T, Tashiro M, Kishino Y, Kido H. Sendai virus infection changes the subcellular localization of tryptase Clara in rat bronchiolar epithelial cells. Eur Respir J 1994;7(4):686–692.

18. Bradburne AF, Bynoe ML, Tyrrell DA. Effects of a "new" human respiratory virus in volunteers. Br Med J 1967;3(568):767–769.

19. Lew TW, Kwek TK, Tai D, et al. Acute respiratory distress syndrome in critically ill patients with severe acute respiratory syndrome. JAMA 2003;290(3):374–380.

20. Cheung CY, Poon LL, Ng IH, et al. Cytokine responses in severe acute respiratory syndrome coronavirus-infected macrophages in vitro: possible relevance to pathogenesis. J Virol 2005;79(12):7819–7826.

21. Donnelly CA, Ghani AC, Leung GM, et al. Epidemiological determinants of spread of causal agent of severe acute respiratory syndrome in Hong Kong. Lancet 2003;361(9371):1761–1766.

22. Lee N, Hui D, Wu A, et al. A major outbreak of severe acute respiratory syndrome in Hong Kong. N Engl J Med 2003;348(20):1986–1994.

23. Myint S, Johnston S, Sanderson G, Simpson H. Evaluation of nested polymerase chain methods for the detection of human coronaviruses 229E and OC43. Mol Cell Probes 1994;8(5):357–364.

24. Lina B, Valette M, Foray S, et al. Surveillance of community-acquired viral infections due to respiratory viruses in Rhone-Alpes (France) during winter 1994 to 1995. J Clin Microbiol 1996;34(12):3007–3011.

25. Peiris JS, Chu CM, Cheng VC, et al. Clinical progression and viral load in a community outbreak of coronavirus-associated SARS pneumonia: a prospective study. Lancet 2003;361(9371):1767–1772.

26. Zhao Z, Zhang F, Xu M, et al. Description and clinical treatment of an early outbreak of severe acute respiratory syndrome (SARS) in Guangzhou, PR China. J Med Microbiol 2003;52(pt 8):715–720.

27. Sydenham T. Classical Descriptions of Disease. London: R. Wellington, 1955.

28. Thomson D, Thomson R. Influenza. New York: Ann Pickett-Thomas Research Labs, 1933.

29. Crosby AW. Epidemic and Peace, 1918. Part IV. Westport, CT: Greenwood Press, 1976.

30. Glezen WP, Keitel WA, Taber LH, Piedra PA, Clover RD, Couch RB. Age distribution of patients with medically-attended illnesses caused by sequential variants of influenza A/H1N1: comparison to age-specific infection rates, 1978–1989. Am J Epidemiol 1991;133(3):296–304.

31. Barker WH, Mullooly JP. Impact of epidemic type A influenza in a defined adult population. Am J Epidemiol 1980;112(6):798–811.

32. Wilson IA, Cox NJ. Structural basis of immune recognition of influenza virus hemagglutinin. Annu Rev Immunol 1990;8:737–771.

33. Alford RH, Kasel JA, Gerone PJ, Knight V. Human influenza resulting from aerosol inhalation. Proc Soc Exp Biol Med 1966;122(3):800–804.

34. Little JW, Douglas RG Jr, Hall WJ, Roth FK. Attenuated influenza produced by experimental intranasal inoculation. J Med Virol 1979;3(3):177–188.

35. Hemmes JH, Winkler KC, Kool SM. Virus survival as a seasonal factor in influenza and poliomyelitis. Nature 1960;188:430–431.

36. Katze MG, Krug RM. Metabolism and expression of RNA polymerase II transcripts in influenza virus-infected cells. Mol Cell Biol 1984;4(10):2198–2206.

37. Katze MG, DeCorato D, Krug RM. Cellular mRNA translation is blocked at both initiation and elongation after infection by influenza virus or adenovirus. J Virol 1986;60(3):1027–1039.

38. Sanz-Ezquerro JJ, de la Luna S, Ortin J, Nieto A. Individual expression of influenza virus PA protein induces degradation of coexpressed proteins. J Virol 1995;69(4):2420–2426.

39. Larson HE, Parry RP, Tyrrell DA. Impaired polymorphonuclear leucocyte chemotaxis after influenza virus infection. Br J Dis Chest 1980;74(1):56–62.

40. Roberts NJ Jr, Steigbigel RT. Effect of in vitro virus infection on response of human monocytes and lymphocytes to mitogen stimulation. J Immunol 1978;121(3):1052–1058.

41. Mozdzanowska K, Furchner M, Zharikova D, Feng J, Gerhard W. Roles of CD4+ T-cell-independent and -dependent antibody responses in the control of influenza virus infection: evidence for noncognate CD4+ T-cell activities that enhance the therapeutic activity of antiviral antibodies. J Virol 2005;79(10):5943–5951.

42. Murphy BR, Nelson DL, Wright PF, Tierney EL, Phelan MA, Chanock RM. Secretory and systemic immunological response in children infected with live attenuated influenza A virus vaccines. Infect Immun 1982;36(3):1102–1108.

43. Murphy BR, Kasel JA, Chanock RM. Association of serum anti-neuraminidase antibody with resistance to influenza in man. N Engl J Med 1972;286(25):1329–1332.

44. Schulman JL, Khakpour M, Kilbourne ED. Protective effects of specific immunity to viral neuraminidase on influenza virus infection of mice. J Virol 1968;2(8):778–786.

45. Kilbourne ED, Laver WG, Schulman JL, Webster RG. Antiviral activity of antiserum specific for an influenza virus neuraminidase. J Virol 1968;2(4):281–288.

46. Wagner DK, Clements ML, Reimer CB, Snyder M, Nelson DL, Murphy BR. Analysis of immunoglobulin G antibody responses after administration of live and inactivated influenza A vaccine indicates that nasal wash immunoglobulin G is a transudate from serum. J Clin Microbiol 1987;25(3):559–562.

47. Lamb JR, Eckels DD, Lake P, Woody JN, Green N. Human T-cell clones recognize chemically synthe-

sized peptides of influenza haemagglutinin. Nature 1982;300(5887):66–69.

48. Lamb JR, Woody JN, Hartzman RJ, Eckels DD. In vitro influenza virus-specific antibody production in man: antigen-specific and HLA-restricted induction of helper activity mediated by cloned human T lymphocytes. J Immunol 1982;129(4):1465–1470.

49. Hashimoto G, Wright PF, Karzon DT. Antibody-dependent cell-mediated cytotoxicity against influenza virus-infected cells. J Infect Dis 1983;148(5):785–794.

50. McMichael AJ, Gotch FM, Noble GR, Beare PA. Cytotoxic T-cell immunity to influenza. N Engl J Med 1983;309(1):13–17.

51. Newton DW, Mellen CF, Baxter BD, Atmar RL, Menegus MA. Practical and sensitive screening strategy for detection of influenza virus. J Clin Microbiol 2002;40(11):4353–4356.

52. Covalciuc KA, Webb KH, Carlson CA. Comparison of four clinical specimen types for detection of influenza A and B viruses by optical immunoassay (FLU OIA test) and cell culture methods. J Clin Microbiol 1999;37(12):3971–3974.

53. Noyola DE, Clark B, O'Donnell FT, Atmar RL, Greer J, Demmler GJ. Comparison of a new neuraminidase detection assay with an enzyme immunoassay, immunofluorescence, and culture for rapid detection of influenza A and B viruses in nasal wash specimens. J Clin Microbiol 2000;38(3):1161–1165.

54. Habib-Bein NF, Beckwith WH 3rd, Mayo D, Landry ML. Comparison of SmartCycler real-time reverse transcription-PCR assay in a public health laboratory with direct immunofluorescence and cell culture assays in a medical center for detection of influenza A virus. J Clin Microbiol 2003;41(8):3597–3601.

55. Landry ML, Ferguson D. Suboptimal detection of influenza virus in adults by the Directigen Flu A+B enzyme immunoassay and correlation of results with the number of antigen-positive cells detected by cytospin immunofluorescence. J Clin Microbiol 2003;41(7):3407–3409.

56. Centers for Disease Control and Prevention. Prevention and control of influenza. Part I: antiviral agents. MMWR 1994;43:1.

57. Hayden FG, Belshe RB, Clover RD, Hay AJ, Oakes MG, Soo W. Emergence and apparent transmission of rimantadine-resistant influenza A virus in families. N Engl J Med 1989;321(25):1696–1702.

58. Belshe RB, Burk B, Newman F, Cerruti RL, Sim IS. Resistance of influenza A virus to amantadine and rimantadine: results of one decade of surveillance. J Infect Dis 1989;159(3):430–435.

59. Ziegler T, Hemphill ML, Ziegler ML, et al. Low incidence of rimantadine resistance in field isolates of influenza A viruses. J Infect Dis 1999;180(4):935–939.

60. Bui M, Whittaker G, Helenius A. Effect of M1 protein and low pH on nuclear transport of influenza virus ribonucleoproteins. J Virol 1996;70(12):8391–8401.

61. Air GM, Ritchie LR, Laver WG, Colman PM. Gene and protein sequence of an influenza neuraminidase with hemagglutinin activity. Virology 1985;145(1):117–122.

62. Colman PM, Ward CW. Structure and diversity of influenza virus neuraminidase. Curr Top Microbiol Immunol 1985;114:177–255.

63. Kruse W. Die Erreger con Husten and Schupfen. Munchen Med Wochenschr 1914;61:1547.

64. Andrewes C. The Common Cold. New York: W.W. Norton, 1965.

65. Hayflick L, Moorhead PS. The serial cultivation of human diploid cell strains. Exp Cell Res 1961;25:585–621.

66. Hamparian VV, Colonno RJ, Cooney MK, et al. A collaborative report: rhinoviruses—extension of the numbering system from 89 to 100. Virology 1987;159(1):191–192.

67. Gwaltney JM Jr, Hendley JO, Simon G, Jordan WS Jr. Rhinovirus infections in an industrial population. I. The occurrence of illness. N Engl J Med 1966;275(23):1261–1268.

68. Halperin SA, Eggleston PA, Hendley JO, Suratt PM, Groschel DH, Gwaltney JM Jr. Pathogenesis of lower respiratory tract symptoms in experimental rhinovirus infection. Am Rev Respir Dis 1983;128(5):806–810.

69. Winther B, Greve JM, Gwaltney JM Jr, et al. Surface expression of intercellular adhesion molecule 1 on epithelial cells in the human adenoid. J Infect Dis 1997;176(2):523–525.

70. Monto AS. A community study of respiratory infections in the tropics. 3. Introduction and transmission of infections within families. Am J Epidemiol 1968;88(1):69–79.

71. Monto AS, Johnson KM. A community study of respiratory infections in the tropics. II. The spread of six rhinovirus isolates within the community. Am J Epidemiol 1968;88(1):55–68.

72. Forsyth BR, Bloom HH, Johnson KM, Chanock RM. Patterns of illness in rhinovirus infections of military personnel. N Engl J Med 1963;269:602–606.

73. Karim YG, Ijaz MK, Sattar SA, Johnson-Lussenburg CM. Effect of relative humidity on the airborne survival of rhinovirus-14. Can J Microbiol 1985;31(11):1058–1061.

74. Makela MJ, Puhakka T, Ruuskanen O, et al. Viruses and bacteria in the etiology of the common cold. J Clin Microbiol 1998;36(2):539–542.

75. Pereira MS, Andrews BE, Gardner SD. A study on the virus aetiology of mild respiratory infections in the primary school child. J Hyg (Lond) 1967;65(4):475–483.

76. Douglas RG Jr, Cate TR, Gerone PJ, Couch RB. Quantitative rhinovirus shedding patterns in volunteers. Am Rev Respir Dis 1966;94(2):159–167.

77. Doyle WJ, Boehm S, Skoner DP. Physiologic responses to intranasal dose-response challenges with histamine, methacholine, bradykinin, and prostaglandin in adult volunteers with and without nasal allergy. J Allergy Clin Immunol 1990;86:924–935.

78. Rossen RD, Douglas G Jr, Cate TR, Couch RB, Butler WT. The sedimentation behavior of rhinovirus neutralizing activity in nasal secretion and serum following the rhinovirus common cold. J Immunol 1966;97(4):532–538.

79. Cate TR, Rossen RD, Douglas RG Jr, Butler WT, Couch RB. The role of nasal secretion and serum antibody in the rhinovirus common cold. Am J Epidemiol 1966;84(2):352–363.

80. Johnston SL, Sanderson G, Pattemore PK, et al. Use of polymerase chain reaction for diagnosis of picornavirus infection in subjects with and without

respiratory symptoms. J Clin Microbiol 1993;31(1):
111–117.

81. Turner RB. New considerations in the treatment and
prevention of rhinovirus infections. Pediatr Ann 2005;
34(1):53–57.

82. Blount RE Jr, Morris JA, Savage RE. Recovery of
cytopathogenic agent from chimpanzees with coryza.
Proc Soc Exp Biol Med 1956;92(3):544–549.

83. Hall CB, Douglas RG Jr. Clinically useful method for
the isolation of respiratory syncytial virus. J Infect Dis
1975;131(1):1–5.

84. Foy HM, Cooney MK, Maletzky AJ, Grayston JT. Incidence and etiology of pneumonia, croup and bronchiolitis in preschool children belonging to a prepaid
medical care group over a four-year period. Am J Epidemiol 1973;97(2):80–92.

85. Peret TC, Hall CB, Hammond GW, et al. Circulation
patterns of group A and B human respiratory syncytial virus genotypes in 5 communities in North
America. J Infect Dis 2000;181(6):1891–1896.

86. Hambling MH. Survival of the respiratory syncytial
virus during storage under various conditions. Br J
Exp Pathol 1964;45:647–655.

87. Shek LP, Lee BW. Epidemiology and seasonality of respiratory tract virus infections in the tropics. Paediatr
Respir Rev 2003;4(2):105–111.

88. Henderson FW, Collier AM, Clyde WA Jr, Denny FW.
Respiratory-syncytial-virus infections, reinfections
and immunity. A prospective, longitudinal study
in young children. N Engl J Med 1979;300(10):530–
534.

89. Holberg CJ, Wright AL, Martinez FD, Ray CG, Taussig
LM, Lebowitz MD. Risk factors for respiratory syncytial virus-associated lower respiratory illnesses in
the first year of life. Am J Epidemiol 1991;133(11):
1135–1151.

90. Knight V, Kapikian AZ, Kravetz HM, et al. Ecology
of a newly recognized common respiratory agent
RS virus. Combined clinical staff conference at the
National Institutes of Health. Ann Intern Med 1961;
55:507–524.

91. Johnson KM, Chanock RM, Rifkind D, Kravetz HM,
Knight V. Respiratory syncytial virus. IV. Correlation
of virus shedding, serologic response, and illness in
adult volunteers. JAMA 1961;176:663–667.

92. Kravetz HM, Knight V, Chanock RM, et al. Respiratory
syncytial virus. III. Production of illness and clinical
observations in adult volunteers. JAMA 1961;176:657–
663.

93. Aherne W, Bird T, Court SD, Gardner PS, McQuillin J.
Pathological changes in virus infections of the lower
respiratory tract in children. J Clin Pathol 1970;
23(1):7–18.

94. Graham BS, Johnson TR, Peebles RS. Immune-
mediated disease pathogenesis in respiratory syncytial virus infection. Immunopharmacology 2000;
48(3):237–247.

95. Wohl ME, Stigol LC, Mead J. Resistance of the total respiratory system in healthy infants and infants with
bronchiolitis. Pediatrics 1969;43(4):495–509.

96. Crowe JE Jr. Immune responses of infants to infection
with respiratory viruses and live attenuated respiratory virus candidate vaccines. Vaccine 1998;16:
1423–1432.

97. Munoz JL, McCarthy CA, Clark ME, Hall CB. Respiratory syncytial virus infection in C57BL/6 mice: clear-

ance of virus from the lungs with virus-specific cytotoxic T cells. J Virol 1991;65(8):4494–4497.

98. Hall CB, Walsh EE, Long CE, Schnabel KC. Immunity
to and frequency of reinfection with respiratory syncytial virus. J Infect Dis 1991;163(4):693–698.

99. Welliver RC. Immunology of respiratory syncytial
virus infection: eosinophils, cytokines, chemokines
and asthma. Pediatr Infect Dis J 2000;19(8):780–783.

100. Hall CB, Powell KR, MacDonald NE, et al. Respiratory
syncytial viral infection in children with compromised immune function. N Engl J Med 1986;315(2):
77–81.

101. Abu-Harb M, Bell F, Finn A, et al. IL-8 and neutrophil
elastase levels in the respiratory tract of infants
with RSV bronchiolitis. Eur Respir J 1999;14(1):139–
143.

102. Fleming DM, Pannell RS, Elliot AJ, Cross KW. Respiratory illness associated with influenza and respiratory
syncytial virus infection. Arch Dis Child 2005;90(7):
741–746.

103. Kotaniemi-Syrjanen A, Laatikainen A, Waris M,
Reijonen TM, Vainionpaa R, Korppi M. Respiratory
syncytial virus infection in children hospitalized for
wheezing: virus-specific studies from infancy to preschool years. Acta Paediatr 2005;94(2):159–165.

104. Isaacs D, Bangham CR, McMichael AJ. Cell-mediated
cytotoxic response to respiratory syncytial virus in
infants with bronchiolitis. Lancet 1987;2(8562):769–
771.

105. Shay DK, Holman RC, Newman RD, Liu LL, Stout JW,
Anderson LJ. Bronchiolitis-associated hospitalizations
among US children, 1980–1996. JAMA 1999;282(15):
1440–1446.

106. Engler HD, Preuss J. Laboratory diagnosis of respiratory virus infections in 24 hours by utilizing shell vial
cultures. J Clin Microbiol 1997;35(8):2165–2167.

107. Abels S, Nadal D, Stroehle A, Bossart W. Reliable detection of respiratory syncytial virus infection in children
for adequate hospital infection control management.
J Clin Microbiol 2001;39(9):3135–3139.

108. Falsey AR, Formica MA, Walsh EE. Diagnosis of respiratory syncytial virus infection: comparison of
reverse transcription-PCR to viral culture and serology in adults with respiratory illness. J Clin Microbiol
2002;40(3):817–820.

109. Kimpen JL, Schaad UB. Treatment of respiratory syn-
cytial virus bronchiolitis: 1995 poll of members of the
European Society for Paediatric Infectious Diseases.
Pediatr Infect Dis J 1997;16(5):479–481.

110. Edell D, Khoshoo V, Ross G, Salter K. Early ribavirin
treatment of bronchiolitis: effect on long-term respiratory morbidity. Chest 2002;122(3):935–939.

111. Khoshoo V, Ross G, Edell D. Effect of interventions
during acute respiratory syncytial virus bronchiolitis
on subsequent long term respiratory morbidity.
Pediatr Infect Dis J 2002;21(5):468–472.

112. American Academy of Pediatrics Committee on Infectious Diseases and Committee of Fetus and Newborn.
Prevention of respiratory syncytial virus infections:
indications for use of palivizumab and update on the
use of RSV-IVIG. Pediatrics 1998;1:1211–1216.

113. Reynolds HY. Defense mechanisms against infections.
Curr Opin Pulmon Med 1999;5(3):136–142.

114. Coonrod JD. Human alveolar lining material and
antibacterial defenses. Am Rev Respir Dis 1986;134(6):
1337.

115. Coonrod JD. The role of extracellular bactericidal factors in pulmonary host defense. Semin Respir Infect 1986;1(2):118–129.
116. Wright JR. Immunomodulatory functions of surfactant. Physiol Rev 1997;77(4):931–962.
117. Lohmann-Matthes ML, Steinmuller C, Franke-Ullmann G. Pulmonary macrophages. Eur Respir J 1994;7(9):1678–1689.
118. Sibille Y, Reynolds HY. Macrophages and polymorphonuclear neutrophils in lung defense and injury. Am Rev Respir Dis 1990;141(2):471–501.
119. MacNee W, Selby C. Neutrophil kinetics in the lungs. Clin Sci 1990;79(2):97–107.
120. Luster AD. Chemokines—chemotactic cytokines that mediate inflammation. N Engl J Med 1998;338(7):436–445.
121. Reynolds H. Normal and defective respiratory host defense mechanisms. In: Pennington JE, ed. Respiratory Infections: Diagnosis and Management, 2nd ed. New York: Raven, 1988:1–33.
122. Agostini C, Chilosi M, Zambello R, Trentin L, Semenzato G. Pulmonary immune cells in health and disease: lymphocytes. Eur Respir J 1993;6(9):1378–1401.
123. Fishman AP, Reynolds, HY, Elias JA, et al. Pulmonary defense mechanisms against infection. In: Fishman AP, ed. Fishman's Pulmonary Diseases and Disorders, 3rd ed. New York: McGraw-Hill, 1998:265–274.

7

Ophthalmic Manifestations of Viral Diseases

Steven Yeh and Mitchell P. Weikert

The ocular manifestations of viruses are diverse and complex, ranging from benign to potentially sight-threatening conditions. Some viruses, such as adenovirus, have direct ocular manifestations after host cell infection. Others require co-infection of the host cell with another virus, for example, cytomegalovirus in patients infected with human immunodeficiency virus (HIV) (1). This chapter reviews the clinical ophthalmic manifestations of common viruses. Other mucocutaneous manifestations of these viruses are contained in previous chapters of this book (Table 7.1). A short summary of viruses with ocular manifestations and their effect on the eye is outlined in Table 7.2.

Herpes Simplex Virus Types 1 and 2

Ocular herpes simplex affects an estimated 400,000 patients in the United States and is generally associated with herpes simplex type 1 (2). Herpes simplex virus type 1 (HSV-1) is transmitted through contact with skin lesions or the saliva of persons with active disease. It may also be passed through a virus-shedding carrier or fomites (3). Ocular disease associated with HSV-2 is transmitted through ocular contact with the genitals of infected individuals (4) and is the cause of 80% of neonatal herpetic infections, acquired during passage through the birth canal. Approximately 20% of neonatal HSV-2 presents with ophthalmic manifestations (5).

Like other forms of herpes, HSV remains latent in the sensory and autonomic ganglia, typically the trigeminal ganglion, and possibly corneal cells (6). The virus and the body's immune response determine its pathogenesis (7). Infections may be primary or recurrent (8). Primary infections commonly appear in facial regions innervated by the maxillary branch of the trigeminal nerve. This is called the "backdoor" approach because the latent virus spreads during primary infection or reactivation through the trigeminal ganglion of the ophthalmic division of the trigeminal nerve. It may also be the reason that primary and recurrent infections often appear at different sites. Ocular infection may even occur without previous mucocutaneous involvement of the ophthalmic branch (9).

Three theories explain how viral reactivation develops. One suggests HSV disease is triggered in the eye or skin when the infected ganglion is stimulated to release virus. Another theory purports that the virus is continuously released in small amounts from the ganglion. The third theory assumes latent HSV resides in peripheral tissue but is reactivated when the ganglion is stimulated. Emotional or physical stress, or a compromised immune system, often precipitates reactivation (10).

Primary Eye Disease

Herpetic involvement of the eye may be a primary infectious disease or, when recurrent, may be related to suppressed immunity (11). The first HSV infection in a nonimmune host is con-

Table 7.1. Additional information on the following viruses can be obtained in the listed chapter

Viral diseases	Chapter
Human herpes virus	8
Cytomegalovirus	7
Epstein-Barr	6
Human papillomavirus	11
Molluscum contagiosum	3
Adenoviruses	
HIV	13
Measles	15
Mumps	
Rubella	21
Picornaviruses	18

sidered primary HSV. Primary infections are usually self-limited and rarely affect the eyes. Ocular involvement occurs in less than 1% of primary infections (12) and is most often subclinical.

Two to 12 days after exposure to an infected carrier, patients may develop pain, foreign body sensation, tearing, and photophobia. This is often accompanied by malaise and fever. Ocular manifestations of HSV include intense periocular blepharitis, follicular and pseudomembranous conjunctivitis, and preauricular adenopathy. Dermatologic lesions in the periocular area generally resolve without scarring. Fifty percent of patients develop keratitis 1 to 2 weeks after the onset of conjunctivitis. Corneal involvement begins as a coarse punctate epithelial keratitis, which may develop into microdendrites. These may then coalesce into large dendrites and geographic ulcers. A linear cell-to-cell spread of the virus gives the keratitis its characteristic dendritic pattern (13). At the base of the dendritic ulcer, fluorescein stain reveals areas where epithelial cells are missing (Fig. 7.1). Rose Bengal stains infected swollen epithelial cells at the edge of the dendrite. Stromal involvement is rarely observed in primary herpetic eye disease. Corneal lesions heal in 2 to 3 weeks with minimal or no scarring. The cornea may experience a loss of sensation after keratitis in primary ocular HSV (14,15). Diagnosis of primary ocular HSV is made by characteristic clinical features

Table 7.2. Summary of ocular viruses

Viral family	Virus	Genetic material	Ocular conditions
Herpesviridae	Herpes simplex virus-1	ds DNA	Keratoconjunctivitis, blepharitis, keratitis, uveitis
	Herpes simplex virus-2	ds DNA	Keratoconjunctivitis, blepharitis, keratitis, uveitis
	Varicella-zoster virus	ds DNA	Keratitis, uveitis, chorioretinitis, optic neuritis, blepharitis, conjunctivitis, canaliculitis, dacryoadenitis, episcleritis, scleritis, glaucoma, extraocular muscle palsies, anterior segment ischemia, vasculitis
	Epstein-Barr virus	ds DNA	Follicular conjunctivitis, keratitis, oculoglandular syndrome papilledema, optic neuritis, multifocal choroiditis, nodular episcleritis, iridocorneal endothelial syndrome, Sjögren's syndrome
	Cytomegalovirus	ds DNA	Keratoconjunctivitis, blepharitis, keratitis, uveitis
	Human herpes virus-8	ds DNA	Eyelid Kaposi's sarcoma, conjunctival Kaposi's sarcoma
Poxviridae	Molluscum contagiosum	ds DNA	Eyelid and conjunctival lesions, keratoconjunctivitis, follicular conjunctivitis
Papovavirus	Papillomaviruses	ds DNA	Benign and malignant tumors of the conjunctiva, lids, and lacrimal sac
Adenoviridae	Adenoviruses	ds DNA	Pharyngoconjunctival fever, epidemic keratoconjunctivitis, nonspecific follicular conjunctivitis
Retroviridae	Human immunodeficiency virus	ds RNA	Cotton wool spots, retinal hemorrhages
Paramyxoviridae	Measles	ss RNA	Conjunctivitis, keratitis, chorioretinitis, retinopathy
	Mumps	ss RNA	Conjunctivitis, keratitis, chorioretinitis, retinopathy, dacryoadenitis, episcleritis, scleritis, optic neuritis, uveitis, extraocular muscle palsies
Togaviridae	Rubella	ss RNA	Cataract, glaucoma, iris atrophy, micro-ophthalmos, microcornea, strabismus, nystagmus, pigmentary retinopathy
Picornaviridae	Coxsackievirus A24 Enterovirus 70	ss RNA	Acute hemorrhagic conjunctivitis, keratitis, optic atrophy

ds, double stranded; ss, single stranded.

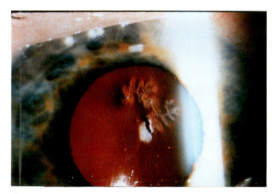

Figure 7.1. Herpes simplex dendrites on the corneal epithelium illuminated by slit beam and retroillumination. The linear branching pattern and "terminal bulbs" are pathognomonic of the herpetic dendrites. Fluorescein stains the base of the dendrite where epithelium is absent. The virus replicates in the epithelial cells of the dendrite margin.

and a history of exposure within 2 weeks. In newborns, primary ocular infections manifest as conjunctivitis and keratitis, accompanied by vesicular skin lesions. A primary infection may be severe and potentially fatal in neonates and immunocompromised individuals (16,17). A life-threatening meningitis or encephalitis may result from primary HSV infection; systemic antiviral therapy is recommended for these patients (18).

Recurrent Eye Disease

Herpes simplex virus keratitis recurs in 25% of patients within 2 years after the initial episode. Additional attacks occur at decreasing intervals (19). The visual morbidity of HSV is primarily due to disease recurrence, and may include epithelial keratitis, stromal keratitis, and anterior uveitis. Recurrent ocular herpes is not associated with fever or malaise. The severity of an episode is dependent on the host's immune response and the virus strain. Diagnosis is made by clinical findings and a known history of previous primary herpetic infection.

In recurrent ocular HSV, epithelial keratitis may be punctate or plaque-like in stellate patterns. Corneal epithelial dendrites may appear in the central or paracentral cornea. Lesions generally heal over a 2-week period, although a few may progress to large geographic ulcers, especially if treated with topical corticosteroids. These, in turn, may develop into trophic sterile ulcers distinguished by gray, heaped-up epithelium with rolled margins. They occur after repeated episodes of keratitis with resulting structural damage of the epithelium. Trophic sterile ulcers may persist for several months and are predisposed to stromal melt and subsequent perforation. Herpes simplex virus may be cultured from dendrites and geographic ulcers.

Stromal keratitis is thought to be an immune-related phenomenon, but a complete understanding of the association between the virus and the host's immune response is unclear (20). Interstitial keratitis, immune Wessley rings, and limbal vasculitis are antigen-antibody-complement–mediated reactions. Diskiform keratitis, a specific subset of stromal keratitis with a well-circumscribed, disk-shaped appearance, is caused by delayed hypersensitivity reactions (21). Both mechanisms of keratitis may present simultaneously. Inflammation, neovascularization, and stromal scars in the visual axis may result in visual loss. Additional episodes compound ocular morbidity. Keratitis may also recur following corneal transplantation.

Recurrent ocular HSV may also lead to iridocyclitis. Uveitis may occur alone or in conjunction with keratitis, and may be immune-related, similar to stromal keratitis (22). Herpes simplex virus has been recovered from the aqueous humor of affected eyes. Patients with herpetic uveitis may present with an abrupt onset of pain, photophobia, and conjunctival injection, and examination may reveal a severe secondary glaucoma (23).

Ocular HSV disease is diagnosed clinically and is supported by laboratory studies. A rise in the serum antibody titer occurs after 1 week and may persist for several weeks (24). There is no rise in serum antibody between recurrences. The virus can be isolated from the skin, as well as corneal and conjunctival lesions in the early phase of the disease. The virus can be recovered only in the early phase of a dendritic ulcer, but not in a geographic ulcer or stromal keratitis. Giemsa stains multinucleated epithelial cells in corneal scrapings. Herpes simplex virus may also be confirmed using the Papanicolaou method to reveal eosinophilic intranuclear inclusions. Additionally, corneal scrapings can be stained with fluorescent antibody to confirm the presence of the HSV antigen. Enzyme immunoassay and polymerase chain reaction (PCR) may also be useful in the rapid diagnosis of HSV keratitis (25,26).

Treatment

Treatment of epithelial keratitis includes topical antiviral agents (Table 7.3). There are three recommended options, each used for 2 to 3 weeks:

Trifluridine (TFT, F3T, Viroptic)	1% drop 8 times/day
Idoxuridine (IDU, Stoxil, Herplex)	0.5% ointment 5 times/day
Vidarabine (Ara-A, Vira-A)	5% ointment 5 times/day

Trifluridine offers the least viral resistance and drug toxicity to the epithelium (27). Although not commercially available in the United States, studies performed in the United Kingdom suggest topical acyclovir ophthalmic ointment is also effective for epithelial keratitis (28). A review of multiple reports of antiviral therapy for herpetic keratitis using meta-analysis models suggests that trifluridine, acyclovir, and vidarabine were more effective than idoxuridine for the treatment of herpetic dendritic and geographic epithelial keratitis. Oral acyclovir is equal to topical antiviral therapy, but does not appear to hasten healing when given with topical antiviral (29).

Adjunctive therapy includes epithelial debridement to decrease the virus and antigen affecting the stroma. Corticosteroids should not be used because of potential increased viral proliferation, healing time, size of the epithelial lesion, and stromal involvement (30,31). In children, the elderly, or disabled patients, oral acyclovir offers an easier alternative to topical antiviral regimens. However, the reported effectiveness of oral acyclovir alone or with topical acyclovir for HSV keratouveitis has been disputed (32–34). According to the Herpetic Eye Disease Study (HEDS), patients who were treated for HSV epithelial keratitis with topical trifluridine within 1 week of onset received no additional benefit from oral acyclovir in the prevention of HSV stromal keratitis or iritis in the following year (35).

Topical prednisolone is indicated for herpetic stromal keratitis, and has been shown to reduce persistence, progression, and duration of stromal inflammation (36). Postponing the initiation of corticosteroids for stromal keratitis delayed its resolution, but did not lead to worsening vision at 6 months follow-up according to HEDS data. The combination of topical corticosteroids and trifluridine limits the duration and progression of stromal keratitis (37). Topical antiviral agents are routinely used with steroids because corticosteroids can potentially intensify viral replication. After the active stromal inflammation has resolved, topical corticosteroids should be slowly tapered to minimize potential rebound inflammation or steroid-induced ocular side effects.

In patients treated with both trifluridine and topical prednisolone, oral acyclovir does not significantly alter the time to treatment failure, proportion of patients who failed treatment, or visual acuity at 6-month follow-up (38). However, a 10-week course of oral acyclovir (400 mg, five times daily) may decrease the incidence of HSV iridocyclitis in patients receiving topical prednisolone and trifluridine. This trend was suggested by HEDS data; however, the sample size did not reach statistical significance (39). Patients who had an episode of HSV eye disease within the preceding year may benefit from suppressive antiviral therapy. In these patients, oral acyclovir 400 mg twice a day reduced the rate of recurrent HSV epithelial and stromal keratitis, as well as orofacial herpes. The benefit was greatest in patients who had experienced prior HSV stromal keratitis (40,41).

Stromal keratitis that progresses to corneal perforation is managed with cyanoacrylate glue and corneal transplantation, along with postoperative topical steroids and antiviral agents. After corneal grafting, recurrent herpetic eye disease occurs in 15% to 32% of eyes within 2 years. Recurrence of keratitis and the risk of graft failure are reduced by postoperative oral acy-

Table 7.3. Diagnosis of herpes simplex virus

Ophthalmic findings	Treatment
Primary disease	
Epithelial keratitis (accompanied by fever and malaise)	Topical trifluridine and cycloplegic
Recurrent disease	
Epithelial keratitis	Topical trifluridine and cycloplegic
Stromal keratitis	Topical steroid + Topical trifluridine or oral acyclovir + cycloplegic
Iridocyclitis	Oral acyclovir 400 mg 5× QD 10 wks + Topical steroid + Topical trifluridine and cycloplegic

clovir (42). Topical cycloplegics are administered as needed to relieve the discomfort of ciliary spasm associated with keratitis or iridocyclitis. Topical antibiotics may be used judiciously to treat corneal ulcers in viral keratitis to prevent secondary bacterial infections.

In postinfectious trophic keratitis, ocular lubricants, patching and bandage contact lens may be used to restore corneal epithelial integrity. Treatment may be lengthy after recurrent episodes of keratitis.

In immunocompromised patients, HSV keratitis is commonly bilateral, severe, atypical with peripheral cornea involvement, more resistant to therapy, and associated with frequent recurrence. Systemic acyclovir is indicated in this group of patients (43).

Varicella-Zoster Virus

Varicella-zoster virus (VZV) is a herpes virus known to infect only humans. Most people are seropositive before age 60. Varicella-zoster virus is transmitted by airborne respiratory secretions or by direct contact with skin lesions. The first exposure to the virus often leads to varicella, commonly known as chickenpox. Following primary infection, VZV establishes latency in the trigeminal or spinal cord ganglia unless reactivated. Herpes zoster, or shingles, occurs when the virus reappears in adults (44). Few eye problems develop from childhood chickenpox. However, severe pain and ocular complications may result from herpes zoster ophthalmicus (HZO) (45).

Ocular Varicella

Ocular varicella may present as primary infectious varicella with the typical cutaneous lesions of chickenpox or in infancy in congenital varicella syndrome. Following host exposure, the virus incubates for 2 weeks and presents as a cutaneous rash with vesicular lesions in various stages of healing, accompanied by fever and malaise. The rash resolves within a week, signaling the end of a patient's contagious phase and the beginning of virus latency. Although most children experience the disease with little or no severity, infants or immunocompromised adults face the prospect of a potentially sight- or life-threatening infection.

Ocular varicella may present as perilimbal epibulbar phlyctenule-like lesions called "pocks." Usually mild and lasting 1 or 2 weeks, these lesions can occur during the infectious period or even months later. The pathophysiology of pocks is unknown, but an immune reaction or live virus is suspected.

Varicella may also result in a punctate or a dendritic epithelial keratitis. It is not uncommon for several episodes of dendrites to occur during the course of infection. Several exam features distinguish herpes simplex dendrites from those of herpes zoster, sometimes termed "pseudo-dendrites." Unlike herpes simplex dendrites, varicella dendrites do not leave an ulcerated base when scraped. The dendrites of herpes zoster are typically elevated, broader, and polymorphous with less distinct branching patterns and fewer terminal bulbs than those of herpes simplex. Herpes simplex dendrites also demonstrate fluorescein staining in the ulcer base and rose Bengal staining along the border of the dendrite.

Both herpes simplex and herpes zoster may cause decreased corneal sensation, so the presence of corneal anesthesia is not helpful in differentiating these entities.

Weeks or months after infection, an immunogenic reaction may result in a diskiform stromal keratitis. Topical corticosteroids may hasten resolution and cycloplegic drops may be used to relieve the discomfort of ciliary spasm. Iritis, chorioretinitis, or optic neuritis may also occur, but are less common. These entities are thought to be immune related and may be treated with corticosteroids unless occurring in the initial phases of infection (46).

Of women of childbearing age, 5% to 16% lack immunity to varicella (47). Congenital varicella syndrome is a more severe systemic and ocular disease than childhood varicella. Chorioretinitis, microphthalmos, optic nerve atrophy or hypoplasia, congenital cataract, and Horner's syndrome may develop in the infant's affected eye (48,49). There is currently no proven treatment for congenital varicella, but the Food and Drug Administration (FDA) approved a vaccine for varicella in 1995, which may reduce occurrences of congenital varicella syndrome.

Varicella-zoster virus in the eye is diagnosed by associated systemic and cutaneous findings along with a rise in serum antibodies during the first 2 weeks of infection.

Herpes Zoster Ophthalmicus

Herpes zoster is relatively common, occurring in 20% of all adults (50). It affects the eye more frequently than does varicella.

Although zoster can affect any of the three divisions of the trigeminal nerve, the ophthalmic division is 20 times more likely to be involved. There are three branches in the ophthalmic division of the trigeminal nerve: frontal, lacrimal, and nasociliary. Zoster involving the ophthalmic branch of the trigeminal nerve is termed herpes zoster ophthalmicus (HZO) regardless of the presence or absence of intraocular involvement (Fig. 7.2). Patients first experience pain and hyperesthesia of the dermatome. This is followed within a few days by vesicular lesions. Ocular complications may be seen concurrently or much later. Zoster sine herpete is a rare condition presenting with HZO-like involvement of the eye, without the classic skin lesions (51). Hutchinson's sign, or zoster lesions on the tip of the nose, indicates involvement of the nasociliary branch of the ophthalmic branch of the trigeminal nerve, the primary sensory nerve of the eye. Of patients with Hutchinson's sign, 50% to 80% develop ocular inflammation (52). However, in the absence of Hutchinson's sign, 61% of patients still have ocular involvement (53).

Possible HZO-related ocular complications include blepharitis, conjunctivitis, canaliculitis, dacryoadenitis, keratitis, keratouveitis, iridocyclitis, secondary cataract, episcleritis, scleritis, glaucoma, vitreitis, retinitis, acute retinal necro-sis, retinal vasculitis, choroiditis, optic neuritis, extraocular muscle palsies, and anterior segment ischemia. These are caused by viral proliferation, immune reaction, inflammatory changes, or occlusive vasculitis. Other possible extraocular manifestations are poliosis, madarosis, trichiasis, cicatricial ectropion or entropion, cicatricial punctal stenosis, and epiphora. Any ocular tissue can be affected, and complications can appear during acute disease or months afterward.

The presentation of conjunctivitis is variable, and may be papillary, pseudomembranous, membranous, or follicular. Different forms of corneal involvement include acute epithelial keratitis, chronic epithelial keratitis, nummular stromal keratitis, interstitial keratitis, diskiform keratitis, and neurotrophic keratitis. The absence of terminal bulbs and the lack of ulcerations can help to distinguish the dendrites of acute epithelial keratitis from those of herpes simplex. The virus can be recovered from the dendrites. Acute epithelial keratitis or zoster skin lesions may be followed by chronic epithelial or nummular stromal keratitis, which are self-limiting and presumed to be immune related. Interstitial keratitis and diskiform keratitis are also immune related. Diskiform keratitis is a delayed hypersensitivity cell-mediated reaction, whereas interstitial keratitis is antigen-antibody-complement related (54). Neurotrophic keratitis occurs after significant corneal damage and may lead to corneal thinning and perforation. Herpes zoster ophthalmicus iritis results from a chronic and recurrent ischemic occlusive vasculitis. Focal or sectoral iris atrophy distinguishes it from the nonischemic, diffuse iris atrophy of herpes simplex iritis. Anterior segment ischemia may occur from extensive perilimbal vasculitis. Laboratory investigation of HZO is similar to varicella.

Prophylactic topical antibiotics, combined with meticulous hygiene may prevent secondary bacteria infections of the HZO skin and lid lesions (Table 7.4). Topical antiviral agents such as idoxuridine, vidarabine, and trifluridine have little effect on ocular lesions. Although 800 mg of oral acyclovir given five times daily for 10 days has been shown to reduce acute pain, the duration of viral shedding, and the formation of new vesicles on the skin, its benefit for ocular complications is unclear (55–57). Some studies support the use of oral acyclovir for treating ocular complications of HZO with decreased

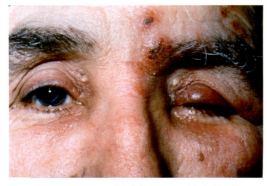

Figure 7.2. Herpes zoster ophthalmicus occurs with viral involvement of the ophthalmic division of the trigeminal nerve. Unilateral vesicles, pustules, and crusted skin lesions are seen in this dermatome. The eyelid is edematous from lesions on the eyelid.

Table 7.4. Diagnosis and treatment of herpes zoster ophthalmicus

Ophthalmic findings	Treatment
Vesicular lesion(s) on the side of the nose tip (Hutchinson's sign). Pain and hyperesthesia of trigeminal nerve ophthalmic branch dermatome. (Begin antiviral within 3 days to lessen incidence of postherpetic neuralgia and ocular complications).	Oral acyclovir 800 mg 5 × per day for 10 days, famciclovir 500 mg TID for 7 days or valacyclovir 1 g TID for 7 days with or without antibiotic ointment to prevent secondary bacterial infection
Stromal keratitis, iridocyclitis	Topical corticosteroid and cycloplegic
Neurotrophic keratitis	Aggressive topical lubrication
Retinitis, cranial nerve involvement	Intravenous acyclovir

incidence and severity (58,59), whereas others have found no benefit (60). Famciclovir, another antiviral agent with better bioavailability (77%) than acyclovir (18%), has been reported to speed resolution of postherpetic neuralgia. Famciclovir 500 mg three times daily demonstrated similar efficacy to acyclovir 800 mg five times daily for the treatment of ophthalmic zoster. No increase in ocular complications was observed with its use (61). Valacyclovir, the prodrug of acyclovir, has a bioavailability of 54% (62), and has demonstrated similar efficacy to acyclovir for herpes zoster in immunocompetent adults. The recommended dosage for valacyclovir is 1 g three times daily (63).

For punctate or dendritic epithelial keratitis, gentle debridement with cotton swabs decreases the amount of virus and antigen in the cornea (64). The efficacy of antivirals in epithelial keratitis is unproven, but antivirals should be considered if the diagnosis is unclear (e.g., HSV epithelial keratitis).

Topical corticosteroids may be used for immune-related ocular complications including iridocyclitis, diskiform keratitis, sclerokeratitis, and keratouveitis. These must be used judiciously and slowly tapered to minimize rebound inflammation. Topical corticosteroids should not be used to treat epithelial keratitis.

Therapeutic options for neurotrophic keratitis include soft contact lenses, lubricants, patching, and tarsorrhaphy to promote epithelial healing. Intravenous acyclovir is administered along with systemic steroids for retinal and optic nerve conditions.

Herpes zoster ophthalmicus has been shown to be an early clinical indicator for AIDS in high-risk young patients (65,66). Features of HZO that may be observed in HIV patients include skin eruption in multiple dermatomes, ocular disease sine herpete, progressive outer retinal necrosis (PORN) syndrome, chronic infectious pseudodendrites, and serious neurologic disease (67).

Because HZO is more severe and prolonged in immunosuppressed individuals, intravenous acyclovir should be used (68) (Fig. 7.3). Progressive outer retinal necrosis, a necrotizing retinopathy described in severely immunocompromised individuals, has been associated with herpes zoster infection. A retrospective study of 38 patients of PORN showed 67% with a history of cutaneous herpes zoster and 41% with HZO (69). Early manifestations of PORN include multifocal deep retinal opacification, which progresses rapidly to total retinal necrosis with retinal detachment and poor visual outcome (70). Treatment with a combination of intravenous agents (e.g., foscarnet and ganciclovir, foscarnet and acyclovir) or an intravenous antiviral and intravitreal ganciclovir agents may arrest progression of retinitis and maintain remission (71,72).

Acute retinal necrosis (ARN) has also been attributed to herpes zoster, although it was originally described in association with herpes simplex (73,74). It may manifest in immunocompromised or healthy patients as a severe peripheral retinitis associated with prominent vitreous inflammation and occlusive retinal vasculitis, which may result in retinal detachment. Papillitis may also be observed. The vitreous

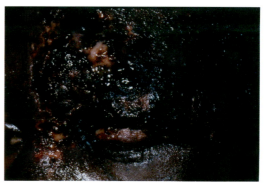

Figure 7.3. Herpes zoster ophthalmicus in an AIDS patient. The condition has a severe and prolonged disease course. Intravenous antiviral treatment is recommended for immunocompromised patients with this condition.

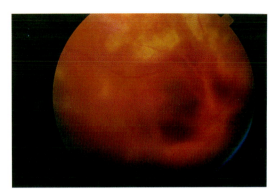

Figure 7.4. This patient has decreased vision from acute retinal necrosis secondary to the herpes zoster virus. Vitreitis and hemorrhages obscure the view of the retina and affected retinal vessels.

inflammation of ARN distinguishes this entity from PORN, as the otherwise healthy patients affected are able to mount a brisk immune response. In addition, visual prognosis tends to be better in ARN than PORN (Fig. 7.4). Herpes zoster, herpes simplex, and cytomegalovirus each have been associated with both ARN and PORN (75–77). The necrotizing herpetic retinopathies may represent a spectrum of disease caused by any one member of the herpes family of virus, with the clinical manifestations dependent on the host immune status (78).

Intravenous acyclovir may prevent progression of acute retinal necrosis; however, its use does not appear to prevent retinal detachment (79). Successful use of intravenous famciclovir has also been reported (80). Some authors have reported successful management of herpetic retinitis with intravitreal ganciclovir as adjunctive therapy following intravenous acyclovir or in combination with two intravenous antiviral medications (81,82). Rhegmatogenous retinal detachment may occur after the onset of inflammation requiring surgical repair.

Epstein-Barr Virus

Epstein-Barr virus (EBV) was discovered in 1964 when Epstein and Barr examined Burkitt's lymphoma tumor cells by electron microscopy. They found viral particles resembling other members of the herpes virus family (83). In vivo, EBV usually infects B lymphocytes and epithelial cells (84), and may establish latent infection of mucosa-associated lymphoid tissue (MALT)

(85). Epstein-Barr virus has been associated with infectious mononucleosis (IM) (86,87), endemic Burkitt's lymphoma (88), nasopharyngeal carcinoma (89), and thymic carcinoma (90). It has also been implicated as a pathogenic agent in Sjögren's syndrome (91,92). Elevated EBV viral capsid antibodies have been observed in patients with iridocorneal endothelial (ICE) syndrome (93).

In the 1970s, studies revealed EBV-specific antibodies were present in 26% to 82% of American college students and military cadets. It has also been reported that 50% to 85% of children in low socioeconomic conditions acquire antibodies by age 4 (94–96). Childhood EBV infection is unremarkable. However, adolescence or adulthood EBV is the cause of IM. The virus is transmitted through upper respiratory droplets. Symptoms and signs of IM include fever, lymphadenopathy, sore throat, hepatitis, pericarditis, polyarthritis, myositis, and atypical lymphocytosis on peripheral blood smear.

A wide range of ocular manifestations have been associated with IM. Reported anterior segment findings include follicular conjunctivitis, dry eye syndrome, nodular episcleritis, iridocyclitis, oculoglandular syndrome, stromal, and epithelial keratitis. Epstein-Barr–associated retinitis (97) and multifocal choroiditis (98) have also been reported. Neuro-ophthalmic manifestations include papilledema, optic neuritis, and cranial nerve palsies (Table 7.5) (99).

Laboratory confirmation of IM is based on a positive heterophile antibody test or rising titers of EBV-specific serologic antibodies (100,101). Because IM is usually self-limiting, treatment is primarily supportive. In the presence of splenomegaly, strenuous physical activity and contact sports should be limited. Epstein-Barr

Table 7.5. Diagnosis and treatment of Epstein-Barr virus

Ophthalmic findings	Treatment
Dry eye syndrome, follicular conjunctivitis, epithelial keratitis	Artificial tears
Stromal keratitis	Mild cases: Artificial tears or observation
	Severe cases: Topical corticosteroids
Retinitis or multifocal choroiditis, papilledema, optic neuritis	No treatment or systemic steroids
Cranial nerve palsies	Monocular occlusion for diplopia

virus–associated ocular diseases are similarly treated with supportive therapy. Topical antiviral therapy is not necessary and the role of systemic acyclovir is unclear. Topical steroids may be considered for ocular inflammation. The treatment for stromal keratitis is dependent on the degree of inflammation present. Artificial tears or no treatment is appropriate in cases of minimal inflammation. Topical steroid drops are recommended for severe stromal keratitis.

Epstein-Barr virus should be considered in the differential diagnosis of any atypical ocular inflammation. Serologic testing of EBV infection should also be considered when the diagnosis is unclear.

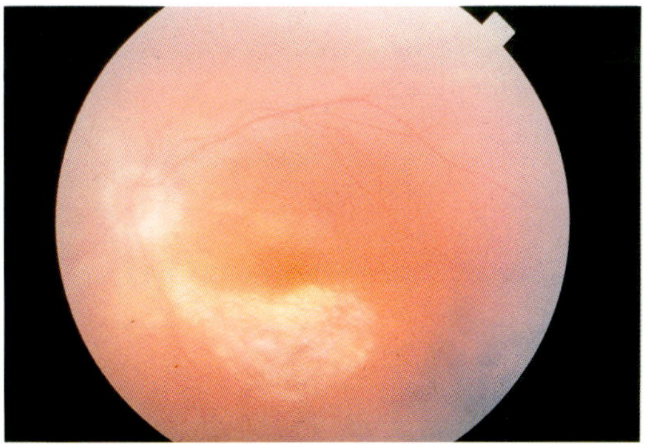

Figure 7.5. Healed atrophic retina in the left eye of an individual on cyclosporine after renal transplant. As the immune status was restored with decreased immunosuppressive medication, the cytomegalovirus (CMV) retinitis resolved.

Cytomegalovirus

Cytomegalovirus (CMV) antibodies can be found in the blood of more than one half of Americans over the age of 50 (102). It is the most common congenital infection, affecting approximately 2% all newborns, but the majority of these infections are subclinical. In affected neonates, systemic manifestations may include hepatosplenomegaly, jaundice, respiratory distress, and intracranial calcifications. The primary ocular manifestation is neonatal chorioretinitis. Cytomegalovirus is among the five intrauterine and perinatal infections collectively known as the TORCHS organisms (*to*xoplasmosis, *ru*bella, *cy*tomegalovirus, *h*erpes simplex, and *s*yphilis (99).

Cytomegalovirus may be transmitted through close contact with infected individuals or, less commonly, by blood transfusions or via infected organ transplants. In most immunocompetent individuals, CMV infections are asymptomatic or may present with symptoms resembling infectious mononucleosis. Cytomegalovirus may result in severe clinical disease with end-organ damage in immunosuppressed individuals (Fig. 7.5).

Cytomegalovirus is a member of the herpes virus family. It is believed that CMV interacts with HIV in a bidirectional manner. Specifically, HIV-1 may enhance productive CMV infection, and co-infection of monocytes with CMV and HIV-1 results in enhanced HIV replication (103).

Although CMV retinitis generally appears in advanced HIV infection, its presence may be the first indicator of AIDS in an estimated 1.8% of HIV-infected persons (104). Cases of opportunistic infection, including CMV retinitis, *Pneumocystis carinii* pneumonia, and *Mycobacterium avium* complex disease, have declined, in the era of highly active antiretroviral therapy (HAART), from 21.9 cases per 100 person-years in 1994 to 3.7 per 100 person-years in mid-1997 (105). However, CMV retinitis remains a common ocular complication in areas with limited economic resources or limited access to HAART. The likelihood of CMV retinitis increases significantly with the decline of CD4 counts below 50 cells/mm^3 (106).

Although the systemic manifestations of CMV such as pneumonitis, gastroenteritis, and encephalitis may be difficult to diagnose, CMV retinitis may identified clinically because of its characteristic funduscopic appearance. Some patients may complain of blurred vision, floaters, photopsias, or blind spots, whereas others may be totally asymptomatic, especially if lesions are small or located in the peripheral retina.

Cytomegalovirus enters the retina via the blood vessels, which may result in a perivascular distribution of retinal lesions. White retinal lesions with adjacent retinal hemorrhages may be observed when the macula is involved and may appear granular when located in the peripheral retina. Lesions are often surrounded by multiple small, round satellite lesions at their posterior border. Cytomegalovirus retinitis may also be associated with vasculitis. Without treatment, CMV retinitis spreads to adjacent areas of healthy retina, leaving the retina atrophic, avascular, and nonfunctioning. Visual field decreases

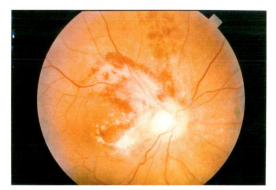

Figure 7.6. Cytomegalovirus retinitis was diagnosed in this individual with AIDS. The patient had a CD4 count of 20 and experienced decreased vision associated with retinitis in the central retina.

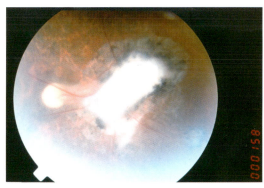

Figure 7.8. Toxoplasmic retinochoroiditis was diagnosed in this HIV patient with CD4 count of 120. A neuroimaging study was performed and confirmed central nervous system involvement. Vision was hand motions only because of macula involvement.

as the area of retinitis increases (Figs. 7.6 and 7.7). Central vision may be preserved unless the macula or optic nerve is involved. Visual loss may also occur from detachment of the necrotic retina.

Small lesions of CMV retinitis in the central retina can resemble HIV retinopathy. The differential diagnosis may also include other viral causes of retinitis such as varicella-zoster or herpes simplex. *Toxoplasma gondii*, *Treponema pallidum*, and intraocular lymphoma may also present with similar fundus findings (Figs. 7.8 to 7.10). A toxoplasmosis retinal lesion may be the first sign of intracranial or disseminated toxoplasmosis (Table 7.6) (107). Neuroimaging studies should be considered when patients present with dense, thick retinitis or an atypical appearance of CMV retinitis. Infiltration of the

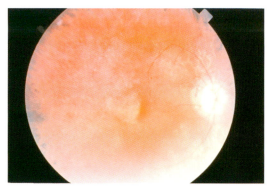

Figure 7.9. Patient with AIDS and extensive retinal pigment epithelial atrophy after treatment with neurosyphilis regimen for syphilitic retinitis in both eyes. The patient's visual acuity was preserved at 20/40 in the right eye and 20/30 in the left.

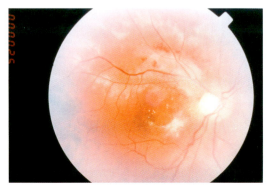

Figure 7.7. This is the same CMV retinitis patient as in Figure 6.6: 6 weeks later. He experienced progressive visual field loss and floaters from progression of the retinitis in the superior retina.

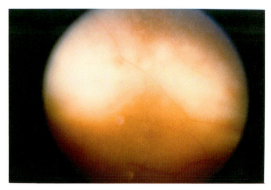

Figure 7.10. Patient with AIDS, mild vitreitis, and retinochoroidal infiltrate. A diagnostic vitrectomy of the eye revealed large cell lymphoma.

Table 7.6. Differential diagnosis of cytomegalovirus (CMV) retinitis

HIV retinopathy: Early CMV retinitis may resemble cotton wool spots. Unlike cotton ool spots, untreated CMV retinitis will progress. A repeat examination in 2 weeks will istinguish the two conditions.

Herpes zoster: Clinical course or herpes zoster retinitis is rapidly progressive and outcome is often poorer than CMV retinitis.

Toxoplasma gondii: Active toxoplasmosis typically presents with a focal area of retinal necrosis and vitreitis. Concomitant central nervous system disease may be present.

Treponema pallidum: Syphilitic lesion may manifest as necrotizing retinitis or as cream-colored posterior placoid chorioretinitis.

Intraocular lymphoma: Intraocular lymphoma associated with AIDS can resemble *Pneumocystis* choroiditis or fungal retinochoroiditis more than CMV retinitis. Differentiation may require a diagnostic vitrectomy or a retinal biopsy in some cases.

retina by lymphoma can also resemble CMV retinitis. In AIDS, intraocular lymphoma may be associated with intracranial disease (108). Central nervous system (CNS) lymphoma in the absence of systemic lymphoma is rare and carries a poor prognosis (109). Ocular syphilis may rarely present as a necrotic retinitis, and should be considered in the differential diagnosis with CMV retinitis. Ocular syphilis may be the initial manifestation of HIV disease and requires a treatment regimen that targets neurosyphilis (110).

Systemic medical treatments for CMV currently include intravenous ganciclovir, foscarnet, and cidofovir (Table 7.7) (111–114). Each requires an induction dose for 2 weeks followed by a maintenance regimen. A periodic reinduction dosage is necessary to slow reactivation or progression. The interval between reinduction treatment decreases as the disease progresses. Valganciclovir, an orally administered prodrug of ganciclovir, has been shown to be as effective as intravenous ganciclovir for induction treatment and may be effective for long-term management of CMV retinitis (115).

Table 7.7. Diagnosis and treatment of CMV retinitis

Ophthalmic findings	Treatment
Retinal opacification and edema associated with adjacent retinal hemorrhages in the central retina, granular opacities when located in the peripheral retina	Intravenous: ganciclovir, foscarnet, cidofovir Intravitreal implant: ganciclovir Intravitreal injection: ganciclovir, foscarnet, fomiversen

In March 1996, local intraocular therapy with a sustained-released ganciclovir pellet (Vitrasert™) was approved (116). This sustained-release intraocular implant was found to decrease the risk of progression of retinitis by three times that of intravenous ganciclovir. However, the risks of CMV disease in the initially uninvolved eye and systemic CMV disease were higher in the patients treated with the intraocular implant alone. Oral ganciclovir administered with a local implant reduces the overall risk of new CMV disease, delays the progression of CMV retinitis in the operated eye, and reduces the risk of Kaposi's sarcoma (117).

Several factors may influence retinitis progression such as subtherapeutic intraocular drug level, development of CMV-resistant strains to virustatic agents, and progressive deterioration in the patient's immunity. Choice of medication may also be influenced by their systemic or ophthalmic side effect profile. Potential dose-limiting toxicities of ganciclovir and valganciclovir include neutropenia, anemia, and thrombocytopenia. Both cidofovir and foscarnet are associated with nephrotoxicity, which requires periodic monitoring of renal function and dosage adjustment.

Cytomegalovirus resistance after prolonged therapy to ganciclovir or foscarnet has been documented and investigated (118). Fomivirsen (Vitravene™), an antisense compound available since 1998, was found to decrease lesion activity in some patients with CMV retinitis not controlled by other anti-CMV drugs (119). Known adverse events of this intravitreally administered drug include anterior chamber inflammation, ocular hypertension, and a reversible bull's-eye maculopathy (120).

With the development of HAART, the combination of protease inhibitors with reverse transcriptase inhibitors, immune recovery uveitis has emerged in HIV patients, changing the incidence and course of CMV retinitis (121,122). While on HAART, AIDS patients with reconstituted immune systems may experience anterior and posterior uveitis following adequate treatment of CMV retinitis, resulting in visual morbidity from vitreitis, cystoid macular edema, epiretinal membrane formation, and papillitis (Fig. 7.11). Mild cases may be observed off therapy, and moderate to severe cases of immune recovery uveitis may respond to periocular corticosteroid injections (123–126). Following restoration of CD4 levels with HAART, discon-

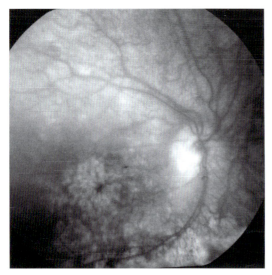

Figure 7.11. Fundus fluorescein angiogram of AIDS patient with decreased vision from cystoid macular edema of immune recovery uveitis. The CMV retinitis was inactive in this eye. CD4 count was 180 after highly active antiretroviral therapy (HAART).

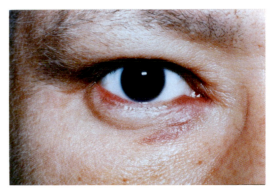

Figure 7.12. This patient with AIDS had painless, violaceous Kaposi's sarcoma lesions along the right lower lid.

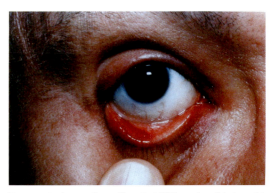

Figure 7.13. Same patient as Figure 7.12. With eversion of the lower lid, these Kaposi's sarcoma lesions were found to involve the adjacent conjunctiva. The lid and visceral lesions regressed after chemotherapy.

tinuation of maintenance CMV therapy may be considered to potentially prevent immune recovery uveitis, provided excellent follow-up can be ensured (127).

An annual dilated funduscopic exam to screen for CMV retinitis is recommended for patients with CD4 counts greater than 100. Patients with CD4 counts between 50 and 100 should be screened every 6 months, and patients with CD4 counts less than 50 cells/mm^3 should be examined at 2- to 3-month intervals. Screening is also advisable if patients are diagnosed with systemic CMV disease. Vigilant screening can increase early detection and treatment, preserving vision (128).

Human Herpes Virus-8

Human herpes virus type 8 DNA sequences were identified by Chang et al. (129) in AIDS-associated Kaposi's sarcoma. The pathogenesis of this virus in malignancy is under investigation (130). On the skin, Kaposi's sarcoma is a marker for HIV (131). The same skin tumor also affects ocular structures in 20% to 30% of AIDS patients (132,133). Although other ocular adnexa can be involved, Kaposi's sarcoma of the eye is usually found on the eyelid or conjunctiva (Figs. 7.12 and 7.13). Ocular Kaposi's sarcoma rarely threat-

ens vision. Conjunctival lesions may resemble subconjunctival hemorrhages in the inferior conjunctival fornix. When lesions are cosmetically unacceptable or associated with entropion, trichiasis, secondary ulceration, or infection, local therapy can be employed. Methods include cryotherapy, surgical excision with or without fluorescein angiography, and localized radiation (133,134). Ocular lesions are also reduced in size when visceral lesions are treated with chemotherapy (Table 7.8).

Table 7.8. Diagnosis and treatment of human herpes virus-8

Ophthalmic findings	Treatments
Asymptomatic violaceous-appearing, red to purple lesions of the eyelid or the conjunctiva.	Chemotherapy
Orbital Kaposi's sarcoma: rare but presence on eyelid may affect visual field	Cryotherapy, surgical excision, or radiation

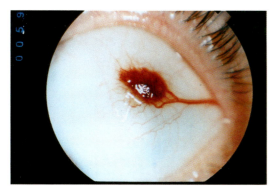

Figure 7.14. An oval, elevated Kaposi's sarcoma lesion in the left eye superior bulbar conjunctiva. The lesion may be mistaken as a subconjunctival hemorrhage.

The differential diagnosis of eyelid or conjunctival lesion includes chalazion, subconjunctival hemorrhage (Fig. 7.14), pyogenic granuloma, lymphoma, and metastatic lymphoma (Table 7.9). Kaposi's sarcoma can be confirmed through biopsy, though it is often unnecessary in AIDS patients (Fig. 7.15).

Table 7.9. Differential diagnosis of human herpes virus-8

Chalazion: Lipogranulomatous inflammation of an obstructed meibomian or sebaceous gland
Subconjunctival hemorrhage: Presence of blood in subconjunctival space, which may be associated with trauma, Valsalva maneuver, or an occult bleeding disorder
Pyogenic granuloma: Pedunculated, deep-red lesion associated with trauma or surgery
Lymphoma: Also seen in AIDS patients and most commonly are of B-cell type in this group; these lesions typically present as gradually enlarging, smooth, salmon-colored lesions

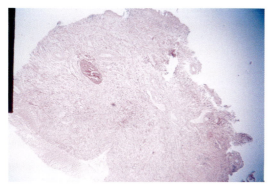

Figure 7.15. Histopathology of a conjunctival Kaposi's sarcoma lesion. Conjunctival surface with goblet cells and underlying mononuclear inflammatory infiltrate. In the deeper subepithelial stroma are dilated vessels and numerous characteristic irregular "jagged" vascular spaces separated by collagen bundles. [Hematoxylin and eosin (H&E), ×10.]

Molluscum Contagiosum

Molluscum contagiosum is a DNA virus of the pox virus family. It proliferates in the skin and mucous membrane epidermis as multiple umbilicated wart-like lesions (135,136). The virus is transmitted through direct contact with lesions or fomites.

Periocular molluscum initially presents as a flesh-colored, smooth, and dome-shaped papule, which may become centrally umbilicated (Fig. 7.16). Lesions may be associated with chronic follicular conjunctivitis and a superficial epithelial keratitis (Fig. 7.17). Corneal pannus may develop in chronic conditions. Lesions are usually asymptomatic; however, pruritic secondary bacterial infections have been described

Figure 7.16. Molluscum contagiosum. Round, waxy, umbilicated lesions on the lower lid, a common site of involvement. In immunocompromised patients, lesions have a high recurrence rate and are more resistant to treatment.

Figure 7.17. Molluscum contagiosum-associated follicular conjunctivitis incited by viral particles from lid lesions. The conjunctivitis will regress when the skin lesions are treated.

(137). Most patients present with fewer than 20 lesions, although more can be present (138).

The diagnosis of molluscum contagiosum is made by clinical examination and confirmed by characteristic pathologic findings (Fig. 7.18). Eosinophilic hyaline cytoplasmic inclusions, or molluscum bodies, are typically observed on biopsy specimens (139). An immunopathologic study of molluscum contagiosum has demonstrated a T-cell–mediated lymphocytic response that is observed in epidermis and dermis adjacent to the molluscum lesions (140).

Periocular molluscum contagiosum involving only the conjunctiva is difficult to diagnose because lesions may be confused with chalazia, ectopic lacrimal gland tissue, granulomas, foreign bodies, or epithelial neoplasms (141).

Molluscum lesions are also found with increased frequency in patients with HIV (141,142). These patients may have numerous confluent lesions, which may recur 6 to 8 weeks after therapy (142). Spontaneous resolution of lesions in HIV patients is unlikely. Keratoconjunctivitis is less common in HIV patients.

Periocular molluscum lesions usually resolve without treatment within a few months. Effective therapies include curettage, chemical cauterization, or cryotherapy. The differential diagnosis of molluscum lid lesions includes hordeolum, seborrheic keratosis, papilloma, nevus, and keratoacanthoma (Table 7.10).

Table 7.10. Differential diagnosis of molluscum contagiosum

Squamous papilloma: Finger-like lesion with a fibrovascular core
Nevus: Pigmented and well-circumscribed lesion of the lids with typical nevus cells in different layers of skin
Keratoacanthoma: Rapidly growing umbilicated lesion that can spontaneously resolve
Hordeolum: Acute onset, well-circumscribed lesion of the eyelids associated with blepharitis, caused by obstruction of meibomian or sebaceous gland
Chalazion: Chronic, lipogranulomatous inflammation of obstructed meibomian or sebaceous gland
Seborrheic keratosis: Greasy and elevated lesion seen in older patients
Pyogenic granuloma: Pedunculated, deep-red lesion that associated with trauma or surgery
Malignant tumors (e.g., cystic basal cell carcinoma): Many epithelial tumors are associated with ulceration and inflammation of normal skin and are diagnosed histologically

Human Papilloma Viruses

Human papilloma virus (HPV) is a double-stranded circular DNA papovavirus that was first identified from genital lesions. Since its discovery, more than 80 subtypes have been identified. Each subtype of virus is site- and cell-type specific. Some subtypes cause warts in specific organs, whereas others are associated with malignancy (143,144).

Human papilloma virus has been associated with tumors of the larynx (145), oral mucosa (146), lacrimal sac epithelium (147), and the conjunctiva (148). Human papilloma virus subtypes 6, 11, and 16 are seen in benign and malignant conjunctival lesions (148,149). Subtypes 16 and 18 have been associated with cervical cancer (150,151). The virus has also been detected in the conjunctiva of patients with cervical dysplasia associated with HPV-16 (152).

Though rare, primary epithelial tumors of the lacrimal sac have been reported. Human papilloma virus type 11 has been associated with benign lacrimal sac tumors. Human papilloma virus type 18 is associated with epithelial malignancies in the lacrimal gland (147), and HPV 16 DNA has been detected in conjunctival epithelial neoplasia (153). Human papilloma virus 16 or 18 DNA and messenger RNA (mRNA) have been detected in conjunctival intraepithelial neoplasia specimens (154). One report failed to detect human papillomavirus DNA in a series of patients with conjunctival epithelial malignancies (155).

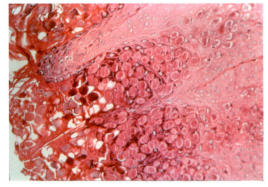

Figure 7.18. Histopathology of molluscum contagiosum showing Henderson-Paterson cytoplasmic eosinophilic inclusion bodies, or so-called molluscum bodies, composed of many viral particles. (H&E, ×64.)

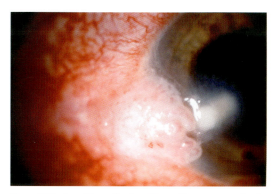

Figure 7.19. A cauliflower-like, irregular papillary mass occupies the inferotemporal conjunctival fornix. Squamous cell conjunctival papilloma is believed to be induced by papovavirus. Lesions may recur after excision.

Table 7.11. Differential diagnosis of human papilloma viruses

Pyogenic granuloma: Pedunculated, deep-red lesion associated with trauma or surgery

Lymphangioma: Multiloculated cystic mass seen before young adulthood

Kaposi's sarcoma: Violaceous red or purple lesions seen in AIDS patients

Lymphoid tumors: Salmon-colored lesion seen in young to middle-aged adults

Amyloid: Smooth and waxy masses seen mainly in lower fornix

Sebaceous cell carcinoma: Should be considered in patients with recurrent chalazion refractory to conventional therapies (i.e. excision, intralesional corticosteroids)

In the eye, HPV-associated papillomas occur more frequently on lid margins than conjunctiva, and are usually seen in children and adolescents. Human papilloma virus is transmitted by autoinoculation or direct contact. The virus has a long incubation period of several months to 2 years. Eyelid papillomas may be single or multiple. They commonly involve the upper eyelids with a subacute, papillary conjunctivitis. Patients may complain of a foreign body sensation and photophobia. In some patients, keratoconjunctivitis may develop. Lid lesions in children tend to be recurrent. Lesions are non-transmissible in older patients, but may rarely undergo malignant transformation.

Conjunctival papillomas are seen with less frequency than eyelid lesions (Figs. 7.19 and 7.20). In an Armed Forces Institute of Pathology review of 1016 epibulbar lesions, 126 papilloma cases were identified (156). The Mayo Clinic reported excising 27 papillomas over a 64-year period (157).

Small asymptomatic ocular lesions may resolve spontaneously, so treatment may consist of observation. Since excision is commonly associated with spread and recurrence, conservative measures should also be considered for lesions in children and adolescents. Treatment is indicated in cases of suspected malignancy, rapid lesion growth, or if vision is affected from eyelid involvement. Treatment may include excision, cryotherapy, or both. Excision of the lesion with a 1-mm margin of adjacent normal tissue is recommended. Other reported treatment methods include carbon dioxide laser, electrodesiccation and curettage, dinitrochlorobenzene, and intralesional injection of α-interferon (158,159). Topical α-interferon has been effective in two patients with papilloma recurrence after excision and cryotherapy (159).

Differential diagnoses of ocular papilloma lesions include both benign and malignant lesions (Table 7.11). Human papilloma virus can be detected by PCR, immunohistochemical staining, and in situ hybridization (160).

Adenoviruses

Adenoviruses are nonenveloped DNA viruses. They replicate entirely inside the nucleus of an infected cell. There are 47 different adenovirus serotypes found throughout the world, causing infections in the upper respiratory tract and the eye. Common manifestations include pharyngoconjunctival fever (PCF), epidemic keratoconjunctivitis (EKC), and acute nonspecific

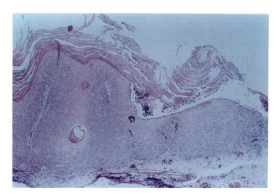

Figure 7.20. Corresponding histopathology of squamous cell conjunctival papilloma. The lesion is composed of squamous epithelium with acanthosis, hyperkeratosis, parakeratosis, and horn cysts. Solar changes and chronic inflammation are present in the stroma, but there is no dysplasia of epithelium. (H&E, ×16.)

Table 7.12. Diagnosis and treatment of adenovirus

Symptoms	Treatment
Pharyngoconjunctivitis fever (PCF): fever, pharyngitis, conjunctivitis, coryza, anterior cervical and preauricular adenopathy, itciness, irritation and tearing, lid edema, ecchymosis, and keratitis	PCF is self-limiting. Proper hygiene is highly effective for prevention; children should be kept from school for 2 weeks.
Epidemic keratoconjunctivitis (EKC): conjunctivitis, keratitis, no systemic symptoms, tearing, foreign body sensation, lid edema, hyperemia, follicular conjunctivitis, chemosis, tender preauricular nodes, pseudomembrane, and symblepharon formation	EK is self-limiting. Prevention is a very important part of management. Supportive treatment includes vasoconstrictors, cold or warm compresses, ocular lubricants, and cycloplegic agents. Topical corticosteroids are used to treat severe keratitis.
Nonspecific follicular conjunctivitis (NFC): seen without any keratitis in children or adults	Nonspecific follicular conjunctivitis is self-limiting with a very mild clinical course.

follicular conjunctivitis (NFC) (Table 7.12). Pharyngoconjunctival fever is primarily associated with types 1, 3, 4, 5, 6, 7, and 14; EKC is associated with 1, 2, 3, 7, 8, 9, 10, 11, and 19; and NFC is associated with many of the serotypes that also cause EKC or PCF (161). The differential diagnosis for PCF, EKC, and NFC is summarized in (Table 7.13).

Pharyngoconjunctival Fever

Pharyngoconjunctival fever is a relatively common syndrome affecting children and young adults. Epidemics often occur within families, schools, or other institutional settings. The virus is transmitted through contact with respiratory droplets, contaminated swimming water, or fomites. The incubation period is 5 to 12 days. Symptoms include fever, pharyngitis, follicular conjunctivitis, hemorrhagic conjunctivitis, coryza, and anterior cervical or preauricular lymphadenopathy (161,162). Initial ocular symptoms may include mild itching and burning, or marked irritation and tearing. These are followed by lid edema and diffuse hyperemia that occur with greater severity in the lower eyelid. There may also be ecchymosis of the lower lid. Presentation is typically bilateral and sequential with one eye symptomatic a few days prior to the fellow eye. Several days after the onset of symptoms, a punctate epithelial keratitis may occur. Keratitis begins with superficial corneal epithelial erosions that stain with fluorescein. The erosions may progress to epithelial and subepithelial focal infiltrates that occupy the central cornea. Infiltrates are thought to be antigen-antibody immune complexes. Although the acute phase of illness can resolve within a few days to a month, the subepithelial infiltrates may persist for several months and may cause glare or diminished vision.

A diagnosis of adenovirus is usually made clinically. Laboratory diagnosis can be determined by direct fluorescent antibody staining and enzyme-linked immunosorbent assay (ELISA) testing. The virus can be cultured in the first 8 to 10 days.

Pharyngoconjunctival fever is a self-limiting condition. Acute symptoms resolve within several days to a month. Patients should be advised regarding contact precautions and strict handwashing to prevent disease spread to others. Pharyngoconjunctival fever is resistant to chlorination, so swimming pools should be avoided. The virus is also resistant to detergent and low pH. Useful disinfectants are phenols, formalin, and 10% household bleach solutions.

Table 7.13. Differential diagnosis of adenoviruses

Other viral conjunctivitis: Viral culture can aid the diagnosis.
Bacterial conjunctivitis: Purulent discharge is a typical symptom. Other signs include chemosis and injection. Bacterial culture is helpful for diagnosis.
Toxic conjunctivitis: Patients have a history of chemical exposure including topical eye drops.
Allergic conjunctivitis: ITCiness with a history of allergy is common.
Ocular cicatricial pemphigoid: Inferior symblepharon with inferior fornix shortening may be observed.

Epidemic Keratoconjunctivitis

More common in adults than children, EKC typically occurs in the fall and winter months. Onset of conjunctivitis is usually subacute. It can be bilateral, lasting for weeks, and followed by keratitis persisting for months. The disease spreads via hand-to-eye contact or from contaminated medical instrument. The most common EKC serotypes are 8 and 19 (163). Type 19 has also been isolated from the cervix and the eye in women with active EKC and cervicitis, suggesting venereal transmission (164). Epidemic keratoconjunctivitis differs from PCF in that there are no systemic symptoms. Following an incubation time of approximately 8 days, patients present with eyelid edema, conjunctival hyperemia, follicular and papillary conjunctivitis and chemosis (161). Subconjunctival hemorrhages and tender preauricular lymph nodes may be present. Patients may complain of persistent tearing, foreign body sensation, or mild photophobia. The fellow eye usually is infected 4 to 5 days later. Patients may form pseudomembranes with eventual conjunctival scarring and symblepharon formation. Active viral replication occurs in corneal epithelial cells. After 2 to 3 weeks, subepithelial corneal infiltrates develop, with decreased visual acuity if the central cornea is affected (Fig. 7.21).

Epidemic keratoconjunctivitis should be suspected in patients with bilateral follicular conjunctivitis. Definitive laboratory confirmation can be obtained by isolation of adenovirus from conjunctival swabs or scrapings in cell culture. The virus is readily isolated during the first week of ocular disease (165). Other tests include serial antibody titer or antigen testing by ELISA (166,167). The differential diagnoses include conjunctivitis from other viruses, bacteria, toxin, and allergens. Ocular cicatricial pemphigoid should be considered if significant pseudomembranes or symblepharon are present.

As with PCF, strict contamination precautions and hand-washing are recommended to prevent

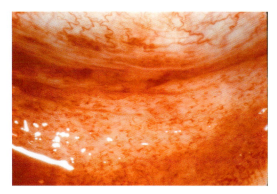

Figure 7.22. Conjunctival petechiae, conjunctival follicles, subconjunctival hemorrhage, swelling of the lids, and profuse tearing characterize adenovirus acute follicular conjunctivitis. The virus spreads easily through unwashed hands and fomites in the acute phase.

spread of the infection. Supportive therapy may be considered to palliate symptoms. Cold or warm compresses and ocular lubrication may be beneficial. Topical vasoconstrictors and a cycloplegic agent may be considered. Topical antibiotics are unnecessary, and acute punctate keratitis resolves spontaneously without treatment. Corneal subepithelial lesions recede gradually over a period of months to years as vision improves. Topical steroids may be used in cases of severe keratitis; however, they may delay viral clearance from the cornea and promote viral shedding (168).

Nonspecific Follicular Conjunctivitis

Nonspecific follicular conjunctivitis presents without keratitis (Fig. 7.22) in either children or adults. Its clinical course is very mild and may be undetected in some cases. Symptoms resolve in 7 to 10 days (169).

Human Immunodeficiency Virus

Human immunodeficiency virus (HIV), a member of the retrovirus family, leads to acquired immune deficiency syndrome (AIDS). In ocular tissue, the virus has been isolated from tears, conjunctiva, cornea, and retina. In the conjunctiva, clinical signs include dilated capillaries and microaneurysms. The most common ophthalmic manifestation of HIV has been observed in the retina. Human immunodeficiency virus retinopathy is seen in more than 50% of AIDS

Figure 7.21. Epidemic keratoconjunctivitis subepithelial infiltrates caused by adenovirus illuminated by a slit beam. Infiltrates may persist beyond the acute phase of the infection and interfere with vision.

patients with a declining CD4 count, and may be a marker for advanced disease (170,171).

Human immunodeficiency virus retinopathy likely reflects retinal ischemia. Clinical signs include focal, ischemic lesions in the superficial central retina, termed cotton wool spots, and peripheral intraretinal hemorrhages. This microvasculopathy may be caused by immune complex deposition in the retinal vessels (172,173) or from infected vascular endothelial cells (174). Ischemic retinopathy from altered blood flow due to hyperviscosity/hypergammaglobulinemia has been suggested as a possible mechanism for HIV retinopathy (175). Acquired immune deficiency syndrome was first recognized as a unique disease in the late 1970s. Before 1978, similar retinal findings were seen in patients with diabetes, hypertension, or collagen vascular disease. Today, HIV retinopathy is an important consideration in the differential diagnosis of cotton wool spots.

Cotton wool spots may resemble early CMV retinitis in the central retina (Fig. 7.23). Human immunodeficiency virus retinopathy can be confirmed by fundus photography of lesions with close follow-up examination in 2 weeks or sooner if necessary. This distinction is important, as HIV retinopathy requires no treatment whereas CMV retinitis may lead to severe visual loss if it is unrecognized and untreated. A clinical appearance of HIV retinopathy in patients with risk factors for HIV testing requires further serologic evaluation.

Opportunistic infections may also affect ocular structures in HIV. Cytomegalovirus

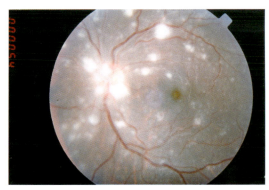

Figure 7.24. Acquired immune deficiency syndrome patient with miliary tuberculosis. Multifocal retinochoroiditis was diagnosed in both eyes. Lesions and other symptoms resolved after tuberculosis medication.

retinitis appears in advanced AIDS. Other viral entities that may lead to visual loss in AIDS patients include herpes simplex and herpes zoster, as discussed above. Other AIDS-related opportunistic eye infections may be caused by bacteria, fungi, and parasites. These include *T. pallidum, Mycobacterium* (Fig. 7.24), *Cryptococcus neoformans, P. carinii,* and *T. gondii* (176). *T. pallidum, T. gondii* retinitis, and intraocular lymphoma may resemble CMV retinal infections. *C. neoformans* and *P. carinii* eye infections involve the choroid with multifocal round lesions (177) (Figs. 7.25 and 7.26). Papilledema is associated with cryptococcal meningitis.

Drug-related ophthalmic side effects have been observed in AIDS patients including rifabutin-associated uveitis (Fig. 7.27), cidofovir-associated uveitis and hypotony (Fig. 7.28)(178), didanosine-associated pigmentary retinopathy

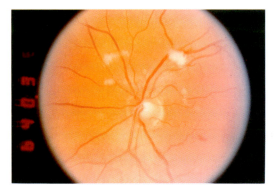

Figure 7.23. Human immunodeficiency virus or AIDS retinopathy indicated by cotton wool spots and intraretinal hemorrhages seen in a HIV-infected young man. This retinal microvasculopathy does not affect vision and lesions disappear within a few weeks. Early CMV retinitis lesions may be mistaken for cotton wool spots. Without treatment, CMV lesions may progress within weeks.

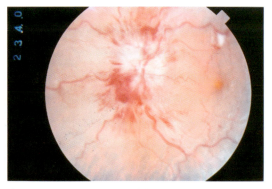

Figure 7.25. Acquired immune deficiency syndrome patient with blurry vision and papilledema, diagnosed with cryptococcal meningitis and a disseminated infection.

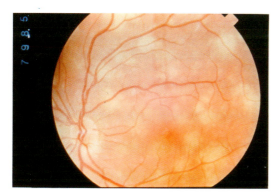

Figure 7.26. Multiple *Pneumocystis carinii* choroidal lesions were found in this AIDS patient. The patient was on aerosolized pentamidine prophylaxis. This opportunistic eye infection is now infrequently seen, as prophylaxis regimens have changed from aerosolized to oral trimethoprim/sulfamethoxazole.

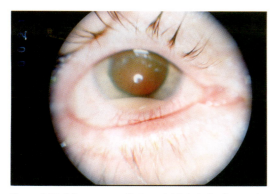

Figure 7.27. Inflammatory cells layered inferiorly in the anterior portion of the eye of an AIDS patient. The patient reported decreased vision while on clarithromycin and rifabutin. The medication-associated toxicity resolved after administration of topical steroid drops and discontinuation of clarithromycin and rifabutin.

Figure 7.28. Cidofovir-related ocular toxicity was diagnosed in this AIDS patient. The patient received the drug for treatment of CMV retinitis. The patient reported sensitivity to light and pain in both eyes. Slit-lamp examination showed anterior chamber inflammatory cells, and the patient's intraocular pressure was low (ocular hypotony).

(179), and fomivirsen-associated uveitis, ocular hypertension, and retinal pigment epithelial toxicity. Familiarity with these conditions helps to distinguish medication-related side effects from acute viral ophthalmic conditions.

Measles

Measles, or rubeola, is caused by the RNA paramyxovirus and is transmitted through the respiratory tract (180,181). Highly contagious, the virus is known to affect only humans. Measles is an acute, febrile exanthematous disease predominantly affecting children and adolescents (182).

Prior to the introduction of a vaccine in 1963, measles occurred epidemically worldwide. Despite a decline in measles after the introduction of its vaccine, the World Health Organization (WHO) estimates that in 2000, measles accounted for approximately 777,000 deaths worldwide with 452,000 (58%) occurring in Africa (183). Measles is estimated to be responsible for 14% to 33% of childhood blindness in Africa. In Zambia, 80% of childhood blindness results from corneal disease, half of which is associated with measles (184).

Although the incidence of measles in the United States has decreased, a significant portion of the population remains susceptible due to lack of immunization or vaccination failure (185–187). Most acquired cases in the U.S. arise from imported cases or import-associated measles strains. During 2001 to 2003, the Centers for Disease Control and Prevention (CDC) reported that no endemic measles strain was circulating in the U.S. (188).

Clinical features of measles include a generalized maculopapular rash lasting 3 or more days, fever, and a triad of cough, coryza, and conjunctivitis. Common ocular manifestations of measles are conjunctivitis and keratitis. Koplik's spots, 1 to 2 mm blue-white spots with a red halo, may involve the conjunctival, as well as the buccal mucosal surfaces. Other ocular findings include bulbar and tarsal conjunctival hyperemia and subconjunctival hemorrhage. Measles keratitis may present with bilateral, symmetric, punctate corneal epithelial and subepithelial lesions, which stain with fluorescein dye (189,190).

In developing countries, bacterial infection superimposed on viral conjunctivitis leads to blindness from corneal perforation, panophthalmitis, and phthisis bulbi (191,192). Visual loss from measles has been reported from chorioretinitis and optic neuritis (193,194). Subacute sclerosing panencephalitis has been reported to cause cortical blindness in the absence of other neurologic signs at presentation (195).

Measles is thought to lead to corneal blindness in developing countries from multiple different mechanisms. Vitamin A deficiency from measles-associated malnutrition leads to xerophthalmia and corneal necrosis, or keratomalacia (196). Bacterial or herpetic keratitis may develop on this ocular surface, leading to corneal scarring or perforation. Harmful traditional remedies are also thought to play a role in developing countries (197).

The diagnosis of measles is made clinically and supported by isolating the paramyxovirus from sputum, blood, and mucous membrane. Humoral antibody response also helps confirm the diagnoses.

Currently, there is no treatment available for the measles virus. Systemic and ocular symptoms are treated supportively. Measles keratitis resolves without sequelae in healthy patients, although corneal epithelial lesions may persist after resolution of patient symptoms. In malnourished patients, corneal lubrication is recommended. Microbial keratitis should be treated with appropriate antibiotics. Vitamin A therapy can reduce ocular and systemic morbidity as well as mortality in malnourished children (198).

Mumps

Mumps is caused by paramyxovirus, another RNA virus. It is a systemic disease occurring mostly in children and occasionally in nonimmune adults. There was a decrease in incidence after the live attenuated mumps vaccine became available in 1967. In 1985, a record low of 2982 cases of mumps was reported nationwide. In 1987, 12,848 cases were reported reflecting an increase among young adults (199). This has been attributed to underimmunization and a vaccine failure rate between 10% to 25% (200).

Infected individuals transmit the virus from saliva to the respiratory tract of others. The illness appears 2 to 3 weeks after exposure (201). Systemic conditions resulting from the viral infection include parotitis, meningitis, deafness,

Table 7.14. Diagnosis and treatment of mumps

Ophthalmic findings	Treatments
Dacryoadenitis, conjunctivitis, scleritis, keratitis, iridocyclitis, optic neuritis, retinitis, and extra ocular muscle palsies	Vaccination to prevent infection Supportive therapy Topical steroids and cycloplegic Systemic steroids to treat neuroretinitis

encephalitis, epididymitis, orchitis, pancreatitis, myocarditis, nephritis, and thyroiditis (202).

Adults without immunity to mumps may develop extrasalivary manifestations. Ocular complications in decreasing order of frequency include dacryoadenitis (called lacrimal mumps), conjunctivitis, scleritis, keratitis, iridocyclitis, optic neuritis, retinitis, and extraocular muscle palsies (203–205). Bilateral dacryoadenitis may take several weeks to completely resolve. A late complication of dacryoadenitis is keratitis sicca syndrome, or severe dry eyes. Diagnosis is based on clinical findings, isolation of the virus from saliva and tears, and a rise in antibody titer. Supportive therapy is recommended (Table 7.14). Cycloplegic eye drops are used to relieve the discomfort of ciliary spasm and topical corticosteroids can be used for intraocular inflammation. Systemic steroids have been used for neuroretinitis. Vaccination remains the best means of controlling disease spread.

Rubella

Rubella, an RNA virus of the Togavirus family, is transmitted via respiratory secretions. This virus usually causes a febrile illness associated with rash, arthralgia, and lymphadenopathy. Serious complications, however, may develop in children of women contracting rubella in early pregnancy (206–208).

In 1941, the virus's teratogenic effects were reported as a combination of congenital heart disease, cataract, and deafness. The most common ocular manifestation is cataract. Other ocular conditions include corneal edema, atrophy of iris stroma, and pigmentary retinopathy, resulting from uneven distribution of pigment in the retinal pigment epithelium. Rubella retinopathy appears as a "salt-and-pepper" fundus and can progress during the first few years of life (Fig. 7.29). Subretinal neovascu-

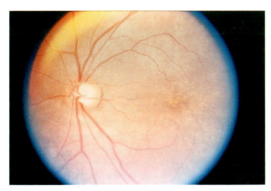

Figure 7.29. Rubella retinopathy in a woman with deafness. The salt-and-pepper retinal pigmentary changes were caused by rubella during the fetal period. Vision is only mildly affected.

larization involving the macula may lead to decreased vision in patients with rubella retinopathy (209–211). Children with congenital rubella who develop glaucoma have poor visual outcomes (212). Other complications include micro-ophthalmos, microcornea, strabismus, nystagmus, and ocular torticollis. Retinitis presumably due to adult rubella has been reported

in a patient exposed to systemic corticosteroids (213).

Diagnosis of rubella is made clinically and supported by antibody titers. Rubella infections are treated with supportive therapy, and ophthalmic disease is treated as necessary (e.g., cataract extraction for congenital cataract). Attenuated live virus vaccination offers protection against the disease. Paresthesias, optic neuritis, and myelitis may rarely be observed following rubella vaccination (214–219). The differential diagnosis of rubella retinopathy is shown in (Table 7.15).

Picornaviruses

Coxsackievirus A24 and enterovirus 70 cause acute hemorrhagic conjunctivitis (AHC). These RNA viruses belong to the picornavirus family (220). Acute hemorrhagic conjunctivitis is known for its highly contagious nature. The mode of transmission is person to person or via infected fomites.

Acute hemorrhagic conjunctivitis was discovered in Ghana, West Africa in 1969 (221). Over 2 million cases have been reported in the Caribbean, the northern part of South American, Central America, and southern Florida between 1980 and 1982 (222). During the spring of 2003, an outbreak of AHC from coxsackievirus A24 began in Brazil, affecting approximately 200,000 individuals, spread to Central America, and eventually led to an outbreak in Puerto Rico, where an estimated 490,000 individuals were affected. School-aged children (i.e., 5 to 18 years of age) and residents of crowded urban areas were at highest risk of acquiring the infection (223).

Following a short incubation of 1 to 2 days, AHC presents with sudden-onset eyelid swelling, conjunctival hemorrhaging, foreign body sensation, photophobia, and ocular pain (Fig. 7.30). Bilateral involvement is typical, with one eye affected 24 hours before the fellow eye (224). The affected eye may develop epithelial keratitis and even a secondary bacterial infection that can lead to visual loss. Systemic features may include malaise, myalgias, fever, headache, depression, upper respiratory tract symptoms with discharge, and preauricular lymphadenopathy. Symptoms may persist for 2 to 3 weeks. Neurologic symptoms can develop 10 to 20 days after the onset of conjunctivitis or as late as 3 months

Table 7.15. Differential diagnosis of rubella

TORCHS syndrome: toxoplasmosis (TO), rubella (R), cytomegalic inclusion disease (C), herpes simplex (H), and syphilis (S)

Toxoplasmosis: The organism has a particular affinity for the central nervous system and the retina. Patient will have white to cream-colored lesions of the retina with vitreous cells. If there is anterior uveitis present and patient is symptomatic, then topical steroid and cycloplegic agent can be used.

Peripheral retinal lesions can be observed in a healthy individual, but lesions close to the optic nerve or macula should be treated with pyrimethamine, folinic acid, and sulfadiazine or clindamycin with sulfadiazine. Another alternative is trimethoprim/sulfamethoxazole with or without clindamycin. Oral prednisone can also be used.

Cytomegalovirus is a virus from the herpes family and can infect a wide array of human organs. The retinochoroiditis is the most common ocular manifestation. The lesions appear to be white to cream color with hemorrhages. The treatment includes use of ganciclovir, foscarnet, cidofovir, fomivirsen, and valganciclovir.

Herpes simplex can cause keratoconjuctivitis, keratitis and uveitis. Keratitis has typical dendrites, which can be treated with topical antiviral agents such as vidarabine, trifluridine, and idoxuridine or debridement.

Syphilis: *Treponema pallidum* is the cause of this infection. This disease is most commonly sexually contracted. Syphilis is diagnosed by use of laboratory testing of VDRL (venereal disease research laboratory) or FTA-ABS (fluorescent treponemal antibody absorption) test and treated with antibiotics such as penicillin.

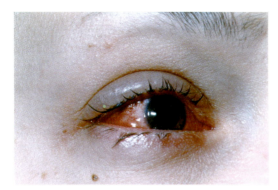

Figure 7.30. Acute hemorrhagic conjunctivitis associated with coxsackievirus and enterovirus. The patient has symptoms of pain, itching, and photophobia. The eye is red from prominent subconjunctival hemorrhage.

after AHC. Neurologic sequelae have been reported to occur in one in 10,000 cases, of which one third may sustain permanent neurologic impairment in the form of facial or lower limb polio-like paralysis, cranial nerve involvement and primary optic atrophy (225). The virus replicates in the epithelial cells of the conjunctiva and cornea resulting in defects on the surface (226).

Diagnosis is suggested by acute, painful, and hemorrhagic conjunctivitis. Confirmation is accomplished by histopathologic identification of either enterovirus 70 or coxsackievirus A24 from conjunctival swabs. Serum assay of immunoglobulin M (IgM) antibody for enterovirus 70 was found to increase by fourfold or greater in 69% of patients during the disease's active phase (227). Differential diagnosis includes ocular trauma, other viral infections, bacterial conjunctivitis, and allergic conjunctivitis.

Treatment of AHC is supportive with cold compresses and appropriate antibiotic coverage if a secondary bacterial infection is suspected. Topical corticosteroids should not be used (228). The importance of strict contamination precautions, frequent hand washing, and avoidance of sharing towels, bedding, and makeup should be reinforced to the patient.

Conclusion

The eye is a common site of viral infection. Viral ophthalmic diseases range from self-limited conditions, such as adenoviral conjunctivitis, to the sight-threatening complications of herpetic eye disease, to potentially life-threatening conditions, such as CMV in immunocompromised patients. Timely diagnosis and appropriate treatment are important to minimize the visually morbidity from viral eye disease. Treatment may require coordination among physicians due to complex nature of some of these viral conditions.

References

1. Faber DW, Wiley CA, Lynn GB, Gross JG, Freeman WR. Role of HIV and CMV in pathogenesis of retinitis and retinal vasculopathy in AIDS patients. Invest Ophthal Vis Sci 1992;33:2345–2353.
2. Leisegang TJ, Melton III LJ, Daly PJ, Ilstrup DM. Epidemiology of ocular herpes simplex. Incidence in Rochester, Minn, 1950 through 1982. Arch Ophthalmol 1989;107:1155–1159.
3. Nahmias A, Roisman B. Infection with herpes simplex viruses I and II. Part III. N Engl J Med 1973;289: 781–789.
4. O'Connor GR. Recurrent herpes simplex uveitis in humans. Surv Ophthalmol 1976;21:165–170.
5. Nahmias AJ, Alford CA, Korones SB. Infection of the newborn with herpesvirus hominis. Adv Pediatr 1970; 17:185–226.
6. Cook SD, Hill JH. Herpes simplex virus: molecular biology and the possibility of corneal latency. Surv Ophthalmol 1991;36:140–148.
7. Pepose JS. Herpes simplex keratitis: role of viral infection vs. immune response. Surv Ophthalmol 1991;35: 345–352.
8. Cook SD. Herpes simplex virus in the eye. Br J Ophthalmol 1992;76:365–366.
9. Liesegang TJ. Biology and molecular aspects of herpes simplex and varicella-zoster virus infections. Ophthalmology 1992;99:781–799.
10. Hyndiuk RA, Glasser. Herpes simplex keratitis. In: Tabbara KF, Hyndiuk RA, eds. Infections of the Eye, 2nd ed. Boston: Little, Brown, 1996:366.
11. Pavan-Langston D. Viral disease of the cornea and external eye. In: Albert DM, Jakobiec FA, eds. Principles and Practice of Ophthalmology. Philadelphia: Saunders, 1994:122.
12. Nahmias A, Roisman B. Infection with herpes simplex viruses I and II. Part III. N Engl J Med 1973;289:781–789.
13. Blaum J. Morphogenesis of the dendritic figure in herpes simplex keratitis. A negative study. Am J Ophthalmol 1970;70:722–724.
14. Paven-Langston D. Major ocular viral infections. In: Galasso G, Whitley R, Merrigan T, eds. Antiviral Agents and Viral Diseases of Man, 3rd ed. New York: Raven Press, 1990:183.
15. Kaufman H, Raefield M. Viral conjunctivitis and keratitis: Herpes simplex virus. In: Kaufman H, Barron B, McDonald M, Waltman S, eds The Cornea. New York: Churchill Livingstone, 1988:299.
16. Ostler HB. Herpes simplex: the primary infection. Surv Ophthalmol 1976;21:91–99.

17. Kimberlin D. Herpes simplex virus, meningitis and encephalitis in neonates. Herpes. 2004;11(suppl 2): 65A–76A.

18. Mommeja-Marin H, Lafaurie M, Scieux C, et al. Herpes simplex virus type 2 as a cause of severe meningitis in immunocompromised adults. Clin Infect Dis 2003;37: 1527–1533.

19. Wilhelmus KR, Coster DJ, Donovan HC, Falcon MG, Jones BR. Prognostic indicators of herpetic keratitis: Analysis of a five-year observation period after corneal ulceration. Arch Ophthalmol 1981;99:1578–1582.

20. Hyndiuk RA, Glasser DB. Herpes simplex keratitis. In: Tabbara KF, Hyndiuk RA, eds. Infections of the Eye, 2nd ed. Boston: Little, Brown, 1996:367.

21. You T, Paven-Langston D. Immune reactions in corneal herpetic disease. Int Ophthal Clin 1996;36:31–39.

22. Abelson M, Pavan-Langston D. Viral uveitis. In: Schlaegel T, ed. Essentials of Uveitis. Int Ophthalmol Clin 1977;17:109–120.

23. O'Connor GR. Recurrent herpes simplex uveitis in humans. Surv Ophthalmol 1976;21:165–170.

24. Meyers JF. Immunology of herpes simplex virus infection. Int Ophthalmol Clin 1975;15:37–47.

25. Sillis M. Clinical evaluation of enzyme immunoassay in rapid diagnosis of herpes simplex infections. J Clin Pathol 1992;45:165–167.

26. Chichili GR, Athmanathan S, Farhatullah S, et al. Multiplex polymerase chain reaction for the detection of herpes simplex virus, varicella-zoster virus, and cytomegalovirus in ocular specimens. Curr Eye Res 2003;27:85–90.

27. O'Day DM. Herpes simplex keratitis. In: Leibowitz H, ed. Corneal Disorders: Clinical Diagnosis and Management. Philadelphia: Saunders, 1984:387.

28. LaLau C, Oosterhuis JA, Versteeg J, et al. Acyclovir and trifluorothymidine in herpetic keratitis—a multicenter trial. Br J Ophthalmol 1982;66:506–508.

29. Wilhelmus KR. The treatment of herpes simplex virus epithelial keratitis. Trans Am Ophthalmol Soc. 2000; 98:505–532.

30. Kimura SJ, Okumoto M. The effect of corticosteroids on experimental herpes simplex keratoconjunctivitis in the rabbit. Am J Ophthalmol 1957;43:131.

31. Ostler HB. Glucocorticoid therapy in ocular herpes simplex. I. Limitations. Surv Ophthalmol 1978;23: 35–43.

32. Sanitato J, Asbell P, Varnell E, Kissling GE, Kaufman HE. Acyclovir in the treatment of herpetic stromal disease. Am J Ophthalmol 1984;98:537–547.

33. Schwab IR. Oral acyclovir in the management of herpes simplex ocular infections. Ophthalmology 1988;95:423–430.

34. Teich SA, Cheung RW, Friedman AH. Systemic antiviral drugs used in ophthalmology. Surv Ophthalmol 1992;37:19–53.

35. The Epithelial Keratitis Trial. The Herpetic Eye Disease Study Group A controlled trial of oral acyclovir for the prevention of stromal keratitis or iritis in patients with herpes simplex virus epithelial keratitis. Arch Ophthalmol 1997;115:703–712.

36. Cohen EJ, Laibson PR. The use of corticosteroids in herpes simplex keratitis. In: Blodi FC, ed. Herpes Simplex Infections of the Eye. New York: Churchill Livingstone, 1984:109–116.

37. Collum LMT, Logan P, Ravenscroft T. Acyclovir (Zovirax) in herpetic disciform keratitis. Br J Ophthalmol 1983;67:115–118.

38. Barron BA, Gee L, Hauck WW, et al. Herpetic Eye Disease Study. A controlled trial of oral acyclovir for herpes simplex stromal keratitis. Ophthalmology 1994;101(12):1871–1882.

39. The Herpetic Eye Disease Study Group. A controlled trial of oral acyclovir for iridocyclitis caused by herpes simplex virus. Arch Ophthalmol 1996;114(9): 1065–1072.

40. Herpetic Eye Disease Study Group. Oral acyclovir for herpes simplex virus eye disease: effect on prevention of epithelial keratitis and stromal keratitis. Arch Ophthalmol 2000;118:1030–1036.

41. Herpetic Eye Disease Study Group. Acyclovir for the prevention of recurrent herpes simplex virus eye disease. N Engl J Med 1998;339(5):300–306.

42. Foster CS, Barney NP. Systemic acyclovir and penetrating keratoplasty for herpes simplex keratitis. Doc Ophthalmol 1992;80:363–369.

43. Young TL, Robin JB, Holland GN, et al. Herpes simplex keratitis in patients with acquired immune deficiency syndrome. Ophthalmology 1989;96:1476–1479.

44. Weller TH. Varicella-herpes Zoster virus. In: Evans AS, ed. Viral Infections of Humans. Epidemiology and Control. New York: Plenum, 1976:457–480.

45. Chang S, De Luise V. Varicella and herpes zoster ophthalmicus. In: Duane's Clinical Ophthalmology, vol 4. Philadelphia: JB Lippincott, 1996:1.

46. Pavan-Langston D. Varicella-zoster ophthalmicus. Int Ophthalmol Clin 1975;15:171–185.

47. Gershon A. Varicella in mother and infant. In: Krugman S, Gershon A, eds. Infections of the Fetus and the Newborn Infant. New York: Alan R Liss, 1975:79.

48. Charles NC, Bennett TW, Margolis S. Ocular pathology of the congenital varicella syndrome. Arch Ophthalmol 1977;95:2034–2037.

49. Lambert S, Taylor D, Kriss A, Holzel H, Heard S. Ocular manifestations of the congenital varicella syndrome. Arch Ophthalmol 1989;107:52–56.

50. Juel-Jensen BE, MacCallum FO. Herpes Simplex, Varicella and Zoster: Clinical Manifestations and Treatment. Philadelphia: JB Lippincott, 1972.

51. Yamamoto S, Tada R, Shimomura Y, Pavan-Langston D, Dunkel EC, Tano Y. Detecting varicella-zoster virus DNA in iridocyclitis using polymerase chain reaction: a case of zoster sine herpete. Arch Ophthalmol 1995;113:1358–1359.

52. Osler HB, Thygeson P. The ocular manifestations of herpes zoster, varicella, infectious mononucleosis and cytomegalovirus disease. Surv Ophthalmol 1976;21: 148–159.

53. Jones DB. Herpes Zoster Ophthalmicus. In: Golden B, ed. Ocular Inflammatory Disease. Springfield, IL: Thomas, 1974:198–209.

54. Liesegang T. Corneal complications from herpes zoster ophthalmicus. Ophthalmology 1985;92:316–324.

55. O'Brien JJ, Campoli-Richards DM. Acyclovir: an updated review of its antiviral activity, pharmacokinetic properties, and therapeutic efficacy. Drugs 1989; 37:233–309.

56. McKendrick MW, McGill JI, White JE, Wood MJ. Oral acyclovir in acute herpes zoster. Br Med J (Clin Res Ed) 1986;293:1529–1532.

57. Cobo LM, Foulks GN, Liesegang T, et al. Oral acyclovir in the therapy of acute herpes zoster ophthalmicus. Ophthalmology 1985;92:1574–1583.

58. Hoang-Xuan T, Buechi ER, Herbort CP, et al. Oral acyclovir for herpes zoster ophthalmicus. An interim report. Ophthalmology 1992;99:1062–1071.

59. Herbort CP, Buechi ER, Piguet B, Zografos L, Fitting P. High-dose oral acyclovir in acute herpes zoster ophthalmicus: the end of the corticosteroid era. Curr Eye Res 1991;10(suppl):171–175.

60 Aylward GW, Claou CMP, Marsh RJ, Yasseem N. Influence of oral acyclovir on ocular complications of herpes zoster ophthalmicus. Eye 1994;8:70–74.

61. Tyring S, Barbarash RN, Nahlik JE, et al. (Collaborative Famciclovir Herpes Zoster Study Group). Famciclovir for the treatment of acute herpes zoster: effects on acute disease and postherpetic neuralgia: a randomized, double-blind, placebo-controlled trial. Ann Intern Med 1995;123:89–96.

62. Acosta EP, Fletcher CV. Valacyclovir Ann Pharmacother 1997;31(2):185–191.

63. Physicians' Desk Reference. 1997:1168.

64. Wilson FM II. Varicella and herpes zoster ophthalmicus. In: Tabbara KF, Hyndiuk RA, eds. Infections of the Eye. Boston: Little, Brown, 1986:369.

65. Sanfor E, Croxson T, Millner A, Mildvan D. Herpes zoster ophthalmicus in patients at risk for AIDS. N Engl J Med 1984;310:1118–1119.

66. Kestelyn P, Stevens AM, Bakkers E, Rouvroy D, Van de Perre P. Severe herpes zoster ophthalmicus in young African adults: a marker for HTLV-III seropositivity. Br J Ophthalmol 1987;71:806–809.

67. Margolis TP, Milner MS, Shama A, Hodge W, Seiff S. Herpes zoster ophthalmicus in patients with human immunodeficiency virus infection. Am J Ophthalmol 1998;125(3):285–291.

68. Seiff SR, Margolis T, Graham SH, O'Donnell JJ. Use of intravenous acyclovir for treatment of herpes zoster ophthalmicus in patients at risk for AIDS. Ann Ophthalmol 1998;20:480–482.

69. Engstrom RE Jr, Holland GN, Margolis TP, et al. The progressive outer retinal necrosis syndrome: a variant of necrotizing herpetic retinopathy in patients with AIDS. Ophthalmology 1994;101:1488–1502.

70. Holland GN. The progressive outer retinal necrosis syndrome. Int Ophthalmol 1994;18:163–165.

71. Perez-Blazquez E, Traspar R, Mendez MI, Montero M. Intravitreal ganciclovir treatment in progressive outer retinal necrosis. Am J Ophthalmol 1997;124:418–421.

72. Ciulla TA, Rutledge BK, Morley MG, Duker JS. The progressive outer retinal necrosis syndrome: successful treatment with combination antiviral therapy. Ophthalmic Surg Lasers 1998;29:198–206.

73. Fisher JP, Lewis ML, Blumenkranz M, et al. The acute retinal necrosis syndrome. Part 1: clinical manifestations. Ophthalmology 1982;89:1309–1316.

74. Ludwiz IH, Zegarra H, Zakov ZN. The acute retinal necrosis syndrome: possible herpes simplex retinitis. Ophthalmology 1984;91:1659–1664.

75. Duker JS, Nielsen JC, Eagle RC, Bosley TM, Granadier R, Benson WE. Rapidly progressive acute retinal necrosis secondary to herpes simplex virus, type 1. Ophthalmology 1990;97(12):1638–1643.

76. Yeo JH, Pepose JS, Stewart JA, et al. Acute retinal necrosis syndrome following herpes zoster dermatitis. Ophthalmology 1986;93:1418–1422.

77. Mitchell SM, Fox JD, Tedder RS, et al. Vitreous fluid sampling and viral genome detection for the diagnosis of viral retinitis in patients with AIDS. J Med Virol 1994;43:336–340.

78. Guex-Crosier Y, Rochat C, Herbort CP. Necrotizing herpetic retinopathies. A spectrum of herpes virus-induced diseases determined by the immune state of the host. Ocul Immunol Inflamm 1997;5:259–265.

79. Blumenkranz MS, Culbertson WW, Clarkson JG, Dix R. Treatment of the acute retinal necrosis syndrome with intravenous acyclovir. Ophthalmology 1986;93:296–300.

80. Figueroa MS, Garabito I, Gutierrez C, Fortun J. Famciclovir for the treatment of acute retinal necrosis (ARN) syndrome. Am J Ophthalmol 1997;123:255–257.

81. Luu KK, Scott IU, Chaudhry NA, et al. Intravitreal antiviral injections as adjunctive therapy in the management of immunocompetent patients with necrotizing herpetic retinopathy. Am J Ophthalmol 2000;129:811–813.

82. Chau Tran TH, Cassoux N, Bodaghi B, Lehoang P. Successful treatment with combination of systemic antiviral drugs and intravitreal ganciclovir injections in the management of severe necrotizing herpetic retinitis. Ocul Immunol Inflamm 2003;11:141–144.

83. Epstein MA, Achong BG, Barr YM. Virus particles in cultured lymphoblasts from Burkitt's lymphoma. Lancet 1964;1:702–703.

84. Rickinson AB. On the biology of Epstein-Barr virus persistence: a reappraisal. In: Lopez C, ed. Immunobiology and Prophylaxis of Herpesvirus Infections. New York: Plenum, 1990:137–146.

85. Klein G. Viral latency and transformation: the strategy of Epstein-Barr virus. Cell 1989;58:5–8.

86. Henle W, Henle G. Epstein-Barr virus and infectious mononucleosis. N Engl J Med 1973;288:263–264.

87. Henle G, Henle W, Diehl V. Relation of Burkitt's tumor-associated herpes-type virus to infectious mononucleosis. Proc Natl Acad Sci USA 1968;59:94–101.

88. Lenoir GM. Role of the virus chromosomal translocations and cellular oncogens in the etiology of Burkitt's lymphoma. In: Epstein MA, Achong BG, eds. The Epstein-Barr Virus: Recent Advances. New York: John Wiley, 1986:184–207.

89. Henle G, Henle W. Epstein-Barr virus-specific IgA serum antibodies as an outstanding feature of nasopharyngeal carcinoma. Int J Cancer 1976;17:1–7.

90. Leyvraz S, Henle W, Chahinian AP, et al. Association of Epstein-Barr virus with thymic carcinoma. N Engl J Med 1985;312:1296–1299.

91. Pflugfelder SC, Roussel TJ, Culbertson WW. Primary Sjogren's syndrome after infectious mononucleosis. JAMA 1987;257:1049–1050.

92. Miyasaka N, Saito I, Haruta J. Possible involvement of Epstein-Barr virus in the pathogenesis of Sjogren's syndrome. Clin Immunol Immuopathol 1994;72:166–170.

93. Alvarado JA, Murphy CG, Juter RP, Hetherington J. Pathogenesis of Chandler's syndrome, essential iris atrophy and the Cogan-Reese syndrome. II: estimate age at disease onset. Invest Ophthal Vis Sci 1986;27:873–882.

94. Evans AS, Niederman JC, Cenabre LC, West B, Richards VA. A prospective evaluation of heterophile and Epstein-Barr virus-specific IgM antibody tests in

clinical and subclinical infectious mononucleosis: specificity and sensitivity of the tests and persistence of antibody. J Infect Dis 1975;132:546–554.

95. Henle W, Henle G. Observations on childhood infections with Epstein-Barr virus. J Infect Dis 1970;121: 303–310.

96. Niederman JC, Evans AS, Subrahmanyan L, McCullom RW. Prevalence, incidence and persistence of EB virus antibody in young adults. N Engl J Med 1970;282: 361–365.

97. Raymond LA, Wilson CA, Linnemann CC, Ward MA, Bernstein DI, Love DC. Punctate outer retinitis in acute Epstein-Barr virus infection. Am J Ophthalmol 1987;104:424–425.

98. Tiedeman JS. Epstein-Barr Viral antibodies in multifocal choroiditis and panuveitis. Am J Ophthalmol 1987;103:659–663.

99. Remington JS, JO Klein, eds. Infectious Diseases of the Newborn Infant. Philadelphia: WB Saunders, 1976.

100. Matoba AY. Ocular disease associated with Epstein-Barr virus infection. Surv Ophthalmol 1990;35:145–150.

101. Pflugfelder SC, Crouse CA, Atherton SS. Ophthalmic manifestations of Epstein-Barr virus infection. Int Ophthalmol Clin 1993;33:95–101.

102. Krech U, Jung M, Jung F. Cytomegalovirus Infections of Man. New York: S Karger, 1971.

103. Skolnik PR, Kosloff BR, Hirsch MS. Bidirectional interactions between human immunodeficiency virus type 1 and cytomegalovirus. J. Infect Dis 1988;157:508–513.

104. Sison RF, Holland GN, MacArthur LJ, Wheeler NC, Gottlieb MS. Cytomegalovirus retinopathy as the initial manifestation of the acquired immunodeficiency syndrome. Am J Ophthalmol 1991;112: 243–249.

105. Palella FJ, Delaney KM, Moorman AC, et al. Declining morbidity and mortality among patients with advanced human immunodeficiency virus infection. N Engl J Med 1998;338:853–860.

106. Drew WL. Cytomegalovirus infection in patients with AIDS. J Infect Dis 1988;158:449–456.

107. Cochereau-Massin I, LeHoang P, Lautier-Frau M, et al. Ocular toxoplasmosis in human immunodeficiency virus-infected patients. Am J Ophthalmol 1992;114: 130–135.

108. Schanzer MC, Font RL, O'Malley RE. Primary ocular malignant lymphoma associated with the acquired immune deficiency syndrome. Ophthalmology 1991; 98:88–91.

109. Levine AM. Epidemiology, clinical characteristics, and management of AIDS-related lymphoma. Hematol Clin North Am 1991;5:331–342.

110. Aldave AJ, King JA, Cunningham ET Jr. Ocular syphilis. Curr Opin Ophthalmol 2001;12:433–441.

111. Jabs DA, Newman C, De Bustros S, Polk BF. Treatment of cytomegalovirus retinitis with ganciclovir. Ophthalmology 1987;94:824–830.

112. Lehoang P, Girard B, Robinet M, et al. Foscarnet in the treatment of cytomegalovirus retinitis in acquired immune deficiency syndrome. Ophthalmology. 1989; 96:865–873; discussion 873–734.

113. Jacobson MA, Drew WL, Feinberg J, et al. Foscarnet therapy for ganciclovir-resistant cytomegalovirus retinitis in patients with AIDS. J Infect Dis. 1991;163: 1348–1351.

114. Lalezari JP, Stagg RJ, Kuppermann BD, et al. Intravenous cidofovir for peripheral cytomegalovirus retinitis in patients with AIDS. A randomized, controlled trial. J Acquir Immune Defic Syndr Hum Retrovirol. 1998;17:339–344.

115. Martin DF, Sierra-Madero J, Walmsley S, et al. A controlled trial of valganciclovir as induction therapy for cytomegalovirus retinitis. N Engl J Med 2002;346: 1119–1126.

116. Musch DC, Martin DF, Gordon JF, et al. Treatment of cytomegalovirus retinitis with a sustained-release ganciclovir implant. N Engl J Med 1997;337:83–90.

117. Martin DF, Kuppermann BD, Wolitz RA, Palestine AG, Li H, Robinson CA. Oral ganciclovir for patients with cytomegalovirus retinitis treated with a ganciclovir implant. N Engl J Med 1999;340:1063–1070.

118. Proceedings on meeting of drug resistance in cytomegalovirus: current knowledge and implications for patient management. J Acquir Immune Defic Syndr Hum Retrovirol 1996;2(suppl 1):1–22.

119. Perry CM, Balfour JA. Fomivirsen. Drugs 1999;57(3): 375–380.

120. Stone TW, Jaffe GJ. Reversible bull's-eye maculopathy associated with intravitreal fomivirsen therapy for cytomegalovirus retinitis. Am J Ophthalmol 2000; 130(2):242–243.

121. Deeks SG, Smith M, Holodniy M, Kahn JO. HIV-1 protease inhibitors. A review for clinicians. JAMA 1997; 277:145–153.

122. Nguyen QD, Kempen JH, Bolton SG, et al. Immune recovery uveitis in patients with AIDS and cytomegalovirus retinitis after highly active antiretroviral therapy. Am J Ophthalmol 2000;129:634–639.

123. Henderson HW, Mitchell SM. Treatment of immune recovery vitreitis with local steroids. Br J Ophthalmol 1999;83:540–545.

124. El-Bradey MH, Cheng L, Song MK, et al. Long-term results of treatment of macular complications in eyes with immune recovery uveitis using a graded treatment approach. Retina 2004;24:376–382.

125. Arevalo JF, Mendoza AJ, Ferretti Y. Immune recovery uveitis in AIDS patients with cytomegalovirus retinitis treated with highly active antiretroviral therapy in Venezuela. Retina 2003;23:495–502.

126. Karavellas MP, Azen SP, MacDonald JC, et al. Immune recovery vitreitis and uveitis in AIDS: clinical predictors, sequelae, and treatment outcomes. Retina 2001;21:1–9.

127. Jabs DA, Bolton SG, Dunn JP, Palestine AG. Discontinuing anticytomegalovirus therapy in patients with immune reconstitution after combination antiretroviral therapy. Am J Ophthalmol 1998;126:817–822.

128. Kupperman BD, Petty JG, Richman DD, et al. Correlation between CD4+ counts and prevalence of cytomegalovirus retinitis and human immunodeficiency virus-related noninfectious retinal vasculopathy in patients with acquired immunodeficiency syndrome. Am J Ophthalmol 1993;115:575–582.

129. Chang Y, Cesarman E, Pessin MS, et al. Identification of herpes virus-like DNA sequences in AIDS-associated sarcoma. Science 1994;266:1865–1869.

130. Levy JA. Three new human herpesviruses (HHV6, 7, and 8). Lancet 1997;349:558–563.

131. Hatcher VA. Mucocutaneous infections in acquired immune deficiency syndrome. In: Friedman-Kien AE,

Laubenstein LJ, eds. AIDS: The Epidemic of Kaposi's Sarcoma and Opportunistic Infections. New York: Masson, 1984:245–251.

132. Dugel PU, Gill PS, Frangieh GT, Rao NA. Ocular adnexal Kaposi's sarcoma in acquired immunodeficiency syndrome. Am J Ophthalmol 1990;110: 500–503.

133. Shuler JD, Holland GN, Miles SA, Miller BJ, Grossman I. Kaposi's sarcoma of the conjunctiva and eyelids associated with the acquired immunodeficiency syndrome. Arch Ophthalmol 1989;107:858–862.

134. Dugel PU, Gill PS, Frangieh GT, Rao NA. Treatment of ocular adnexal Kaposi's sarcoma in acquired immune deficiency syndrome. Ophthalmology 1992;99:1127–1132.

135. Epstein WL. Molluscum contagiosum. Semin Dermatol 1992;11:184–189.

136. Charteris DG, Bonshek RE, Tullo AB. Ophthalmic molluscum contagiosum: clinical and immunopathological features. Br J Ophthalmol 1995;79:476–481.

137. Deoreo GA, Johnson HH Jr, Binkey GW. An eczemous reaction associated with molluscum contagiosum. Arch Dermatol 1956;74:344–348.

138. Margo C, Katz NNK. Management of periocular molluscum contagiosum in children. J. Pediatric Ophthalmol Strabismus 1993;20:19–21.

139. Pepose JS, Esposito JJ. Molluscum contagiosum, orf, and vaccinia. In: Pepose JS, Holland GN, Wilhelmus KR, eds. Ocular Infection and Immunity. St. Louis: Mosby, 1986:846–856.

140. Charteris DG, Bonshek RE, Tullo AB. Ophthalmic molluscum contagiosum: clinical and immunopathological features. Br J Ophthalmol 1995;79(5):476–481.

141. Charles NC, Friedberg DN. Epibulbar molluscum contagiosum in acquired immune deficiency syndrome, case report and review of the literature. Ophthalmology 1992;99:1123–1126.

142. Robinson MR, Udell IJ, Garber PF, Perry HD, Streeten BW. Molluscum contagiosum of the eyelids in patients with acquired immune deficiency syndrome. Ophthalmology 1992;99:1745–1747.

143. Miller DM, Bredell RT, Levine MR. The conjunctival wart: report of a case and review of treatment options. Ophthalmic Surgery 1994;25(8):545–548.

144. de Villiers EM. Heterogeneity of the human papilloma virus group. J Virol 1989;63:4898–4903.

145. Quick CA, Watts SL, Krizyzek RA, Faras AJ. Relationship between condylomata and laryngeal papilloma: clinical and molecular biological evidence. Ann Otol Rhinol Laryngol 1980;89:467–471.

146. de Villiers EM, Weidauer H, Otto H, Zur Hausen H. Papillomavirus DNA in human tongue carcinomas. Int J Cancer 1985;36:575–578.

147. Madreperla SA, Green WR, Daniel R, Shah KV. Human papillomavirus in primary epithelial tumors of the lacrimal sac. Ophthalmology 1993;100(4):569–573.

148. McDonnell PJ, McDonnell JM, Kessis T, Green WR, Shah KV. Detection of human papillomavirus type 6/11 DNA in conjunctival papillomas by in situ hybridization with radioactive probes. Hum Pathol 1987;18:1115–1119.

149. McDonnell JM, Mayr AJ, Martin WJ. DNA of human papillomavirus Type 16 in dysplastic and malignant lesions of the conjunctiva and cornea. N Engl J Med 1989;320:1442–1446.

150. Brescia RJ, Jenson AB, Lancaster WD, Kurman RJ. The role of human papillomaviruses in the pathogenesis and histological classification of precancerous lesions of the cervix. Hum Pathol 1986;17:552–559.

151. Bonfiglio TA, Stoler MH. Human papillomavirus and cancer of the uterine cervix. Hum Pathol 1988;19: 621–622.

152. McDonnell JM, Wagner D, Bernstein G, Sun YY. Human papillomavirus type 16 DNA in ocular and cervical swabs of women with genital tract condylomata. Am J Ophthalmol 1991;112:61–66.

153. McDonnell JM, McDonnell PK, Sun YY. Human papillomavirus DNA in tissues and ocular surface swabs of patients with conjunctival epithelial neoplasia. Invest Ophthalmol Vis Sci 1992;33:184–189.

154. Scott IU, Karp CL, Nuovo GJ. Human papilloma virus 16 and 18 expression in conjunctival intraepithelial neoplasia. Ophthalmology 2000;109:542–547.

155. Eng HL, Lin TM, Chen SY, Wu SM, Chen WJ. Failure to detect human papillomavirus DNA in malignant epithelial neoplasms of conjunctiva by polymerase chain reaction. Am J Clin Pathol 2002;117:429–436.

156. Ash JE. Epibulbar tumors. Am J Ophthalmol 1950;33: 1203.

157. Erie JC, Campbell RJ, Liesegang TJ. Conjunctival and corneal intraepithelial and invasive neoplasia. Ophthalmology 1986;93:176–183.

158. Petersen CS, Nurnberg BM. Carbon dioxide laser vaporization combined with perilesionally injected interferon alfa-2b in the treatment of a hyperkeratotic verruca vulgaris on the upper eyelid. Arch Dermatol 1994;130:1369–1370.

159. de Keizer RJ, de Wolff-Rouendaal D. Topical alpha-interferon in recurrent conjunctival papilloma. Acta Ophthalmol Scand 2003;81(2):193–196.

160. Nakamura Y, Mashima Y, Kameyama K, Mukai M, Oguchi Y. Detection of human papillomavirus infection in squamous tumours of the conjunctiva and lacrimal sac by immunohistochemistry, in situ hybridization, and polymerase chain reaction. Br J Ophthalmol 1997;81:308–313.

161. Pavan-Langston D. Viral disease of the cornea and external eye. In: Albert DM, Jakobiec FA, eds. Principles and Practice of Ophthalmology, Clinical Practice, vol. 1. Philadelphia: WB Saunders, 1994.

162. Duke-Elder S. The adenoviruses. In: Duke-Elder S, ed. System of Ophthalmology, vol. 8. St. Louis: CV Mosby, 1965:348.

163. O'Day D, Guyer B, Hierholzer J. Clinical and laboratory evaluation of epidemic keratoconjunctivitis due to adenoviruses type 8 and 19. Am J Ophthalmol 1976; 81:207–215.

164. Harnett, Newnham W. Isolation of adenovirus type 19 from the male and female genital tracts. Br J Vener Dis 1981;57:55–57.

165. Gibson J, Darougar D, McSwiggan D, Thaker U. Comparative sensitivity of a cultural test and the complement fixation test in the diagnosis of adenovirus ocular infection. Br J Ophthalmol 1979;63:617–620.

166. Schwartz H, Vastine D, Yamashiroya H, West CE. Immunofluorescent detection of adenovirus antigen in EKC. Invest Ophthalmol 1976;15:199–207.

167. Rodrigues M, Lennette D, Arentsen J, Thompson C. Methods for rapid detection of human ocular viral infections. Ophthalmology 1979;86:452–464.

168. Romanowski EG, Yates KA, Gordon YJ. Topical corticosteroids of limited potency promote adenovirus replication in the Ad5/NZW rabbit ocular model. Cornea 2002;21:289–291.

169. Vastine V. Adenoviruses and miscellaneous viral infections. In: Smolin G, Thoft R, eds. The Cornea, 2nd ed. Boston: Little, Brown, 1987:266.

170. Freeman WR, Chen A, Henderly DE, et al. Prevalence and significance of acquired immunodeficiency syndrome-related retinal microvasculopathy. Am J Ophthalmol 1989;107:229–235.

171. Jabs DA, Green WR, Fox R, Polk BF, Bartlett JG. Ocular manifestations of acquired immune deficiency syndrome. Ophthalmology 1989;96:1092–1099.

172. Newsome DA. Microvascular aspects of acquired immune deficiency syndrome retinopathy. Am J Ophthalmol 1984;98:590–601.

173. Pepose JS, Holland GN, Nestor MS, Cochran AJ, Foos RY. Acquired immune deficiency syndrome. Pathogenic mechanisms of ocular disease. Ophthalmology 1985;92:472–484.

174. Pomerantz RJ, Kuritzkes DR, de la Monte SM, et al. Infection of the retina by human immunodeficiency virus type 1. N Engl J Med 1987;317:1643–1647.

175. Engstrom RE, Holland GN, Hardy WD, Meiselman HJ. Hemorrhagic abnormalities in patients with human immunodeficiency virus infection and ophthalmic microvasculopathy. Am J Ophthalmol 1990;109:153–161.

176. Jabs DA, Green WR, Fox R, Polk BF, Bartlett JG. Ocular manifestations of acquired immune deficiency syndrome. Ophthalmology 1989;96:1092–1099.

177. Rosenblatt MA, Cunningham C, Teich SS, Freidman AH. Choroidal lesions in patients with AIDS. Br J Ophthalmol 1990;74:610–614.

178. Fraunfelder FW, Rosenbaum JT. Drug induced uveitis. Drug Safety 1997;17(3):197–207, 199.

179. Whitcup SM, Dastgheib K, Nussenblatt RB, Walton RC, Pizzo PA, Chan CC. A clinical pathologic report of the retinal lesion associated with didanosine. Arch Ophthalmol 1994;112(12):1594–1598.

180. Ray CG. Measles (rubeola). In: Thorn GW, et al., eds. Harrison's Principles of Internal Medicine, 8th ed., vol. 1. New York: McGraw-Hill, 1977.

181. Katz SL. Measles. In: Rudolph AM, Barnett HL, Einhord AH, eds. Pediatrics, 16th ed. New York: Appleton-Century-Crofts, 1977.

182. Bergstrom TJ. Measles infection of the eye. In: Darrell RW, ed. Viral Diseases of the Eye. Philadelphia: Lea & Febiger, 1985:233–238.

183. Measles mortality reduction—West Africa, 1996–2002. MMWR 2004;53:28–30.

184. Awdrey PN, Cobb B, Adams PCG. Blindness in the Luapula Valley. Central Afr J Med 1967;13:197–201.

185. Hinman AR, Brandling-Bennett AD, Nieburg PI. The opportunity and obligation to eliminate measles from the United States. JAMA 1979;242:1157–1162.

186. Hinman AR, Brandling-Bennett AD, Bernier RH, Kirby CD, Eddins DL. Current features of measles in the United States. Epidemiol Rev 1980;2:153–170.

187. Measles—United States, 1977–1980. MMWR 1980;29:598–599.

188. Epidemiology of measles—United States, 2001–2003. MMWR 2004;53:713–716.

189. Kayikcioglu O, Kir E, Soyler M, Guler C, Irkec M. Ocular findings in a measles epidemic among young adults. Ocul Immunol Inflamm 2000;8:59–62.

190. Deckard PS, Bergstrom TJ. Rubeola keratitis. Ophthalmology 1980;88:810–813.

191. Furgiuele FP, Hiles DA, Cignetti FE. Measles. In: Harley RD, ed. Pediatric Ophthalmology. Philadelphia: WB Saunders, 1975.

192. Fedukowicz HB. Measles In External Infections of the Eye, 2nd ed. New York: Appleton-Century-Crofts, 1978.

193. Caruso JM, Robbins-Tien D, Brown WD, Antony JH, Gascon GG. Atypical chorioretinitis as an early presentation of subacute sclerosing panencephalitis. J Pediatr Ophthalmol Strabismus 2000;37:119–122.

194. Azuma M, Morimura Y, Kawahara S, Okada AA. Bilateral anterior optic neuritis in adult measles infection without encephalomyelitis. Am J Ophthalmol 2002; 134:768–769.

195. Senbil N, Aydin OF, Orer H, Gurer YK. Subacute sclerosing panencephalitis: a cause of acute vision loss. Pediatr Neurol 2004;31:214–217.

196. Sommer A. Xerophthalmia, keratomalacia and nutritional blindness. Int Ophthalmol 1990;14:195–199.

197. Gilbert C, Awan H. Blindness in children. BMJ 2003; 327:760–761.

198. Hussey CD, Klein M. Measles-induced vitamin A deficiency. Ann NY Acad Sci 1992;669:188–194.

199. Centers For Disease Control. Mumps—United States, 1985–1988. Leads from the MMWR. JAMA 1989;261: 1702.

200. Kim-Farley R, Bart S, Stetler H, et al. Clinical mumps vaccine efficacy. Am J Epidemiol 1985;121:593.

201. Wilfert CM. Mumps. In: Joklik WK, ed. Principles of Animal Virology. New York: Appleton-Century-Crofts, 1980.

202. Foster ER, Lowder CY, Meisler DM, Kosmorsky GS, Baetz-Greenwalt B. Mumps neuroretinitis in an adolescent. Am J Ophthalmol 1990;110:91–93.

203. Katavisto M. Ocular complications of epidemic parotitis. Acta Ophthalmol 1956;34:208.

204. Bang HO, Bang J. Involvement of the central nervous system in mumps. Acta Med Scand 1943;113:487.

205. Al-Rashid RA, Cress C. Mumps: uveitis complication the course of acute leukemia. J Peidatr Ophthal 1977;14(2):100–102.

206. Yanoff M. The retina in rubella. In: Tasman W, ed. Retinal Disease in Children. New York: Harper and Row, 1971:223–232.

207. Walff SM. The ocular manifestations of congenital rubella. A prospective study of 328 cases of congenital rubella. J Pediatr Ophthalmol 1973;10:101–141.

208. Franceschetti A, Francois J, Babel J. Chorioretinal Heredodegenerations. Springfield, IL: Charles C. Thomas, 1974:1069.

209. Slusher MM, Tyler ME. Rubella retinopathy and subretinal neovascularization. Ann Ophthalmol 1982; 14(3):292–294.

210. Orth DH, Fishman GA, Segall M, Bhatt A, Yassur Y. Rubella maculopathy. Br J Ophthalmol 1980;64: 201–205.

211. Frank KE, Purnell EW. Subretinal neovascularization following rubella retinopathy. Am J Ophthalmol 1978;86:462–466.

212. Wolff SM. The ocular manifestations of congenital rubella. Trans Am Ophthal Soc 1972;70:577–614.

213. Hayashi M, Yoshimura N, Kondo T. Acute rubella retinal pigment epitheliitis in an adult. Am J Ophthalmol 1982;93:285–288.

214. Speier JE. Complications of rubella vaccination. JAMA 1970;213:2272.

215. Kilroy AW, Schaffner W, Fleet WF, Lefkowitz LB Jr, Karzon DT, Fenichel GM. Two syndromes following rubella immunization. JAMA 1970;214:2287–2292.

216. Gilmartin RC, Jabbour JR, Duenas DA. Rubella vaccine myeloradiculoneuritis. Pediatrics 1992;80:406–412.

217. Rubella Surveillance Report. Atlanta: Center for Disease Control, August 1970.

218. Rubella Surveillance Report. Atlanta: Centers for Disease Control, August 1976.

219. Kline LB, Margulies SL, Oh SJ. Optic neuritis and myelitis following rubella vaccination. Arch Neurol 1982;39:443–444.

220. Patriarca PA, Onorato IM, Sklar VEF, et al. Acute hemorrhagic conjunctivitis. investigation of a large-scale community outbreak in Dade County, Florida. JAMA 1983;249:1283–1289.

221. Wright PW, Strauss GH, Langford MP. Acute hemorrhagic conjunctivitis. Am Fam Physician 1992;45:173–178.

222. Patriarca PA. Clinical experience with acute hemorrhagic conjunctivitis in the United States. In: Uchida Y, Ishii K, Migamura K, Yamazaki S, eds. Acute Hem-

orrhagic Conjunctivitis: Etiology, Epidemiology, and Clinical Manifestation. New York: Karger Press, 1989:49–56.

223. Acute hemorrhagic conjunctivitis outbreak caused by Coxsackievirus A24—Puerto Rico, 2003. MMWR 2004;53:632–634.

224. Uchida Y. Clinical features of acute hemorrhagic conjunctivitis due to enterovirus 70. In: Uchida Y, Ishii K, Miyamura K, Yamazaki S, eds. Acute Hemorrhagic Conjunctivitis: Etiology, Epidemiology, and Clinical Manifestation. New York: Karger Press, 1989:213–224.

225. Chopra JS, Sawhney IM, Dhand UK, et al. Neurological complications of acute haemorrhagic conjunctivitis. J Neurol Sci 1986;73:177–191.

226. Langford MP, Yin-Murphy M, Barber JC, Heard HK, Stanton GJ. Conjunctivitis in rabbits caused by enterovirus type 70 (EV 70). Invest Ophthalmol Vis Sci 1986;27:915–920.

227. Wulff H, Anderson LJ, Pallansch MA, de Souza Carvalho RP. Diagnosis of enterovirus 70 infection by demonstration of IgM antibodies. J Med Virol 1987;21:321–327.

228. Sklar VE, Patriarca PA, Onorato IM, et al. clinical findings and results of treatment in an outbreak of acute hemorrhagic conjunctivitis in southern Florida. Am J Ophthalmol 1983;95:45–54.

Index